Mineral Resources and the Environment

A report prepared by the
Committee on Mineral Resources and the Environment (COMRATE)
Commission on Natural Resources, National Research Council

NATIONAL ACADEMY OF SCIENCES
WASHINGTON, D.C. 1975

NOTICE: The project that is the subject of this report was approved by the Governing Board of the National Research Council, acting in behalf of the National Academy of Sciences. Such approval reflects the Board's judgement that the project is of national importance and appropriate with respect to both the purposes and resources of the National Research Council.

The members of the committee selected to undertake this project and prepare this report were chosen for recognized scholarly competence and with due consideration for the balance of disciplines appropriate to the project. Responsibility for the detailed aspects of this report rests with that committee.

Each report issuing from a study committee of the National Research Council is reviewed by an independent group of qualified individuals according to procedures established and monitored by the Report Review Committee of the National Academy of Sciences. Distribution of the report is approved, by the President of the Academy, upon satisfactory completion of the review process.

This study was supported in part by the
Department of the Interior,
The National Science Foundation
and The Population Council.

Library of Congress Catalog Card Number: 75-4176
ISBN 0-309-02343-2

Available from
Printing and Publishing Office
National Academy of Sciences
2101 Constitution Avenue, N.W.
Washington, D.C. 20418

Printed in the United States of America

TABLE OF CONTENTS

SECTION III: THE IMPLICATIONS OF MINERAL PRODUCTION FOR HEALTH AND THE ENVIRONMENT: THE CASE OF COAL

SECTION IV: DEMAND FOR FUEL AND MINERAL RESOURCES

NATIONAL ACADEMY OF SCIENCES - NATIONAL RESEARCH COUNCIL

COMMISSION ON NATURAL RESOURCES

COMMITTEE ON MINERAL RESOURCES AND THE ENVIRONMENT

Arnold J. Silverman
Department of Geology
University of Montana

M. Gordon Wolman
Department of Geography and Environmental Engineering
The Johns Hopkins University

PANEL ROSTERS

PANEL ON MATERIALS CONSERVATION THROUGH TECHNOLOGY

Chairman: Alan G. Chynoweth
Bell Laboratories

Co-Chairman: Herbert H. Kellogg
Henry Krumb School of Mines
Columbia University

MEMBERS

Nathaniel Arbiter
Primary Metals Division
The Anaconda Company

Franklin P. Huddle
Science Policy
Library of Congress

David J. Rose
Department of Nuclear Engineering
Massachusetts Institute of Technology

Konrad T. Semrau
Stanford Research Institute

Theodore B. Taylor
International Research and Technology Corporation

PANEL ON THE ESTIMATION OF MINERAL RESERVES AND RESOURCES

Chairman: Kenneth O. Emery
Woods Hole Oceanographic Institution

Co-Chairman: Brian J. Skinner
Department of Geology & Geophysics
Yale University

MEMBERS

Paul Bailly
Occidental Minerals Corporation

Randolph W. Bromery
Chancellor
University of Massachusetts

Eugene N. Cameron
Department of Geology & Geophysics
University of Wisconsin

John C. Dunlap
Consulting Geologist
Dallas, Texas

Nicholas Georgescu-Roegen
Department of Economics and Business Administration
Vanderbilt University

John D. Moody
Mobil Oil Corporation

PANEL ON THE IMPLICATIONS OF MINERAL PRODUCTION FOR HEALTH AND THE ENVIRONMENT

Chairman: N. Robert Frank
Department of Environmental Health
University of Washington

Co-Chairman: Arnold J. Silverman
Department of Geology
University of Montana

MEMBERS

Thomas D. Brock
Department of Bacteriology
University of Wisconsin

John C. Crowell
Department of Geological Sciences
University of California, Santa Barbara

Frank J. Laird, Jr.
Environmental Engineering
The Anaconda Company

H. Crane Miller
Alvord & Alvord

Hubert E. Risser
U.S. Geological Survey

M. Gordon Wolman
Department of Geography and Environmental Engineering
The Johns Hopkins University

PANEL ON DEMAND FOR FUEL AND MINERAL RESOURCES

Chairman: Richard R. Doell
U.S. Geological Survey

Co-Chairman: Judith Blake Davis
Graduate School of Public Policy
University of California, Berkeley

MEMBERS

William Alonso
Princeton, New Jersey

Herman E. Daly
Professor of Economics
Louisiana State University

Tomas Frejka
The Population Council

Robert B. Fulton, III
E.I. DuPont de Nemours & Company

Denis Hayes
Illinois State Fuel Energy Office

Wilfred Malenbaum
Department of Economics
The Wharton School
University of Pennsylvania

Burkhard Strumpel
Institute for Social Research
University of Michigan

COMRATE STAFF

Francois J. Lampietti
Executive Secretary

Philippa Shepherd
Coordinating Editor

Farlan Speer
Staff Consultant

Mildred Lewis
Administrative Assistant

Joanne Keller
Secretary

Acknowledgments

In addition to my own appreciation of committee and panel members' generous contribution of time and expertise to this study, COMRATE's collective thanks are due to the numerous conferees and observers from academia, government and industry whose advice and assistance have been valuable inputs at all stages of the study.

The Committee wishes to express its sincere gratitude to all who, through individual contributions, insights and suggestions, or administrative effort, have helped in the production of this report.

Brian Skinner
Chairman
Committee on Mineral Resources
and the Environment

NOTE

Changes made to two references to M. King Hubbert on pages 88 and 89 of the report, in accordance with a suggestion from him made after the first printing, may warrant further clarification: References made to 1967 estimates of Company B in the text refer to Hubbert's 1967 estimates which, though developed while he was with Company B, actually appeared when he was with the U.S.G.S. In the table on page 89, consequently, Company B has been deleted and the estimates appear under the U.S.G.S. category as "Hubbert, 1967."

COMRATE HISTORY

The concept of COMRATE as a standing Committee of the NAS-NRC Division of Earth Sciences to provide an ongoing, balanced, long-term review of problems affecting mineral resources and the environment was approved by the NAS Governing Board in September 1971.

The first meeting of the COMRATE nucleus was in October 1972 under the Chairmanship of Preston Cloud who had also headed the Academy's prior study Resources and Man published in 1968.

Starting in early 1973 COMRATE divided itself into four panels each of which undertook separate investigations. With initial support by the Department of the Interior, U.S. Bureau of Mines and U.S. Geological Survey, later supplemented by a grant from the Population Council and from the National Science Foundation, Division of Materials Research, COMRATE and its panels each met approximately five times between January 1973 and June 1974. Characteristically the panels convened workshops to which experts and representatives of government, industry and academia were invited to present their point of view and to contribute to the panel investigations. Together with its panels COMRATE also conducted three field trips: (1) to observe the offshore oil production operations and natural oil and gas seeps offshore at Santa Barbara, California in June 1973; (2) to observe the steel making complex at Sparrows Point, Maryland in September 1973; and (3) to view various open pit, block caving and solution mining operations for copper in the vicinity of Tucson, Arizona in December 1973.

Upon Preston Cloud's resignation in August 1973, responsibility for the COMRATE chairmanship was assumed by Brian J. Skinner and COMRATE itself was administratively transferred to the newly-formed NAS-NRC Commission on Natural Resources. Also in August 1973 COMRATE convened a special Symposium on Mineral Policy which was attended by representatives of many branches of the federal government.

EXECUTIVE SUMMARY

EXECUTIVE SUMMARY
INTRODUCTION

TWO POINTS OF VIEW

Minerals are the staff of civilized living, and growing concern with their continued availability and efficient use was the impetus for this report. In the course of its study, the Committee on Mineral Resources and the Environment (COMRATE) recognized that two increasingly polarized schools of thought are becoming entrenched concerning the future adequacy of the world's mineral resources and the environmental costs of winning them. The study of the general issues involved--demand, supply, technology, and environmental impact of production--by the four COMRATE panels has resulted in findings which may go some way towards reconciling the two extremes.

The "doomsters" see a future in which catastrophic exhaustion of resources is inevitable unless drastic measures are taken to reduce economic growth. In opposition, the "cornucopian" view maintains that mineral resources are economically, and, for any future that may concern us, physically infinite. This unresolved conflict in economic thought was represented within COMRATE panels as well, and is illustrated by the dissenting opinion attached to the Report of the Panel on Demand for Mineral Resources (Section IV).

There are fallacious assumptions and potentially dangerous consequences inherent in both extremes. The "doomsters" pay too little attention to the adjustment potential of the market mechanism, and generally fail to understand the distinction between "reserves" and "resources." Their gloomy outlook is based on a "fixed" supply of materials and fails to recognize that the supply available changes as price rises and technical advances make lower grade resources economically and physically more accessible. The danger of this approach lies in its encouragement of alarmist overreaction on the part of policymakers, which may in turn have unnecessarily disruptive effects on the economy and society as a whole.

The "cornucopians," on the other hand, rely too heavily on the market mechanism for inducing the transformation of "infinite" resources into almost infinite reserves, and on the technological miracle for providing the physical wherewithal. Their hypothesis insufficiently represents the increasingly large capital costs of technological advance, the long lead times involved, the "net energy" factor (the energy cost involved in the technology of increasing production), and the fact that although technology has always come up with an answer in the past, its solutions have always had their social, environmental, or economic costs. These costs can no longer be ignored and are in fact setting a practical limit to the economic/technologic transformation of resources into reserves. More importantly, the economic/technological basis of the cornucopian argument is derived from the very assumption its adherents are concerned to disprove: it is shortages and public awareness of shortages which provide the incentive for increased production, technological solutions, and increased efficiency of use. The paradoxical result of the cornucopian message may thus be the fulfillment of the Cassandras' prophecies: in the relaxed climate fostered by anticipation of plenty, there will be no apparent urgency for setting in motion the economic and technological machinery for maintaining that plenty. This is a particularly important problem for the United States where maintenance or attainment of self-sufficiency in mineral resources is concerned. COMRATE believes that the United States will face serious difficulties in attempting to increase some supplies of energy and mineral raw materials from domestic sources. Indeed, COMRATE believes it is doubtful whether even current levels of supply can be maintained for all materials.

To view the problem moderately, we must draw together the valid arguments of both schools of thought. The overall conclusion that has emerged from this study is by no means a counsel of despair. But separate consideration of the complex problems involved underscores the need (1) to husband resources, (2) to generate information in areas where it is inadequate, and (3) to tackle immediately problems where there is adequate information to form a basis for new action or for augmenting existing efforts. Such actions should always be designed to conserve resources and increase efficiency in their use.

THE THEME OF INTERDEPENDENCE

In all aspects of the study, whether during detailed consideration of a specific mineral or process or in the wider arena of resource supply and demand, the theme of interdependence has insistently claimed attention. It has become increasingly clear that no aspect of materials policy can be considered in isolation. The complex interactions within the materials cycle, the two-way street of energy/minerals production and the effects of both on environmental considerations, the need for reconciling the interests of government, industry and the consumer, and the interrelationships of national and international policy have emerged as essential ingredients at every level of policy formulation. Efforts to increase supply should be made concurrently with policy aimed at decreasing demand; technological progress in substitution and recycling should be stimulated along with, not independently of, the encouragement of a conservation ethic.

The National Research Council (NAS/NAE) has been aware of and responsive to the urgent need for study of these and related issues. In the two years of their existence, COMRATE's panels have attacked selected problems in mineral resources, amplifying points raised in earlier reports, such as those by the National Commission on Materials Policy* and the Environmental Studies Board.**

GENERAL CONCLUSIONS

Some general conclusions emerged from the panels' separate deliberations. Although the time span for this first COMRATE report has not allowed all-encompassing conclusions to be drawn, those conclusions that have been drawn have many implications for policymakers.

1. Mineral resources become available for man's use by a complex and lengthy process which, on a worldwide scale, relates intimately (a) natural process, (b) man's knowledge and technological ingenuity, and (c) man's economic, social, and ethical concerns. Efficiency in use and avoidance of waste in both mineral resources and their end products are essential to alleviate immediate economic strains, and are even more essential if we are to avoid preempting the resources needed for future generations. Policymaking at all levels should recognize interdependencies within the materials cycle, among nations, and among the various users of mineral commodities. But, above all, we should adopt a conservation ethic that has at its heart avoidance of waste and more efficient use of materials.

2. Widely divergent methodologies, based largely on individual judgment, are used both in forecasting demand for, and in estimating supplies of, mineral resources. There are currently no standardized techniques for making either long-term demand forecasts or resource estimates nor are means available to assess adequately the accuracy of the existing methods.

3. Reliable data on mineral resources are difficult to obtain because of their proprietary or international nature. This affects supply estimates. In the U.S. much improvement is still needed in the work of the U.S. Bureau of Mines and the U.S. Geological Survey in the collection, coordination, standardization and dissemination of mineral resource data.

4. Definitional vagueness and numerical imprecision afflict discussion and publications, both inside and outside government, concerning mineral

*Material Needs and the Environment Today and Tomorrow: Final Report of the National Commission on Materials Policy, Washington, D.C., U.S. Government Printing Office, June 1973.

**Man, Materials and the Environment: A Report to the National Commission on Materials Policy by the Environmental Studies Board of the NAS/NAE, Washington, D.C., March 1973.

resources. The notions of reserves and resources used in this report (Figure 1), which were recently formulated for use by the United States Bureau of Mines and the United States Geological Survey, could also serve as a common frame of reference for all resource discussions. Units of measure used for resources are diverse and cumbersome. Recognizing the interdependence of resources, common scales for units and for quantities should apply to resource estimates. In addition, the gathering of resource data should be refined and systematized so that standard error or confidence level appreciations can be applied as elsewhere in the physical sciences.

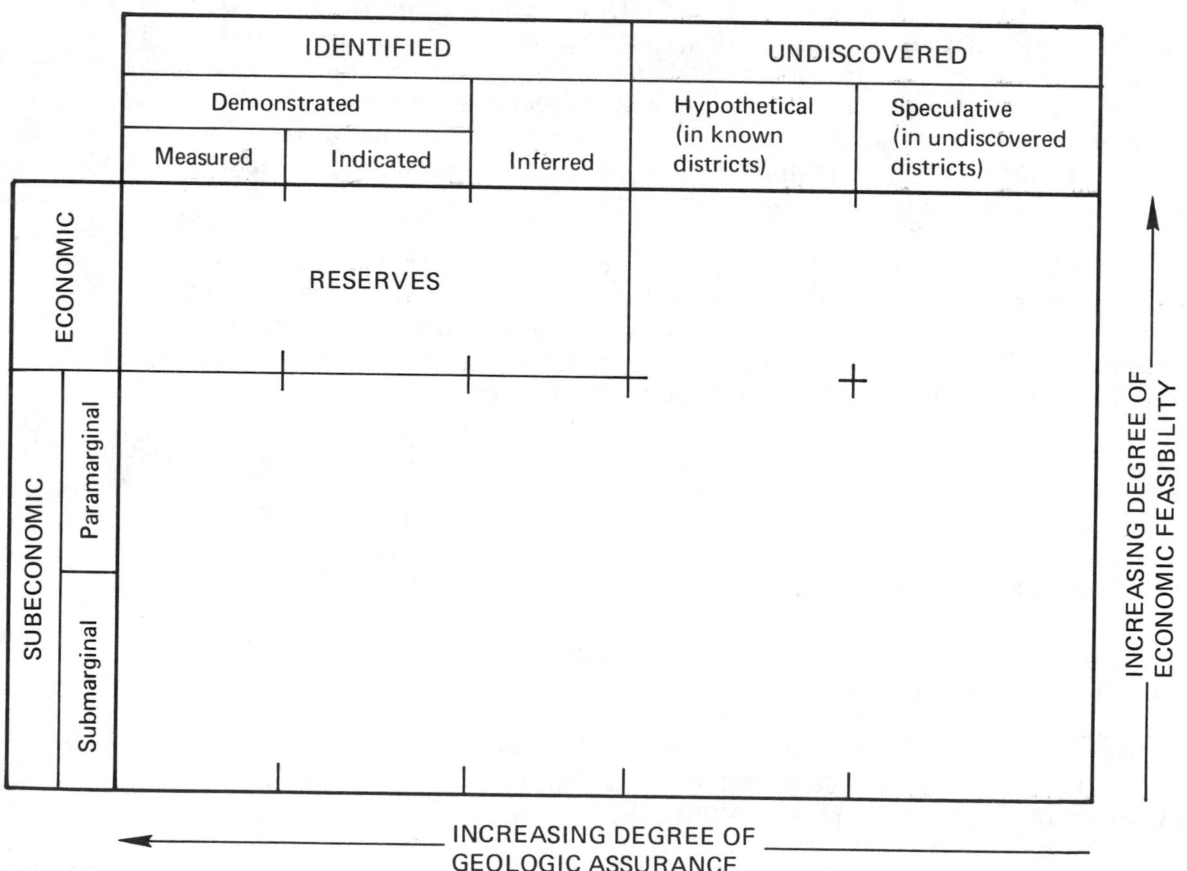

SOURCE: After USBM (1973).

FIGURE 1: Classification of Mineral Resources

SUMMARY OF PANEL FINDINGS AND RECOMMENDATIONS

GENERAL

Within the vast area of mineral resources and the environment covered by its terms of reference, COMRATE carved out four provinces for study by separate panels: technology, supply, the environment and demand. In addressing their allotted topics, the four panels adopted approaches that differ in some important respects.

Section I, the report of the technology panel, entitled Materials Conservation through Technology, is a broad coverage of the entire area under consideration. To solve the problems of space and readability involved, the panel presented an overview summary of the large body of contributed essays on specific aspects of the topic. The contributed essays have been collected in a separate Appendix volume which, while it does not necessarily reflect ultimate panel consensus, nevertheless provides extensive documentation of the panel's findings. The Appendix volume, for which a table of contents appears on page 346, is available on request.

Sections II and III--the reports of the supply and environment panels, entitled respectively Estimation of Mineral Reserves and Resources, and Implications of Mineral Production for Health and the Environment--are case studies. Section II discusses supplies of the fossil fuels and of copper as major representatives of two categories of mineral resources. For Section III, coal was selected for exclusive study because it is a material whose prospective vastly increased exploitation will raise important environmental questions. Both Sections II and III are thus illustrative of the general issues involved in mineral production and use.

The report of the Demand panel, entitled Demand for Fuel and Mineral Resources (Section IV) adopts the more generalized approach. It sets the specific questions of technology, supply and environment addressed in the preceding sections in the larger socio-economic perspective of demand for energy, and of the implications of demand forecasting for both supply and demand for minerals.

Sections II, III and IV, like Section I, are augmented by separate Appendix volumes, which again do not necessarily reflect COMRATE or panel opinion. These Appendices, unlike the substantial information-base Appendix to Section I, contain only selected inputs, which expand or document certain aspects of the main report, and which may be of use to readers with a special interest in following up on background data. All four Appendices are available, separately or together, on request. A contents list of material contained in each Appendix appears at the end of the report. (Annex III).

REPORT OF THE PANEL ON MATERIALS CONSERVATION THROUGH TECHNOLOGY (Section I)

Contents and Conclusions

The report of the Panel on Materials Conservation Through Technology (hereafter called the Technology Panel) concludes that technology will not always be capable of closing the growing gaps between rising demands and limited supplies of mineral resources. While technology can do much towards conserving resources, it cannot do everything. There are limits to efficiency, to substitution, and to other technological responses. The Panel concludes that the federal government should proclaim and deliberately pursue a national policy of conservation of materials, energy, and environmental resources.

Following an introductory chapter outlining the main theme and listing recommendations, the first part of the panel's report (Chapter II) provides a general philosophical and policy framework which is essential to technical considerations. Interdependency is the basic theme: interaction between social and technological issues predicating a concerted attack by social scientists and technologists; interdependencies within the materials cycle which preclude the study of any sector or component in isolation; and chain reactions in international competition and cooperation. Overall conclusions and recommendations center on a "project interdependence" involving a conservation ethic that specifically includes efforts at efficient stockpiling of both materials and processing capabilities.

Chapter II also outlines the limitations on technology imposed by constraints of capital, manpower, physical and technical considerations, and time. An enormous amount of capital investment will be needed for energy supplies--through 1985, about 35 percent of all available investment capital, as compared with the current 22 percent--distorting the whole investment picture and denying needed capital to other industries, including the mineral industry. Lack of adequate manpower could be offset, not so much by an effort to increase the total stock, as by deployment of existing manpower and acceleration of automation. Despite an abundance of alternative energy technologies, physical constraints on progress may eventually be imposed by environmental considerations. The Panel therefore considers it important for the Government "to develop, publish, and update data on the energy and pollution consequences of recovering values from listed reserves and resources." The time constraints, the long lead-times involved in effecting technological substitution, lead to the conclusion that in general it is dangerous to rely on the prospect of immediate technological "fix." R&D should be undertaken to provide long-term answers; in the short-term, the need is for redeployment and rearrangement of existing technology.

The report goes on (Chapter III) to discuss technological opportunities in the materials cycle to offset shortages. The chapter reviews the tactics for science and technology consistent with the theme developed in Chapter II of the necessity for conservation in the materials-energy-environment system.

Under the headings of Energy Sources, Pollution Control, Mineral Exploration, Mining, Benefication, Primary Metal Production, Recycling, Product Design and Substitute Materials, technical activities already being pursued to this conservational end are endorsed and new or intensified efforts are recommended. The identification within each topic of specific technological opportunities to counteract shortages is amplified by detailed treatment in the Appendix to this section. (See Table of Contents of Appendix to Section I, p. 346.)

Recommendations

1. The Technology Panel recognized needs leading to 12 policy recommendations, primarily for Federal Government action (see p. 17-18). The major emphasis of these recommendations is that conservation of materials through all their cycles, involving extraction, use, reuse, and ultimate discard, is essential, as is careful consideration of energy needs and environmental consequences of these cycles. Examples of these detailed recommendations are: (a) to establish equitable international agreements for sharing of natural resources; (b) to create national stockpiles of vital materials; (c) to provide federal support for manpower training and education and for automation in the materials industry as a means of increasing efficient production; (d) to establish a capacity for long-range forecasting of materials resource supplies and for predicting the energy needs and environmental and social consequences of developing these resources.

2. The Panel has also made a set of 40 recommendations in nine specific areas of minerals and materials technology, reflecting the policy recommendations and applying them to all aspects of energy and materials production and use from exploration through production, application, and final disposal. (See p. 20-23.) The recommendations concern the possible applications of present and future technology both to more efficient use of known supplies and to the reduction of demand through substitution, conservation and recycling.

REPORT OF THE PANEL ON ESTIMATION OF MINERAL RESERVES AND RESOURCES (Section II)

Contents and Conclusions

The Panel on Estimation of Mineral Reserves and Resources (hereafter called the Supply Panel) addressed itself to current methods of estimating mineral reserves and resources and the quality of current resource estimates. At a time when the adequacy of mineral supplies to meet future needs is being questioned and concern is growing over the increasing dependence of the

United States on sources of minerals abroad, it is essential that the best possible estimates of reserves and resources of minerals be available. Such estimates are a first and necessary basis for national policy with regard to minerals.

The panel reached two general conclusions on the necessity of improving estimation methods (see p. 77-78):

1. A major problem confronting attempts to improve estimates is the difficulty of bringing into the public domain proprietary information gathered by private organizations. Bringing such information together with that available from government files, without negating proprietary values, would enormously facilitate the maintenance of a running inventory of resources, and greatly increase the accuracy of resource estimates.

2. A second deficiency of current resource estimates is their inadequate recognition of the economics and possible rates of mineral production. Estimation of resources in the ground should be supplemented by engineering and economic studies of factors influencing rates of production and processing.

Reserves and Resources of Fossil Fuels

World resources of coal are large relative to current energy requirements. Resources appear adequate for hundreds of years, even at considerably expanded rates of production, but world distribution of coal is uneven, the principal deposits being in China, the U.S.S.R., and the United States. World resources of petroleum and natural gas are more limited and will be substantially consumed by the first quarter of the twenty-first century if world trends of production and consumption continue.

Undiscovered recoverable resources of oil and natural gas, onshore and offshore, in the United States including Alaska, are considerably smaller than indicated by figures currently accepted within government circles. In the judgment of the Panel, the figure approximates 15 billion metric tons (tonnes) (113 billion barrels) for crude oil and Natural Gas Liquids (NGL) and 15 x 10^{12} cubic meters (530 x 10^{12} cubic feet) for gas.* (See pp. 87-91.) A large increase in rates of United States production of petroleum and natural gas is extremely unlikely.

Both world and United States resources of petroleum in oil shales and tar sands are large, but many technical, economic, and environmental problems must be resolved before significant rates of production from these resources can be achieved.

---

*U.S. consumption of oil and NGL for 1973 was 6 billion barrels, and for natural gas was 23.4x10^{12} cubic feet.

The report recognizes the enormous value of petroleum as a source of energy in the medium term, in view of the lead times of a decade or more required to develop alternate sources of energy and hydrocarbons. The report also stresses the importance of petroleum as a long-term source of chemical raw materials. Conclusions based on these findings are that U.S. independence from external sources is essentially impossible on the basis of increased production of petroleum during the next decade, and that a strong emphasis on conservation is necessary.

Reserves and Resources of Copper

The report concludes that an ore-grade of about 0.1 percent copper represents a point beyond which mining and recovery of copper is unlikely to proceed. The world's copper resources above this grade are on the order of 1.4×10^{10} tons of metal, and approximately 3 percent of this total has already been mined or blocked out as mineable reserves in currently worked deposits.

United States reserves of copper are at least 80×10^{6} tons of metal, approximately 40 times current annual consumption of new copper, but additional discoveries at a significant rate will be necessary if the United States production rate is to be maintained until the end of the century. The Panel's recommendation for copper, therefore, concentrates on the need to aid and stimulate effective domestic mineral exploration and production. World reserves of copper are large, and world supply for the remainder of the century will depend primarily on creation of additional mining and smelting capacity.

Recommendations

Reserves and Resources of Fossil Fuels

1. Estimates of proved reserves and undiscovered resources of petroleum and natural gas indicate that a large increase in annual production from conventional domestic sources is extremely unlikely. In view of this and the importance of petroleum as a source of both energy and chemical raw materials, and because a decade or more will be required for large-scale development of alternative sources of energy and hydrocarbons, we recommend that policy toward the use of petroleum and natural gas place a strong emphasis on conservation.

2. A very large untapped domestic petroleum resource, which could be as large as 100 billion barrels, lies in partly depleted oil reservoirs, and reservoirs no longer capable of production using present technology. We therefore recommend that research and development to increase the recovery of petroleum from oil reservoirs be pursued vigorously.

Reserves and Resources of Copper

1. The availability of copper from domestic deposits during the remainder of the century depends to a significant extent on success in discovery of new domestic copper deposits. We recommend that exploration for copper be stimulated, and that means of improving the effectiveness of exploration be investigated and supported.

2. Copper resources in manganese nodules on the floors of the deep seas are apparently as large as developed reserves in conventional deposits on land. Because there is uncertainty as to the ability of the United States to meet demands for copper from domestic sources, we recommend that developing the recovery of copper and associated metals from these nodules be encouraged with due regard to the potential impact of undersea mining on the environment.

REPORT OF THE PANEL ON THE IMPLICATIONS OF MINERAL PRODUCTION FOR HEALTH AND THE ENVIRONMENT--THE CASE OF COAL (Section III)

Contents and Conclusions

The probable key role of coal in the future energy needs of the nation is well established. Consequently, the large increase in coal production to be expected makes consideration of potential environmental impact imperative. The Panel on the Implications of Mineral Production for Health and the Environment (hereafter called the Environmental Panel) has therefore concentrated on some important environmental problems associated with coal production and use, focussing on human health and safety and on certain impacts on the ecological system as a whole. No precise dollar values can be placed on the damage of such impacts, but it is clear that they are potentially large and important both in themselves and in the general issues they raise for minerals policy.

Two important health effects were considered: Coal Worker's Pneumoconiosis, with its enormous attendant costs, and the pulmonary effect of air pollution on the public at large. Particular attention was paid to recent evidence on emissions of sulfur oxides and their subsequent conversion to particulate sulfates. The conclusion was reached that current analysis and clean-up methods, in concentrating on larger particles, may not be approaching the problem correctly. There is evidence that smaller particles may not only be more harmful to health but are also an important contribution to the oxidation of SO_2 to sulfate, and, further, to acid rain, a matter of increasing ecological concern where further research is warranted.

A general conclusion of the report is that monitoring is inadequate, and that funds for the study of environmental impacts of fossil fuels should be increased.

Recommendations

Mine Health and Safety

1. The expected large increase in numbers of miners makes urgent the need to overhaul methods of detection, diagnosis, and compensation for Pneumoconiosis in the nation's Black Lung Disease program.

2. Because most coal mining accidents involve inexperienced miners, it is recommended that a comprehensive pre-entry training program be mandatory for the influx of miners expected as a result of expanded production, supported by ongoing training of working miners.

Public Health

In view of the effects of sulfur pollutants, and especially of the associated acid rain, on the ecology and health of the nation, the Panel strongly recommends:

1. studies (aided by a national monitoring network) of the cycles, interactions, and chemistry of sulfur and its compounds from emission to absorption;

2. the redesign of protection standards and attendant sampling systems because current sampling systems do not properly reflect how sulfur oxides and sulfates do their damage; and

3. immediate development and installation of sulfur oxides removal equipment in current sulfur oxides emitting operations, development of cleaner fuels, and the acceleration of more efficient processes that will reduce the overall emission of sulfur oxides.

REPORT OF THE PANEL ON DEMAND FOR FUEL AND MINERAL RESOURCES (Section IV)

Contents and Conclusions

The report of the Panel on Demand for Fuel and Mineral Resources (hereafter called the Demand Panel) addresses the problem of demand for mineral resources and the significance of demand forecasts. The report points out that national obsession with expanding supplies arises from the dramatic gap between projected demand and projected supply. But acceptance of the existence of this gap and of the need to close it through expanding supply presupposes both that the demand projections are valid and immutable and that demand can be neither stabilized nor reduced. As a consequence of such a supposition, demand projections can become self-fulfilling.

The report challenges the credibility of the projections and the assumption that they are not susceptible to change, and strongly emphasizes the potential for influencing demand through the market mechanism, through policy, and through technology. This potential, in the Panel's view, has been neglected relative to that for influencing supply.

The report first reviews the state of the art in demand forecasting for mineral resources. This is important because assumptions about future demand are a basis for policy decisions; and, since demand forecasts can be self-fulfilling as noted above, it becomes essential that their conclusions be substantiated. The panel therefore commissioned a bibliographic review of forecasts for energy demand as a starting point for its own general analysis of the kind of assumptions and methodologies that have been influencing policy. (The bibliographic review is presented in the separate Appendix volume to this Section.)

General conclusions reached on the state of the art (Chapters XI and XII) were that forecasting techniques are, in general, relatively primitive and that there is need for improvement in terms of definition, aggregation, correlation with supply forecasts, and specification of data. The demand panel has concluded that inadequacies in these respects have apparently created a tendency toward upward bias that exaggerates demand. This contrasts with forecasts of a decade or so ago that tended to <u>underestimate</u> demand because they did not appreciate the effects on demand of relative declining prices. Such relatively declining prices, however, are unlikely to recur.

The bulk of the report (Chapters XII-XVI) examines sources and components of demand structures, first, to provide material for improving forecasting, and second, to identify areas of demand susceptible to policy. Specific components considered are: GNP, population, consumer response to shortages and price changes, independent social changes, and the accounting of hidden costs in the price of minerals. Chapter XVII suggests applications of the findings of the study as a whole to developing designs for future forecasting, stressing the immediate need for short-term forecasting on the one hand and for research into long-term projecting on the other.

The panel's conclusion that study of demand for resources can readily be neglected because of historical emphasis on increasing supplies, led to its recommendation in August 1974 that the Commission on Natural Resources of the National Research Council establish a Resources Demand Board to focus research on this demand component as a policy variable. The contribution of this panel's report has been a delineation of the major problems facing demand policy, in terms of inadequate information, inadequate development of forecasting techniques and inadequate consideration of means for influencing demand. The scope and magnitude of the issue of demand for resources has confined the panel to identification of the problem and to a very general indication of potential directions for further study and, ultimately, policy. It is hoped that these general findings will provide a foundation for the detailed research and specific recommendations of a Resources Demand Board, on such issues as means for improving forecasting, the effects of price changes, governmental regulation, and secondary effects of changing market responses to restrictions. The establishment of such a Board is currently under consideration.

In the meantime, the panel's own deliberations have enabled it to come up with some general recommendations which could be followed up immediately. These are given in detail in the concluding chapter of Section IV, which endorses the need for a conservation ethic emphasized by the Technology and Supply Panels, and lists recommendations for integration and analysis of information and for policy directed towards reducing demand. These are summarized below.

Recommendations

Needed Research and Its Use

1. *Collection and analysis of information on demand response to price, income, shortage and similar matters, should be undertaken.*

2. *Technical ability to make more accurate short- and long-term forecasts using this information must be improved.*

3. *A governmental instrumentality should be established to encourage and make use of the studies recommended in 1. and 2.*

National Policy

1. *All federal policy actions and legislation, existing and proposed, should be modified or designed to move resource use demand downwards, mainly through the market mechanism.*

2. Procedures for compensation to segments of the society under adverse effects of resource shortages or price increases, or both, should be designed to alleviate these problems.

3. Education, publicity, and similar campaigns should be instigated to lead the nation towards a strong "no waste ethic."

SECTION I

MATERIALS CONSERVATION THROUGH TECHNOLOGY

CHAPTER I
INTRODUCTION TO SECTION I

Faced with growing strains between the supply of mineral resources and the demand for them, society reacts in two general ways: (1) it assumes that economists and legislators will be able to manipulate appropriately the socio-technical environment and (2) it assumes that technology will be able to relieve the strains. This Section is concerned primarily with the second reaction. The problem of imbalances between supply and demand is addressed in the context of the total materials cycle, and ways are sought in which technology can increase the supply of needed materials and reduce the demand for them, or both. The theme is how technology can contribute to conservation of the earth's resources of materials and energy and maintain its environment. The Panel offers no universal panaceas that will banish the problems, but it proposes steps that should move the nation in the right direction. Though technology can do much to conserve resources, it cannot do everything; and furthermore, what it can do takes time and involves many steps.

While Section I is basically concerned with technological issues, these cannot be divorced from some broader concerns which are reviewed in Chapter II, and which lead to certain policy recommendations that are principally nontechnical. Chapter III covers the recommendations for action by the materials technology community and its sponsors, and addresses salient issues at various stages of the materials cycle. The whole of this panel's report, both technical and nontechnical, summarizes more detailed studies and inputs from panel members and conferees. Many of these are included in an Appendix to this report, a separate document available on request to those with a special interest in following up the conclusions presented here. A Table of Contents for this Appendix is given on page 346.

SYNOPSIS OF RECOMMENDATIONS

Policy Recommendations

(The recommendations are not rank ordered, but the keynote is given in policy recommendation 5. The federal agencies, departments and other units to which we believe these policy recommendations are most pertinent are indicated in Table I.)

We recommend that government bodies and other units concerned with matters pertaining to materials and other natural resources take the following steps:

1. *obtain the concerted advice of technologists and social scientists*

2. *take a systems approach to the materials cycle and the materials-energy-environment system*

3. *seek equitable international agreements concerning the exploitation and sharing of material resources*

4. *practice stockpiling of materials and of the technology for acquiring them from domestic sources*

5. *pursue a national policy of conservation of materials, energy and environmental resources*

6. *facilitate industrial cooperation to minimize waste of natural resources*

7. *find ways to share the scale-up costs of materials technology in areas of high risk or delayed payoff*

8. *endeavour to increase the amount of trained manpower available to the materials industries*

9. *support an intensive program to develop automation for the materials industries*

10. *sponsor intensive studies of the effects of waste heat and pollutants on climates and ecosystems*

11. *develop data giving the energy and pollution consequences of recovering values from ores*

12. *create an institutional capability for long-range forecasting of materials issues and taking appropriate actions.*

TABLE 1 Policy Recommendations

Prime Lead Agencies, etc.	1	2	3	4	5	6	7	8	9	10	11	12
Congress	x				x	x	x	x				x
White House	x		x		x							
Department of Agriculture	x	x		x	x					x		x
Atomic Energy Commission	x				x		x	x		x		
Department of Commerce			x	x	x	x	x		x			x
Department of Defense				x	x							
Federal Energy Office	x	x	x		x							
Environmental Protection Agency		x			x					x	x	
General Services Administration				x	x							
Department of Health, Education and Welfare	x	x						x		x	x	
Department of Housing and Urban Development	x	x				x	x			x		
Department of the Interior	x	x		x	x			x	x	x	x	x
Justice Department					x	x						
Labor Department								x				
Office of Management and Budget	x	x										
National Science Foundation				x	x	x	x	x	x	x		x
State Department	x		x									
Office of Technology Assessment		x			x					x	x	x
Department of Transport	x	x				x	x					

Technological Recommendations

Energy

1. *Prime emphasis should be on ways to conserve energy.*

2. *Attention should be paid to assuring adequate supplies of certain materials critical to power generation technology.*

Pollution Control

Work should be supported on:

1. *developing bodies of generalized design information for pollution control systems*

2. *developing basic control technology for nitrogen oxides emitted from combustion processes*

3. *finding uses for sulfur and sulfur by-products*

4. *determining the effects of pollutants on biological systems and developing instrumentation necessary for such studies*

5. *investigating the effects of concentrated, large heat and moisture releases on local and regional climates.*

Mineral Exploration

1. *Research should be intensified on finding improved methods for mineral exploration.*

Mining

Needs for new or intensified research include:

1. *the development of improved methods of rock cutting and tunnel boring*

2. *the development of automated mining techniques*

3. *determining the environmental effects of deep-sea mining*

4. exploration of the capabilities of in situ leaching as a mining process.

Beneficiation

Intensified research is needed on:

1. ways to obtain greater energy efficiency in rock crushing and grinding

2. finding methods for improved mineral recovery from fine powders.

Primary Metal Production

1. Research should be intensified to find extraction processes with greater energy efficiency and better pollution control.

2. More research is needed on processes for more complete recovery of metals from complex and refractory ores, and of metal by-products now lost to waste streams.

3. The symbiotic advantages of industrial complexes for materials should be explored.

4. There should be more emphasis on finding methods of direct processing from melt to stock shapes.

Recycling

Aspects calling for new or intensified effort include:

1. developing economic uses for mine wastes and tailings

2. developing better automated methods for collecting, sorting and segregating municipal wastes

3. recovering material values from buildings at the end of their useful lives

4. methods of separating various classes of materials joined to each other in junk equipment

5. processes for separating intimate mixtures of metals into their component elements

6. means of repurification of degraded alloys as well as finding uses for which their properties are suitable

7. finding satisfactory methods of reclaiming values from composite materials

8. methods for sorting plastics and recovering material and monomer values from them

9. methods for recovering values from glass and ceramic wastes

10. finding incentives to encourage the collection and processing of wastes.

Product Design

1. A set of principles of product design for facilitating material recovery at the end of the product life should be developed.

2. Research aimed at understanding and improving the performance-limiting properties of materials must be adequately and steadily supported.

3. Products and structures should be designed as far as possible to performance and conservation specifications.

Substitute Materials

1. Substitutes need to be found, in their major use contexts, for the following unique materials:

Helium

Mercury

Asbestos

2. Substitutes need to be found, in their major use contexts, for the following specialty metals:

Chromium

Gold

Palladium

Tin

3. The need for substitutes should be assessed for:

Antimony

Tungsten

Vanadium

Silver

Zinc

4. Ways need to be found for extending the range of useful ceramic and glass materials and processes for making them.

5. Research to extend the range of performance properties of plastics needs to be intensified.

6. Economic processes for obtaining feedstocks for plastics from forest and vegetable products need to be found.

7. Research needs to be extended on finding ways to grow more uniform, defect-free wood, uses for wood scraps, and improving the durability of wood products.

8. The range of industrial composites, particularly those made from more abundant materials, should be extended.

FEDERAL ROLES IN STIMULATING TECHNOLOGY

> *"...it is recommended that the Federal Government proclaim and deliberately pursue a national policy of conservation of material, energy, and environmental resources..."*

Many of the conclusions and recommendations which follow implicitly or explicitly involve the Federal Government in the role of stimulating materials technology. While we believe that the private sector should retain the main responsibility for developing technology, federal participation is often not only inevitable but necessary, perhaps even desirable. The effectiveness of such participation, however, depends critically upon its mechanics and, perhaps even more, on the spirit with which it is undertaken. Four especially important roles for the Federal Government are discussed below.

Broad Guidance of the National Economy from an Era of Material Growth to One of Material Conservation

Growth in the material sense has been so long a bedrock of conventional economic wisdom that it may seem heretical to suggest that it will be less valid in the future. But our study strongly suggests that, for purely physical reasons, at least the material consumption part of the economy will increasingly encounter limits to growth. Material economics follow S-shaped curves more nearly than rising exponentials. In our view, national policies exhorting, and based on, ever greater growth rates in the material economy will become increasingly short-sighted and inconsistent with human aspirations. A traditional driving force for the national growth ethic is to keep ahead of foreign competitors, just as it is traditional for domestic competitors to use growth as a measure of their relative success. But it must be recognized that other countries are on S-shaped curves too, and that, in comparison, the net abundance of resources enjoyed by the U.S. should ensure for it a continued strong position in world-wide economies.

Guardianship of the Environment and Stimulation of More Effective Utilization of Materials and Energy

The Federal Government is already moving strongly in a number of ways to implement its role of guardian of the environment in which we live and work. Regulation and the setting of standards are key elements in the federal role and are likely to become even more so as that role is extended to embrace energy and materials. As recent events have shown, however, the methods of technology forecasting and assessing the consequences of technological developments leave much to be desired. Clumsy, hasty or over-zealous regulation or standard-setting can prove more injurious to the economy, the environment, and the standard of living of the individual consumer than the condition they were intended to correct. Often these unintended injuries appear to stem from political or economic decisions taken on the basis of too little consultation with the technical community and too myopic a view of the intrinsic time scales. We see it as vital for future prosperity for such decisions to evolve out of effective collaboration among government, physical scientists, engineers and social scientists. At the same time it is imperative for these various disciplines to learn to work together.

Stimulation of Transfer of Technology Out of the R & D Phase and into Commercial Use

There is little evidence that the federal government can manage an industry more efficiently or wisely than the private sector which is familiar with competition in the commercial market-place and in meeting the needs and wants of consumers. Given adequate support of R & D itself, the biggest need for new federal help is in the sensitive, risky stages of technology transfer from the R & D phase into commercial production. The help need not be direct financial aid but incentives that will make it more attractive to private industries to accept the risks. A variety of fiscal, legislative, and administrative incentives should be experimented with, but this subject falls outside the scope of this report. An important area of concern if resources are to be conserved through more efficient industrial organization concerns the freedom of different industries to cooperate, share risks, and do systems planning in an orderly way. It must be recognized that the more complex and ramified the field of technology the more the need for a systems approach which in turn requires unified responsibility.

Regulation of Industry in the Public Interest

If the energy and environmental fields are any guide, increased federal intervention can be anticipated in the materials field not only because of the problems of materials supply and demand themselves but also because of the ways in which materials interact with energy and the environment. It is quite likely, also, that consumer demand will require some federal regulation in the future. To be successful federal intervention, whether it be by regulation, legislation, or whatever, should aim at creating opportunities within a broad framework for individual and corporate enterprise, not stifling them. Such matters as policies, major technological objectives, and performance specifications need to be worked out more cooperatively with the affected parties in full cognizance of the time taken to develop new techniques. Persistence with the current trend which tends to keep apart industries that would make natural partners for increasing the effective use of natural resources and reducing insults to the environment, and which often casts the Federal Government in an adversary role against each of the major industrial organizations, is likely to lead to wasteful and duplicative rather than constructive competition, and to squandering rather than conservation of natural resources. The traditional picture of the lone inventor, the courageous entrepreneur, may still hold true for relatively discrete, simple products but the whole concept of entrepreneurship requires considerable modification when applied to a complex system such as the materials cycle and its interactions with energy and the environment. Again, as recent events have shown, industries must be allowed adequate profits in order to invest in R & D and to attract the venture capital needed for successful technology transfer. Price regulation (which usually means keeping prices unnaturally, even punitively low) may buy consumer goodwill in the short term but at the expense of long-term ability to provide the goods and services that consumers will continue to expect.

CHAPTER II
BROAD CONCERNS

Until very recently, materials, the stuff that things are made with, have been very much taken for granted by the average citizen. He has tended to accept without question that for whatever gadget, device, structure, or machine he wanted to have, or construct, the necessary materials would be available at an acceptable price. This cozy state of confidence has now been shaken by several recent events which have had universally felt effects--the oil shortage, electricity brownouts, beef and wheat shortages and, it is not inappropriate to add, incidences of overloading of service sectors such as telephone communications and roadway capacity. Nowadays the average citizen asks whether future supplies will be adequate to meet his needs for steel, for water, for energy, for concrete, in fact for every material.

Some of the economists and other analysts attempting forecasts of future material supply and demand situations also seem to have fallen into the same state of complacency; this is particularly true of those who would base their predictions of the future very much on extrapolations, no matter how elegant, of the past. Such projections work best when based on the assumption that there is essentially an inexhaustible source of materials and other factors--population, usage patterns, fashion, product development, etc.--are the prime determinants of the scale of consumption. But this pays no heed to the fact that the earth's resources are finite. The abundances of the materials available in the earth vary enormously but man's pattern of consumption rates bears little resemblance to the abundance pattern. In consequence, man faces the prospect of a series of shocks of varying severity as shortages occur in one material after another, with the first real shortages perhaps only a matter of a few years away. From a practical viewpoint the minerals man uses will not all be exhausted at once. Their occurrence in the earth's crust is subject to almost infinite variation of concentration and accessibility. Man's needs vary in quantity and substance from generation to generation, but the trend is always upward. In contrast, the rate at which new discoveries are made, and the reserves remaining in the ground, in the course of time, trend downward.

As the prospect of materials shortages gradually dawns on man he tends to assume, traditionally, (1) that technology will find the answers, and (2) that economists and legislators will be able to manipulate the socio-technical environment so as to alleviate any strain that the shortages would otherwise cause. Faith in both expedients is weakening. Technology often seems to be as much blamed for present predicaments as regarded as a potential saviour. And under the relentless pressure of inflation, the average citizen doubts the abilities of economists and politicians to manage the economy.

In a free economy supply and demand are supposed to track each other closely, with price being the mediator--the classic picture of a free market economy. And because materials are so world-wide in their ramifications, then in the absence of effective world-wide controls the price mechanism of a free market economy will operate globally. No industrial supplier of materials is going to increase his plant capacity beyond what he believes the market will need, if he is managing his business at all efficiently. Conversely, if demand is exceeding current plant capacity to supply he will try to increase his plant. So, with such feedback systems in operation, supply and demand over a sufficiently large, commercial base and over a not-too-short time scale will track each other closely provided that the basic resource, the amount of material available, does not become the limiting factor. Under these free market economics, any shortages that occur drive the price up until demand and supply are again brought into equilibrium, principally by tapping new, marginal sources.

However, demand for materials tends to be highly inflexible because large investment in capital equipment and labor force to convert materials into products requires a stable supply and cannot tolerate a shortage, while inventory building to create a reserve is prohibitively expensive--especially when interest rates are high. Then, rising prices tend to motivate the opening up of lower grade reserves by marginal companies. But moving to lower grade ores tends to require moving larger tonnages so that productive capacity rises disproportionately with demand, driving prices down. Unfortunately, also, there is a substantial time lag in the opening up of such new marginal deposits and usually considerable investment has to take place. Then, when relative stability has been achieved some event may cause a drastic downward change in the demand pattern.

Thus, the high flexibility plus time lag of raw material production coupled with the low flexibility of industrial demand and the cost of money make materials highly unstable as to price. The free market economy balances materials supply with demand but only on a long time scale within which many people can get "hurt."

In practice, attempts are often made to cushion the more severe results of free market economics through various forms of price control. But, the effects of price controls fall unequally on different commodities and different life styles. In a world-wide market system, any price control or any additional taxes (perhaps for environmental purposes) applied locally, as by one nation, for example, can only generate, sooner or later, compensating reactions elsewhere in the system--black markets can develop as a reaction to prohibition, boycotts, or other legally-imposed controls. Many industries can shift to other countries with lower costs if taxes or labor rates become excessive. The occurrence of compensating reactions is a reminder, as will be apparent repeatedly in this report, that an overall systems approach has to be taken with the materials field.

The approach taken by this panel is to start by looking more closely at the physical limits to materials supply and demand and to derive some conclusions as to how technology can best be deployed to alleviate foreseeable difficulties. Technology can be appealed to for increasing the materials supply and reducing the demand and our interest in this report is to attempt an overview of where technology can be expected to help in both of these directions. But as we shall indicate, we believe there are some physical and practical limits which may well, in time, determine the standard of living available to the world's population.

> Recommendation 1. *It is recommended that agencies, departments, and other government bodies concerned with legislation, administration, and regulation affecting these areas ensure at all opportunities that they seek and consider the concerted advice of technologists and social scientists. This is important particularly because of the close interdependence of technological and societal issues relating to natural resources.*

THE MATERIALS CYCLE AND THE MATERIALS-ENERGY ENVIRONMENT SYSTEM

The COSMAT Report (NAS, 1973) has described how all materials move in a total materials cycle (see Figure 1):

> "From the earth and its atmosphere man takes ores, hydrocarbons, wood, oxygen and other substances in crude form and extracts, refines, purifies, and converts them into simple metals, chemicals, and other basic raw materials. He modifies these raw materials to alloys, ceramics, electronic materials, polymers, composites, and other compositions to meet performance requirements; from the modified materials he makes shapes or parts for assembly into products. The product, when its useful life is ended, returns to the earth or atmosphere as waste. Or it may be dismantled to recover basic materials that reenter the cycle.
>
> The materials cycle is a global system whose operation includes strong three-way interactions among materials, the environment, and energy supply and demand. The condition of the environment depends in large degree on how carefully man moves materials through the cycle, at each stage of which impacts occur."

Materials traversing the cycle incur energy costs at each stage. Indeed, the intimate interaction of the materials cycle with energy warrants urgent attention in view of the present reality of energy shortage and the future prospect of materials shortages. As we proceed around the materials cycle we see energy inputs required to win metals from ores and plastics from crude oil, to work and shape metals, ceramics and plastics, to assemble components and systems, to transport goods at all stages of production, and to operate the final product in the hands of the consumer. Then we begin to see energy dissipated as metals corrode or rust, as plastics degrade, as the product is discarded, and as the trash in the sanitary landfill returns to the low-energy "natural" state.

It cannot be emphasized too strongly that because of the intricate web of interconnections in the materials cycle and in the materials-energy-environment system, an event in one part of the cycle or system can have repercussions throughout the cycle or system. Research and development can open new paths around the cycle with concomitant effects on energy and the environment. For example, development of a magnetically-levitated transportation system could increase considerably the demand for the metals that might be used in the necessary superconducting (e.g. niobium) or magnetic (e.g. cobalt) alloys. Widespread use of nuclear power could alter sharply the consumption patterns of fossil fuels and the related pressures on transportation systems.

The materials cycle can also be perturbed by external factors such as legislation. For example, environmental legislation requires extensive recovery of sulfur from fuels and from smelter and stack gases; by the end of the century the tonnage recovered annually could be twice the domestic demand unless, as is most likely, new markets are found for the sulfur as surpluses drive its price down. Again, safety legislation has made it necessary to install stronger doors, crash bumpers, and other equipment on automobiles, thereby increasing their weight sharply. This weight increase, in turn, has caused a sharp increase in the consumption of gasoline, a significant contributing factor to the current gasoline shortage. One response by the automobile designers is to seek ways of saving weight in the car, one obvious path being to develop aluminum blocks to replace the traditional steel engine blocks. But this substitution, in turn, would require a significant increase in plant capacity of the aluminum industries and associated increase in the electric power required to refine the aluminum. And so on, to say nothing of the related changes in employment and other personal as well as economic aspects. The perturbations resulting from legislation clearly can lead to instability in the economy; good legislation should tend to moderate rather than increase instability.

But in addition to recognizing the various repercussions that can result from seemingly straightforward developments in technology or well-meant legislation, it is equally important to recognize that these repercussions take time to propagate. The "incubation period" for a new material, product, or technology, during which an industrial effort is trying to establish for it an above-critical share (e.g. 5 percent or more) of the market, is usually at least 10 years. The subsequent growth period in which the substitution is becoming more and more complete may take another 10 or 20 years or more. It is worth noting that the desirability or otherwise of the new technology

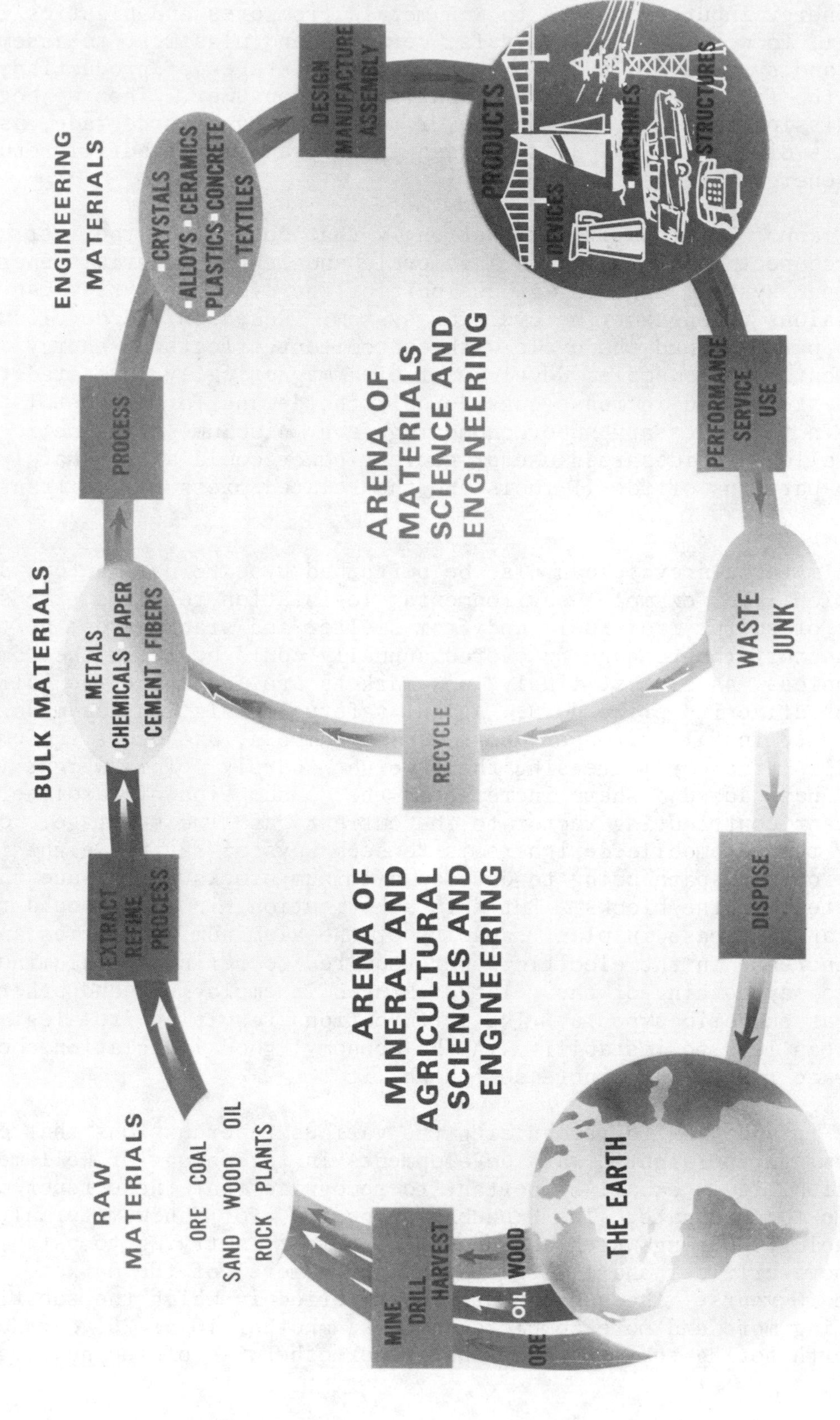
THE TOTAL MATERIALS CYCLE
RAW MATERIALS
ORE COAL
SAND WOOD OIL
ROCK PLANTS
EXTRACT REFINE PROCESS
BULK MATERIALS
METALS
CHEMICALS PAPER
CEMENT FIBERS
PROCESS
ENGINEERING MATERIALS
CRYSTALS
ALLOYS CERAMICS
PLASTICS CONCRETE
TEXTILES
DESIGN MANUFACTURE ASSEMBLY
PRODUCTS
DEVICES
MACHINES
STRUCTURES
PERFORMANCE SERVICE USE
WASTE JUNK
RECYCLE
DISPOSE
THE EARTH
MINE DRILL HARVEST
ORE
OIL
WOOD
ARENA OF MINERAL AND AGRICULTURAL SCIENCES AND ENGINEERING
ARENA OF MATERIALS SCIENCE AND ENGINEERING

will generally show up in this growth period. What is needed is a reliable method of predicting these consequences during the earlier incubation period.

Thus, there is a vital need to develop the methodologies of systems approaches for the materials cycle and the materials-energy-environment system to take into account not only the various levels of repercussion but the time factors as well. Such approaches are starting as witnessed by the econometric studies of "Limits to Growth" and the setting up of a Congressional Office of Technology Assessment, to name but two highly visible indicators. But there is a need for considerable refinement of the tools and techniques employed, including putting the various interactions, both direct and indirect, on as quantitative and objective a basis as possible. The methodologies of technology forecasting and assessment are still in their infancy but they will play an increasingly-important role in otpimizing man's use of the materials-energy-environment system. But always it must be realized that technology can do much but not everything, and what it can do takes time. Hence, not only forecasting forward but the ability to "forecast" backwards in time from future objectives is necessary in order to decide what must be done now to meet man's material needs 20 years and more from now.

Experiences with wartime shortages have shown conclusively that governments under stress can make things happen sooner. But the social costs of living as if in a state of war are not to be invited lightly. Institutions to foresee and prepare in advance for necessary technological changes and adaptations on an easy, voluntary basis seem much more attractive.

> Recommendation 2. *It is recommended that relevant governmental bodies and industrial organizations should use a systems approach and analysis to ensure that their actions are acceptable in the context of the materials cycle and the materials-energy-environment system, as judged by contemporary standards. Further, better methodologies need to be developed for forecasting over a suitable time scale the effects of technological, administrative, legislative and regulatory actions. These matters are important because actions taken in any part of the materials cycle or the materials-energy-environment system can have repercussions at other parts of the cycle or system.*

RAW MATERIALS AND INTERDEPENDENCE

Overall, any country must maintain a balance of payments if the international value of its currency is to remain steady. Short-term fluctuations in the balance of payments can be tolerated but in the long term a close balance must hold; international trade is the major factor in determining the payment balance. The principal ingredients of international trade are: food and agricultural products; materials, including energy materials; technological products, military and civilian; and services. It is not necessary, of course, for there to be a trade balance in each separate category. Countries such as Great Britain export technological products and services to pay for the needed imports of food and raw materials. The Arab oil countries, until recently, have maintained a lifestyle that needed minimal imports to support it and income from oil exports often went to purchasing military hardware. Some countries are so devoid of any known assets for export that they remain the world's poorest nations, requiring foreign aid for mere subsistence.

Fortunately for the United States, at present it is rich in every trade category except several important raw materials. Thus it must export food, technological products, and services to pay for the raw materials to maintain its standard of living. How long the world will accept uneven distributions of resources among nations and what adjustments, some of which are already happening, in international relationships and mechanisms will take place remains to be seen, but these issues are outside the scope of this report.

Supplies of materials for the United States are likely to be forthcoming from other countries as long as these other countries require goods and services from the United States. As we have seen recently, however, this situation may not always obtain. The OPEC countries felt sufficiently free from dependence on U.S. exports, at least for the short term, to dare to embargo oil exports to the United States. The effect of this action has led this country to realize that it cannot afford to be critically dependent on outside sources of such a vital material as oil and as a result, ambitious programs are being embarked upon to achieve an increased level of self-sufficiency in energy materials by 1980. Naturally, this event has triggered similar worries about the Nation's dependence on other countries for other materials and the question has been raised whether the United States should aim for a higher degree of self-sufficiency in all materials. This is an historic turnaround: many of the raw material-producing countries are poor and have hitherto very much depended on the rich, advanced countries for improvements in their standard of living; but now the rich countries are seen also to be dependent on the poor to avoid lowering their own standards of living.

Again it is relevant to witness what is happening in the Arab oil countries. They are now beginning to look for ways to spend the rapidly accumulating wealth resulting from their strong bargaining position for the price of oil. While a good part may still end up as personal wealth (much even invested in the industrialized nations) there are signs that the oil countries would like to acquire a bigger share of the oil and other industrial "action." Why should profits from shipping the crude oil go to foreign companies? Already some Arab countries are investigating the possible purchase of a tanker fleet. And why ship crude oil to be refined elsewhere? Why not erect refineries at home and thereby keep the refinery profits as well? And why not use some of the profits to build up other industries--as Iran is doing for steel? And so it goes on.

At the base of the oil embargo there lies another fact--that the "enormous" Middle East reserves, at present and prospective rates of withdrawal will be used up in perhaps 25-30 years. The Arab countries are entitled to ask themselves (and us) what kind of economy and culture they will have achieved by the time this transient bounty runs out. Will they subside into a poor, thinly-populated pastoral and nomadic state or will they be able to convert this time-limited resource into the basis of a more modern and sustainable, technologically-supported economy?

We can envisage similar situations developing in a number of mineral industries as well. For instance, as a group the bauxite-producing nations are asking for capital to install reduction plants, rolling mills, and extrusion facilities to enable them to ship not low-value ore but high-value materials. The trend is logical, even necessary, as a stage in the development process. As the chief instigator of this process the United States should logically accept its implications. But it poses for this country the alternatives of a lowered domestic standard of living or an expansion of exports in other areas, such as agricultural products, to pay the heightened costs of the imports it needs. This same trend is evident in the move from fire-refined copper to wire bars, from chromite to ferrochrome, from cobalt concentrates to balls and rondels.

As the raw material countries acquire wealth they will build up their industrial base and in this sense, as more and more of the materials "action" builds up overseas the United States becomes more and more dependent on other countries not only for raw materials but for refined products as well. (This trend underscores the increasingly-important role of multinational corporations which by and large will distribute their activities among the countries in which they operate so as to optimize the whole operation.) Since this process raises the standard of living in the poorer countries it can perhaps be viewed sympathetically. However, as these countries industrialize and develop they, in turn, become more and more dependent on other countries in order to sustain their technology and standard of living. As they "level-up" they will become more dependent on the advanced countries for more sophisticated technology and, like everyone else, on the whole range of countries that supply the raw materials for *their* technology. Thus the world trend is towards *interdependence*, though along the way there will be frequent fluctuations of the Arab oil embargo sort. We therefore believe that a long-range objective of overall materials independence or self-sufficiency for the United States is not only unwarranted, it is not even feasible. This is not to say, however, that precautions should not be taken against *temporary* difficulties caused by transient cessations of exports of certain raw materials.

There are various situations which can result in the supply of raw materials to the United States being vulnerable to the policies and practices of other countries. These include:

1. where most of the world's resources are held by a single country, e.g. tungsten (China), mercury (Spain), palladium (Soviet Union).

2. where most of the world's resources are held by a very few countries of similar political persuasion, e.g. chromium (Rhodesia and S. Africa).

3. where the transport routes between sources of supply and the United States are particularly vulnerable to hostile action, such as when the source countries are geographically close together and remote from this country (e.g. tin from Southeast Asia).

4. where various countries, even with widely-varying ideologies, can group together with the common interest of raising the price of the commodity they share (e.g. the bauxite-producing countries).

5. where substitute materials cannot readily be found or developed (e.g. as for iron, chromium, and manganese), or where nations that are sources of the prime commodity and nations which are sources of the most likely substitutes (e.g. the copper and aluminum nations) join together to manipulate the price of both commodities.

For the industrialized nations, the problems of future international trade in materials may be focused not so much on their controlling the rate at which they import materials from the raw material countries but with inducing relaxation of export controls that the latter place on their commodities. In the light of these vulnerabilities it seems prudent for the United States to pursue a policy of stockpiling--

a. "critical" materials--i.e. highly essential and highly vulnerable; and

b. technological capability to develop alternative sources.

The stockpiling of materials of strategic importance to the military has been practiced since World War II. The object has been to assure the United States adequate supplies to feed the U.S. armament requirements for many months, and in some cases three to five years, in the event that this country was cut off from its normal sources of supply. Clearly, similar considerations can also be applied to protection of the U.S. economy and the essential needs of the civilian sector. Materials stockpiles can smooth out the jolts caused by interruptions in supplies from overseas. The size of a stockpile should be such as to keep a significant fraction of U.S. needs satisfied during the period needed for alternative sources of supply, domestic or foreign, to be brought on stream, or until other adaptations can be devised and implemented.

Adjustments to supply difficulties might be made by rearrangements of trading patterns, albeit at some cost because, presumably, the original source was the cheapest. However, for those materials where rearranging trade patterns is less feasible, the United States should undertake research programs to establish a <u>capability</u> of supplying its needs, as far as possible, from domestic resources. This is not to advocate indiscriminate funding of materials research in the belief (or hope) that something useful will eventually come of it. Instead, there should be a deliberate selection of research programs aimed at lessening U.S. vulnerability to overseas suppliers. These include:

a. mining and extraction techniques that will raise the efficiency of processes for obtaining vital materials from lower grade domestic deposits (e.g. aluminum-containing ores);

b. developing substitute materials and exploring their effects on product designs and performance (e.g. silicon nitride ceramics for automobile engines).

Such research need not yield economic processes or products in the contemporary economies; the important fact is that the United States would have a technology available which could be scaled up and brought on stream if normal supplies were in jeopardy for an indefinite period.

Recommendation 3. *Because of the world-wide implications of many actions affecting natural and environmental resources, and because of the inevitable trend towards interdependence between nations, it is recommended that the Federal Government should make strong and continuing efforts to achieve equitable international agreements concerning the exploitation and sharing of these resources.*

The State Department is the lead agency in this area, with the Department of Commerce also playing an important role. Besides pursuing bilateral and multinational agreements there is a particular need for the Federal Government to work through the United Nations and to strengthen the latter's participation in regulating the consumption of the world's resources.

Recommendation 4. *Because of the vulnerability of the USA to shortages of certain materials necessary to sustain minimum standards in the civilian as well as the military sectors, it is recommended that a deliverate policy should be pursued of: (a) stockpiling critical materials in sufficient quantity to provide time until developments are made that will overcome an import shortage by means of domestic production, alternate sources, substitutes, development of recycling systems, and other expedients, and (b) undertaking research programs to establish a technical capability, which could be relatively rapidly scaled up when necessary, for supplying U.S. needs, as far as possible, from domestic resources.*

This recommendation is particularly germane to the Departments of Commerce and the Interior.

CONSERVATION ETHIC

Whenever the balance between supply and demand is threatened, whatever the reason, whether it be with materials of domestic or foreign origin, society can respond by endeavouring to increase the supply, reduce the demand, or both. Hitherto the pressure has been generally placed on the supply sector to meet demand but this alternative threatens to become less rewarding. World competition for available supplies and exploitation of the seller's market by producing nations makes it chancy. Moreover, increasing the supply of material available ususally calls for increased capital investment in mining and transport equipment, the mobilization and training of additional manpower, investment in additional pollution control equipment, and additional consumption of energy. All of these requirements may impose severe constraints and in addition there is the uncertain but perhaps increasing force of societal value judgments together with moral judgments on man's right to deny materials to later generations.

Reliance of the United States on materials from less fortunate producing nations is threatened in several ways: by unwillingness of such nations to permit continued disparity of living standards by supplying a rich country with the means to its continued enrichment, by the competition from other developed and consuming nations for these materials from the poor countries, thereby creating a "seller's market," and unwillingness to forfeit future richness to, in many cases, meagre present gain. Why should we, the Australians plausibly ask, ship our uranium to the United States for dollars today when we can use it in the centuries ahead to energize the development of our own "Outback"?

As a result of all these pressures, increasing emphasis is being placed on reducing demand to preserve the supply-demand equilibrium. At present we are in a transitional phase in which society's reaction to material shortages is to try both to increase supply and to reduce demand for new material at the same time; reducing demand through various conservation measures may well be the primary path of the future. We see a conservation ethic which urges throwing as little away as possible, which implies squeezing every bit of use possible out of energy and the less abundant materials to reduce demands on primary resources. Clearly, the transition from an ethic of material growth to one of material conservation will have significant repercussions throughout the nation's economy and welfare; this argues again for a systems approach being taken in the management of the nation's affairs at the highest levels of government.

Every operation in the materials cycle uses energy. Every operation must, therefore, be made more energy-efficient--mining, beneficiation, extraction, primary metal production, manufacturing methods, product use, and product disposal or recycling.

Material scrap and wastes from material operations must also be minimized at every stage of the materials cycle and any scrap or waste that is generated at any stage must be recycled or reused as much as possible, or secondary uses found for it. Recovery levels from ores must be maximized and, whenever possible, uses found for the overburden and tailings that result from mining and smelting operations. Often these wastes can first be reworked to recover residual metal values as leaner ores become more economic. Scrap from manufacturing operations must be fed back into the materials cycle. Maximum recovery from used products must be practiced. Secondary uses must be found for materials degraded in recycling processes--in fact, materials may have to be put through a whole cascade of successively less demanding applications until they are good only for fuel or landfill.

Products must be designed so as to minimize their use of the scarcer materials. Tactics include miniaturization; designing to closer tolerances; simplification and standardization; the use of more abundant materials in place of scarcer ones and of renewable ones instead of nonrenewable; designing for flexibility in material selection through less demanding performance specifications; designing for minimum dispersal of materials, for longer service life, ease of maintenance and repair, to facilitate dismantling and reclaiming materials, to minimize materials wastage in manufacturing, to make use of reclaimed materials; and the employment of composite materials to perform total functions.

To sum up, all the aforementioned tactics for materials and energy conservation need to be pursued but perhaps the most important themes are:

--reducing energy consumption in materials operations and by-products,

--increasing the durability and maintainability of products,

--development of substitute materials, and

--reclamation and recycling.

The overall aim is to maximize the effectiveness of materials and energy usage. But while there is considerable room for improvement of current practices, there are limits on the extent to which benefits can be obtained from technology.

> Recommendation 5. *Because of limits to natural resources as well as to means for alleviating these limits it is recommended that the Federal Government proclaim and deliberately pursue a national policy of conservation of material, energy and environmental resources, informing the public and the private sectors fully about the needs and techniques for reducing energy consumption, the development of substitute materials, increasing the durability and maintainability of products, and reclamation and recycling.*

INDUSTRIAL SYMBIOSIS

We have mentioned the increasing necessity to take a systems approach to the materials cycle and to the secondary cycles associated with it. There is need for better coordination between the materials supply and materials demand sectors of the economy, especially with the aim of minimizing materials wastage, energy consumption, and environmental pollution. Inasmuch as each of these three categories incur real costs (internalizing the externalities) it might be expected that most industries will tend to move in the desired directions. But the materials cycle is often quite highly divided among a number of industries--e.g. mining and smelting, manufacturing, service, and scrap metal reclamation. Each of these separate industrial segments of the cycle may be individually optimized but this does not necessarily imply that the whole cycle operation is optimal. Systems people have long recognized the dysfunctional consequences of optimizing a subsystem (suboptimization). The dysfunctional consequences of suboptimization are evident in many kinds of systems: political, administrative, educational, medical, transport, utility, economic, etc. Adam Smith in his "Wealth of Nations" hypothesized that the sum of all suboptimizations (for "profit") of individual entrepreneurs resulted in optimization of the total national economy. Today we are just beginning to suspect that the Smith hypothesis is invalid. But we have not yet devised a way of encouraging entrepreneurs "to behave properly" like subsystems to optimize the operation of the total economic system.

A manufacturing industry aiming at producing products at lowest costs may pay little attention to the maintenance costs and recycling costs of its products if these are the concerns of other organizations. This diffused effort illustrates a gap in our national materials system that generates perhaps the most pressing need for a systems approach--the need to design products to minimize total life costs, not just the manufacturing costs. The feasibility of closing this gap by a more systematic approach increases the more the total life cycle of a product is the responsibility of a single organization or a few, closely collaborating organizations.

Another way of expressing the theme is that sometimes the waste generated by one organization can be a valuable input of another, so that by establishing close physical, economic, and managerial relationships between the two industries the costs of both can be reduced. (Here waste is used in a very general sense--it includes industrial waste and effluents from processes and manufacturing, scrap equipment, products and junk, generally, and even municipal waste.) Such collaborations between or among industries often raise the spectre of price-fixing and other monopolistic practices but antitrust laws and regulatory measures must be administered in ways that ensure that the public's best interests are served. It is of paramount importance that any monitoring body be fully cognizant of, and pay due attention to, the advantages of economies of scale and systems approaches for conserving natural resources. The more the materials cycle and major segments of it are fragmented among various industries, the less successful they are likely to be in providing at lowest possible costs the products and services that society needs, and also in minimizing the material, energy, and environmental costs. Fragmented industries may offer the consumer more choice but the price can well be greater demands on resources as well as higher-priced products.

> Recommendation 6. *It is recommended that the Federal Government seek ways to facilitate cooperation between fragmented industries, industries at different parts of the materials cycle, and institutions managing wastes and garbage, to maximize the effective utilization of materials and minimize harm to the environment while assuring that there is no disservice to the public interest from such industrial concentration.*

CAPITAL CONSTRAINTS

The time factors alluded to earlier can in large part be related to the problems of raising capital, training manpower, and the momentum and institutional infra-structure associated with the older product or process that is being displaced. Technological breakthroughs at the research stage are always exciting and conjure up visions of new products and new dimensions to our way of life. But the scaling up from the research into the development and engineering stages in order to arrive at a profitable commercial base usually takes a lot of capital in addition to engineering skill. This is especially true in the basic materials industries, often described as capital-intensive. New basic oxygen furnaces, drilling rigs, smelters, represent enormous investments which can control the financial conditions of materials industries. The risks are high; the consequences of failure, dire.

Yet faced with potential materials shortages these industries must find new investment money. For example, the suggested goal of tripling coal production by 1980 implies at least approximately tripling the investment in coal-mining machinery unless there is a dramatic improvement in productivity caused by the development of new techniques. Or again, just to keep a steady output of certain materials will call for increased mining and milling capacity as we have to turn to leaner and leaner ores.

Capital is needed not only for expansion of plant capacity but for modernization and pollution control equipment. Modernization is necessary for industries to be competitive with foreign producers who, in many of the materials industries have installed or built many modern plants since World War II. Pollution control equipment is usually, though not necessarily, a nonproductive capital burden to industries which tends to make industries less competitive with those in other countries where such equipment may not yet be required. However, most advanced industrial nations are experiencing rather similar pollution problems to those of the U.S. and are having to tackle them in equivalent ways: forces are thus at work tending to blunt the comparative advantages that would otherwise obtain. Overall, spending money on pollution control seems preferable to spending money, in some way or other, for the privilege of polluting, even though the latter charges, if high enough, encourage expenditures for the former.

The capital required for expansion, modernization, and pollution control has to come from somewhere but it must be recognized that there is very little slack in the capital market. Other industrial sectors are making their demands too: energy (much of it relevant to materials technology), transport, communications, schools, and so on. Some capital might come from that share of the GNP that lies in government sectors and in personal consumption, but the traditional remedy is to counteract these pressures as much as possible by increasing productivity, and increasing it at a rate greater than or equal to the rate of increase of wages. But increasing productivity implying (as it does for a given industry size) increasing production runs counter to the uncertainty of materials supplies. Increasing production (and consumption) is hardly the overall answer for the future. The traditional workings of a growth economy seem incompatible with a society that is moving out of a growth phase in the material sense and into a conservation era, from an era characterized by rising exponential functions to one better described by the S-shaped sigmoid curve or even a bell-shaped curve. In fact, the proportion of the GNP going into consumption, federal and personal, may well have to decrease in order to release investment money to sustain the plant capacity that will be needed to create the materials and products needed to maintain an acceptable standard of living.

In this scenario we conclude that new approaches to financing materials industries will have to be found. As Robert C. Holland has said (his remarks were directed at the energy sector but they apply equally to the materials industries): "Because of the large amounts of money that will be required, even major firms may find that they individually lack adequate financial strength to support all needed debt financing. Under these circumstances it is possible that multicompany guarantees may be required to permit the necessary funds to be raised. Such guarantee networks will be soundest if the guaranteeing firms are themselves in the direct line from energy production to energy consumption, and are sustained in their guaranteeing ability by their own, direct market knowledge and customer contacts. The pressures to provide these and other credit arrangements to meet energy and other needs will probably test the flexibility and responsiveness of our financial system as it has seldom been tested before." (Holland, 1973) Government advice and consent will be necessary in most multicompany arrangements but what is needed is a government-industry collaboration in the public interest, in spirit and fact, not an adversary situation. There is also, probably, a general preference for avoiding any direct management of industry by the government.

Holland's remarks are echoed in a study (Little, 1973) of barriers to innovation in industry that was carried out recently. While surveying a broader cross-section of industry than the materials industries alone, the survey found that financial barriers that were rather universally regarded as inhibiting were: (a) lack of funds, and (b) uncertainty about level of investment, i.e. financial risk too great, lack of critical information needed for decision. The report also notes that "the most significant single observation developed from (the study) is that the crucial bottleneck in the innovative venture-development process is the shortage of seed capital to support the earliest stages of venture formulation and start-up."

The report concludes that in view of the high start-up costs of high technology innovation, particularly in capital-intensive industries, or if current public policies (fiscal) are not sufficient incentives for innovations targeted in the public interest, the feasibility could be explored of fiscal incentives that satisfy the following criteria: Target on main barriers of startup costs; reimburse costs and risks incurred (in the past) rather than subsidize (current or) future costs; and maintain fair competition. As a footnote, there is a need for the public to be better informed about capital constraints, the effects of limited earnings, and that choices have to be made, directly or indirectly, as to how the finite amount of capital will be allocated among competing technico-social areas. If out of the new capital available each year for investing, the proportion going to the energy sector must rise from 22 percent (as it has been) to around 35 percent through 1985 (as some have estimated) this means that the proportion of the total capital available for investing in all the other goods and services that society wants--health, education, communication, transport, housing, and so on--must fall from 78 percent to 65 percent.

Recommendation 7. *Because the prime roadblock to technology transfer from the research phase to commercialization in the materials industries is the capital (seed money) needed for prototype development of product or process, it is recommended that the Federal Government find ways to share the burden of such developments with the private sector in areas of high risk or delayed pay-off.*

MANPOWER CONSTRAINTS

As we have noted, even to sustain society's present standard of living will require an increase of activity in materials technology, and one of the constraints on the rate at which the level of materials technology can be increased is the availability of capital. But another is manpower, both brawn and brain.

Except to the extent that productivity can be increased the pressures for materials will require increasing the manpower employed in operating the materials system. Increasing the manpower in mining and ore processing, recycling and other bulk material operations implies reducing, at least, the percentage of the labor force in other industrial and/or service sectors, an event that does not happen overnight. It takes time to train and retrain workers, though perhaps much can be done through company-sponsored training courses at junior colleges, for example. Such investment would be good both for the people involved and for the economy. It takes time for other sectors to become less manpower-intensive, or to deploy their manpower in the different ways needed to implement conservation rather than consumption.

Perhaps more importantly, it takes time for societal and personal attitudes and customs to adjust to different priorities and emphases. The process can be influenced somewhat by financial incentives but this again runs up against the same inelasticities noted in the previous section--most industrial and societal sectors are competing for capital and operating funds, reflecting societal pressures for energy, transportation, education, etc. To a first approximation then, it is prudent for the basic materials industries in the United States to assume that though they will have to engage in more and more activity in order to meet society's needs they will have to do so without an equivalent increase in manpower.

The traditional answer lies, of course, in increasing productivity and this, in turn, often implies developing new or improved technology, particularly automation. Unfortunately, the United States has allowed itself to become woefully weak in basic materials technology. Since World War II, if not earlier, the emphasis in the United States has been on the so-called, science-intensive, high technologies such as aerospace, nuclear energy, and electronics. Mining, extraction, transport engineering, and other "older" technological areas have been sadly neglected. Only about 100 B.S. level graduates in mineral and metal processing are produced in the United States per year. This contrasts with the numbers of chemists (11,500), physicists (5,282), and engineers (44,790) produced in 1971-72. The total senior class in metallurgical engineering in the U.S. and Canadian universities has fluctuated around 800 ± 100 over the last fifteen years. In a recent report (Dresher, 1974) it is concluded that at "current production rates of B.S. engineers, the (metallurgical engineering) profession probably will not be able to fully replace those in the profession who are lost due to death and retirement in the next 10 to 15 years. The metallurgical engineering profession has not kept pace with other physical science-based professions in its growth during the last decade." Furthermore, within the category of metallurgical engineering, the proportion of graduates in the mining and extractive branches has fallen relative to the proportion in other branches of metallurgy. There is clearly an urgent need to modify this distribution of educational effort, a process which again will take a good many years. In the meantime, much might be done to attract scientists and engineers from other disciplines into the basic materials activities which hitherto have suffered by not projecting an image of variety of opportunity in contrast to the science-intensive industries. In fact, the fresh and novel ideas and views that could be brought into the basic materials field by individuals trained in other disciplines might well lead to impressive new advances. The problems are no less intellectually challenging than those of the high technologies of the last quarter century. The need is to turn the hitherto low technology areas into high ones, a process that could be much hastened by engaging the interests and enthusiasms of those engaged in a wide range of physical and engineering sciences. Automation in particular offers novel challenges to the talents and skills of the engineering and information communities.

Recommendation 8. *Because for the foreseeable future there will be growing pressure on basic materials industries to meet demand and because the manpower in these industries will consequently have to be increased, it is recommended that the Federal Government: (1) explore ways to facilitate the training and re-training of manpower including those from other industries; (2) emphasize support in the schools and universities of education and research pertinent to the basic materials industries.*

Recommendation 9. *It is recommended that an intensive program be undertaken to develop suitable automation for the basic materials industries in view of the urgent need to increase their productivity. Such programs may call particularly for increased cooperation between the basic materials industries and the computer and other science-intensive industries.*

PHYSICAL AND TECHNICAL CONSTRAINTS

Wise use of technology will certainly ameliorate materials shortages, but there are limits: limits to the resources we can capture, limits to the possibility of indefinite recycling, limits to the developments of substitutes. The issues are not simple, and often excite discussion unworthy of the topic. Public expectations of research and development have often been over-optimistic (R & D oversold) for the short-term but over-pessimistic for the long term (through lack of imagination).

First, are there upper limits to the amount of material that can be made available? At present rates of consumption the world will be facing within a few decades very severe shortages of certain materials unless accessible new deposits of adequate quality are discovered. Some such discoveries will undoubtedly be made from time to time but notwithstanding, shortages will appear sooner or later. To offset these shortages technology must be developed to increase recovery efficiencies from ores, to make the treatment of leaner ores profitable, and to mine less and less accessible ores. But recovery efficiencies are already quite high--often higher than 80%--so there is not much room for improvement there. This means that greater emphasis will have to be placed on working with leaner ores and less accessible ores. But the leaner the ore the more useless rock has to be handled and broken up to win the desired material, an activity that requires (in general) more energy and produces more environmental degradation.

This trend raises the question of whether energy can be used in even-larger amounts to accomplish such tasks. The question concerns first, how much energy can be provided for this use, and second, whether there are limits to its use.

Regarding the first, there are in principle enormous reserves of energy materials in the world. While difficulties of the sort generated by the Arab oil embargo can be expected from time to time, causing temporary shortages, the reserves of oil shale, coal and tar sands are enough to give grounds for thinking that once technology comes on stream there will be no energy shortage for hundreds of years. Furthermore, there are considerable resources for nuclear fission power and if nuclear fission power becomes a reality the energy resources could become almost inexhaustible. In addition there is the prospect of tapping the significant inputs of solar energy and lesser energy sources, such as geothermal, as well. Abundant energy does indeed seem attainable in due course.

But the second part of the question posts a warning sign....

Whatever the way in which fuel materials are converted into energy this energy eventually ends up as waste heat. It is easy to estimate that present total energy use on earth amounts to about 1/25,000 of what falls onto the earth continuously from the sun and that this energy generation would have to be increased something like a hundred-fold before the earth warmed up enough, by waste heat alone, to constitute a _global_ environmental danger. That will not happen with fossil fuels, because if facilities were built sufficient to generate heat at this rate, the fuel would last only a few years--say 10 or 20 at most; no civilization would embark on a course of that sort. It _is_ possible to get into such "heat death" with the relatively "inexhaustible" sources, nuclear (fission or fusion) and solar power. This latter may appear as a surprise, but it should not be so: one tints an area of the earth dark to make solar collectors, whence more heat is absorbed on earth instead of being reflected.

The cure for the "heat death" problem appears as a corollary to the previous sentence: one can adjust the reflectivity of other parts of the earth's surface, as a counter to heat production via solar, nuclear or any other resource. However, the idea is aesthetically unattractive, and a civilization that exists by such application of _force majeure_ is probably in more serious trouble from other consequences of indiscriminate energy use.

The most obvious difficulty arises from the fact that the heat generation is usually concentrated in relatively small areas, and usually in the same general areas where people live. Thus local and regional, not global, effects will dominate, as they already do in some areas: on the average, the atmospheric temperatures of cities are 1° to 2°C above those of surrounding rural areas, but in some instances the temperature difference may reach 10°C.

More subtle and probably more urgent difficulties arise not from the crude production of heat, but from other aspects of energy production. Burning fossil fuels has increased the levels of atmospheric carbon dioxide globally, and of sulfur oxides and nitrogen oxides over continental areas. The effects of these atmospheric changes on the earth's radiation reflectivity and emissivity and on the earth's climate are not well known. Other effects come from what the energy is used for--in industry, transport, heating and air conditioning, and more--which pollutes the atmosphere, alters surface conditions, and so forth. This increasing use of energy, even if it never polluted _per se._, would have to be accomplished by increasingly strict controls over its use.

Then there are other man-made pollutants which hardly exist in nature or not at all. These include hydrocarbons, various particulate pollutants, and pesticides and herbicides. The subtle and not-so-subtle effects that these can have on man's environment are only beginning to be appreciated. Many of these pollutants may well accumulate to disturb the natural balance rather than be assimilated harmlessly into natural cycles.

We thus have two classes of pollutants: (a) those which take part in natural cycles and can be assimilated if their levels are trivial compared with the background levels of the natural cycles and (b) those which do not take part in natural cycles and therefore accumulate injuriously. For the first type, technology must ensure that man-made levels do not amount to more than an extremely small fraction (though depending on circumstances), of natural levels. For the second type the only answer, other than finding ways to do without the materials in the first place, seems to be for technology to devise ways of containing the pollutants, completely preventing their escape into the biosphere. And for both classes of pollutant, as is being increasingly recognized, action cannot be delayed.

Pollution of a different sort stems from piles of industrial and municipal junk and wastes. Effective reclamation and recycling of wastes serves two major purposes. It not only ameliorates the visual pollution but it also can make a major contribution to conserving materials and energy. The closed cycling of materials is one of today's most urgent needs and it should be an omnipresent attitude on the parts of product designers and industrial and municipal operators to use every feasible technique to prolong the life of materials through as many use cycles as possible.

It is impossible, of course, ever to get 100 percent efficient recycling. Some material will always be irretrievably lost on each cycle but the smaller this proportion the smaller the demand that is placed on prime sources of raw materials and on storage space for discarded wastes. A recycling efficiency of 90 percent though rarely attainable, would mean approximately a tenfold reduction in the rate at which fresh material would need to be mined, not counting any additional material needed for growth. The scarcer the material the higher the need for efficient recycling and all usages and management operations that tend to make salvage operations difficult should be kept to an absolute minimum. It is important to remember that over a period in which materials consumption rate is rising, the proportion of new needs that can be met by recycled materials can fall far short of 100 percent. This is because the total amount of material available to be recycled, even with 100 percent collection efficiency, can only be that which was produced an average product-lifetime earlier. Thus for cars, the absolute maximum number of cars that could be recycled today is approximately the number of cars that was produced 10 years ago, a number which might be quite a bit smaller than today's demand level. Thus, while it is essential to make every effort to make the efficiency of utilization of materials approach as closely as possible to 100 percent, recycling and related tactics such as cascaded uses can never be a nearly complete solution to materials shortages until the consumption rate becomes constant.

Faced with shortages of certain materials some optimists would have us believe that technology can always come up with a substitute. This is not necessarily so. For some applications substitutes may be found quite readily but the more technically complex the application or equipment the material is used in, the more difficult it becomes to find a substitute, perhaps impossible. Products like nuclear reactors, jet engines, and integrated circuits are systems of highly interdependent materials, each carefully adapted to its role in the total structure. The unavailability of a particular material may require a complete and prohibitively costly redesign of the equipment to accept another material with slightly different properties, even if such redesign is technically possible. Palladium is about the only material that has been found to give adequate performance in electrical relay contacts in the telephone system. Other metals which could be substituted for palladium would so drastically degrade contact performance that telephone communications would become much less reliable. In this case completely new telephone systems might have to be built around completely different and perhaps more expensive technology, such as solid state switches.

Substitutions do and will play an important role in relatively simple applications, such as construction materials, household utensils, and fabrics. Substitutes are generally much harder to come by, if not impossible, when the materials are used in an enormous number of different applications (e.g. chromium), or when they are used in such vast quantity that no substitute can be found that has sufficient abundance and is adequately capitalized. Also, when materials are inadequately available for making the more complex mechanical and electrical equipment that man has come to regard as essential for his way of life, it may often be necessary to find alternative ways of performing the desired functions.

Man's ingenuity, stirred by circumstances and nourished by deep knowledge and understanding of materials, will probably rise to these occasions for substitutions for a long time to come. The area of functional substitution will perhaps be the prime determinant of what man's future way of life will be like, but it is an area where hard facts are particularly scarce. One aspect that is frequently overlooked is that while substitution may allow the function to be preserved it usually does so at increased cost (since otherwise, the "substitute" would have been the preferred course to begin with). But increasing the cost means that fewer users will be able to afford it. Thus, to those at the lower end of the economic scale, substitution does not mean that the function will still be available to them. There is clearly a need to sharpen our methodologies for predicting future technology, particularly so that its potential effect on the materials-energy-environment system can be assessed.

As man's technological environment gets more and more sophisticated and complex another technical constraint may begin to set in--reliability. As any homeowner knows, the more "labor-saving" devices he owns the more time and effort he has to devote to maintenance, repair, and replacement. So too with industrial and commercial equipment. Unfortunately, hardly anything can be made absolutely trouble-free, especially things like automobiles, stereo sets, computers, aircraft, and so on. The wonder is that they perform as reliably as they do, though if consumers were to demand greater durability and reliability and were prepared to pay for it there could undoubtedly be fewer defects and improved performance.

Much can be done to reduce causes of trouble to acceptable levels but the fact that failures will occur may present a practical limit to how much technology it makes sense to have. For example, integrated circuits for critical applications can nowadays be made with failure rates of around 10 FIT in the medium term, where the unit, 1 FIT, represents a failure in one billion operating hours, or about 10,000 years, accumulated by a sufficiently large batch of test samples. At first sight this might seem to be overdoing reliability but it must be remembered that complex systems, such as computers, control systems for anti-ballistic missiles, nationwide communication networks, can require the reliable operation simultaneously of many thousands of integrated circuits. A figure of 10 FIT means that, statistically, a system consisting of a thousand integrated circuits will exhibit a system failure rate of about once a year. It is not obvious that much improvement can be made in this figure; we may be reaching a practical limit to reliability in integrated circuits. If so, this circumstance will tend to set an upper limit to the practical size of a complete system. To get around this limit, we may need to resort to alternative, redundant, or backup systems, as with blind landing control equipment for aircraft.

Another potential limit to the diversity and complexity of equipment, besides physical reliability, is the technical training required by those whose job it is to repair the equipment when it does fail. For most people, automobiles and television sets already exceed the point at which simple repairs can be carried out at home. Nowadays the services of specialists are more likely to be required though some relief can be provided by using modular construction. And so it is with industrial and commercial equipment. It is hard to quantify these effects of technological complexity but they are real and may well pose some practical limits to the man-made environment.

In this section we have assembled a rather gloomy catalog of limits to what can be achieved by technical means to alleviate the threatened pressures on materials, energy and the environment. The theses that technology is infinitely improvable and that substitution is infinitely possible, we feel, are highly suspect. They can be accepted only with considerable qualifications. So what can be done? Is there a brighter side? We believe there is. That by focusing attention on the difficulties and liabilities man tends to overlook the opportunities and assets that are available to him.

First, energy. There is perhaps scope for considerable increase yet in the rate at which the <u>world</u> consumes energy (totalling all its forms). However, the United States has to face up to the question of whether it has used up its "share" of energy consumption. To go on increasing could lead quite soon to unacceptable environmental disturbance as well as international recriminations. Instead, the opportunities for the United States have to be sought in using energy more efficiently. Fortunately, there is still much scope here. Through ingenious engineering and the application of scientific knowledge we can perhaps achieve a several-fold increase in the effectiveness of our nation-wide usage of energy.

As for materials, while we may experience shortages of various metals and certain other substances we will never run out of materials _per se_. There is a limitless supply of refractory oxide materials and ways should be found of using these in a greater variety of applications as substitutes for scarcer materials. Then there are the forests and other vegetation which represent a vast renewable resource for organic materials; these, with prudent husbandry, could sustain a steady supply of useful materials at acceptalbe levels of ecological impact, including avoiding depleting the amount of organic carbon in the biosphere.

Finally, there is information technology--the technologies of communications, computers, and control based on solid state electronics. Information technology is not materials- or energy-intensive and so is not likely to run into these limitations though it may well experience other limitations, such as capital requirements and reliability. Some specialty metals may pose supply problems in the future but overall, information technology, in its broadest sense, may well ease the pressures on materials, energy and the environment. Functional substitutions such as facsimile transmission instead of mail systems, video communications instead of personal travel, and computer, control and automation generally instead of human operators are all expedients by which we can reduce the drain on energy, materials, and the environment. Information technology has made such rapid advances in recent years that already we can foresee being able to provide society with a vast smorgasbord of services, from remote health checkups to bank and shipping transactions, from traffic monitoring and control to robotic mining machinery. The exploitation of such wizardry will require wise selection and guidance and knowledgeable acceptance by society.

Recommendation 10. *Because the biosphere is facing severe strains from the effects of the global levels of energy consumption and pollutant emission, intensified research is urgently needed on the effects of waste heat and pollutants on climates and ecosystems.*

Recommendation 11. *Because of the close relationships between materials production, the energy consumed, and the wastes and pollutants caused thereby, it is recommended that cognizant federal bodies, develop, publish, and keep updated data giving the energy and pollution consequences of recovering values from listed reserves and resources.*

TIME CONSTRAINTS

The time taken for new scientific knowledge to reach full commercial implementation is frequently underestimated. As Pry has noted: (See Appendix of Source Materials)

"Three distinct stages are apparent in the birth, growth, and diffusion of a science-technology area. First, the stage from scientific discovery of a new physical phenomenon or a new theoretical concept to the broad acceptance of its applicability to technology. This gestation period involves a broadening of the concept, probing the depth and breadth of the experimental phenomenon, and, subsequently, assimilating the findings into a common technical-knowledge base. The time period for this process varies greatly, has been shortening considerably over the last century, but still may take 40 to 50 years to complete (e.g. electromagnetism in the Nineteenth Century; superconductivity in the Twentieth).

"The second stage consists of the time between the recognition of a scientific solution to a technological need and the time a first viable product emerges utilizing the solution. Several studies undertaken in the U.S. in the last 20 years show this time to be between six and thirty years, with the average time between 10 and 15 years.

"The third stage consists of the time required for the new product or service to grow and diffuse through the technology and society... The span of time required for this process of technological and/or product substitution to occur has not shortened in any discernible way over the last 60 or 70 years. The examination of a number of cases of product substitution led to the intuitive conclusion that the time required for the completion of this stage did vary with the breadth of the product impact, the capital requirements, social changes required, marketing and distribution patterns, as well as product superiority or other technically-related factors, but not with the specific dates of the substitution."

Thus, depending on the nature of the product or process substitution, the total time taken from the initial pertinent scientific discovery to dominance of the market, that is the sum of the three stages which may be termed the basic research, development, and commercialization stages, lies typically in the range from a minimum of 50-60 years to a maximum of 100-150 years. Wars, economic booms and disasters, government and statutory controls and regulations appear to have remarkably little effect on these substitution times. The major conclusions that can be drawn from these observations are:

1. Research and development is not the answer to <u>immediate</u> technical problems. These have to be met by redeployment and rearrangement of today's existing technology, using engineering capability to the full.

2. Problems likely to become serious within the next 10 to 20 years will generally have to be solved, and probably can be, by <u>development</u> of scientific knowledge <u>that we already possess</u>.

3. Continued investment in research is essential to build up the knowledge base that will be necessary for meeting the technico-social problems of 20 years or more from now.

4. Any attempts at "instant" cures or immediate technological "fixes" of technico-social problems through tough, inflexible legislative or administrative measures are likely to be ineffective or worse except insofar as it may be possible to prevent outright poisoning by various forms of pollution. A patient, understanding, long-term partnership between industry and society, with government as intermediary and cognizant of the long time scales that are involved, is more likely to lead to mutually-acceptable progress in the face of the global pressures that affect all mankind. In particular, the ability of the Federal Government to provide substantial and sustained support for the development of techniques needed in the long range future should be recognized as a means to supplement the free market in meeting such needs.

Recommendation 12. *The Federal Government should equip itself with a capability for long-range forecasting of materials supply questions, for conducting analyses of future problems, for devising courses for corrective action, and for obtaining their implementation in timely fashion.*

CHAPTER III
TECHNOLOGICAL OPPORTUNITIES IN THE MATERIALS CYCLE TO OFFSET SHORTAGES

Note: The preceding chapter reviews the factors that have led us to the conclusion that the major theme in the future must be one of conservation in the materials-energy-environment system. The next chapter reviews the tactics for science and technology consistent with this theme. We recognize from the outset that many of the technical activities recommended are already being pursued by various federal agencies, industrial organizations, and others. In such cases our recommendations may be regarded as endorsements or, in some instances, calls for intensified effort. The important point, we feel, is that we have gathered together those technical activities which are consistent with our major recommendation of a conservation policy.

A. ENERGY

Although we have in earlier sections outlined the impossibility of increasing energy use indefinitely (or use of anything else, for that matter), nevertheless a great deal of energy will continue to be required in societies like those in the U.S., in order to accomplish needed tasks. Without energy there would be no materials but without certain materials there would be no energy. It is therefore a task of first priority to ensure adequate supplies of: (a) energy materials--fossil fuels and uranium; and (b) special materials needed in energy technology--e.g. helium, and materials for superconductors, materials for first walls in thermo-nuclear fusion devices, and light absorptive and transmitting materials for solar energy converters.

The list includes, more specifically, chromium and manganese (for steels, including stainless steel), fluorspar (for steelmaking), nickel and cobalt for various alloys, tungsten for tool alloys, platinum for catalysts, and looking further ahead, probably niobium for superconductors, and lithium and helium, especially if fusion power comes into its own, besides a host of other materials whose supply poses less problems. Helium is a special case: it is available relatively cheaply only from some natural gas fields now wasting away. Its stockpiling has been looked upon principally from a narrow economic point of view, but the problem of helium availability for adequate energy supplies (for high temperature gas-cooled fission reactors, for controlled fusion, for superconducting transmission lines) in the 21st century is not a narrow economic problem. Nor is it a trivial technological problem.

A detailed study of the materials research problems pertinent to developing energy technology falls outside the scope of this report and is receiving attention elsewhere. Also, other panels on COMRATE are, or will be, addressing the mineralogical and technological problems associated with the availability of fuel materials. Nevertheless, it is important to remark here that the Committee strongly supports conservation and frugal use of energy resources. To the engineer and scientist, our principle takes the more specific form of minimum free energy change in overall process cycles, or minimum entropy increase. In everyday operating terms, the principle comes to not using energy where it need not be used, and accomplishing desired results with as little energy use as is practical. Natural tradeoffs often exist between using materials and using energy to accomplish a given purpose; the balance must be decided in each case on the peculiar merits.

Recommendation A.1. *Because of the serious environmental and international implications of continued increase in the generation of energy by the United States, conservation of energy resources and their rational utilization deserve a major increase in research attention.*

Recommendation A.2. *As present plans to make energy available from new sources will fail unless there are adequate supplies of certain critical materials it is recommended that the Federal Government should make strong efforts to ensure the continued availability of these materials by facilitating new sources of supply and by tempering current demand through such means as encouraging the search for, and development of, substitutes for these materials in their major use contexts. Materials requiring particular attention include manganese, chromium, fluorspar, nickel, cobalt, tungsten, platinum, niobium and helium.*

B. POLLUTION CONTROL

Lack of available technology is not the principal barrier to control of pollution of air and water by materials operations. The basic technology is known for control of most pollution problems, even though engineering applications may be poorly developed. Pollution control is primarily an economic and political problem. It is often developed only with reluctance since it is hard to make profitable.

The quantity of a pollutant emitted into the atmosphere or into streams or bodies of water can be reduced by physical or chemical treatment of the effluent stream of contaminated gas or liquid being discharged from the pollution source. However, the cost of such treatment is determined primarily by the volume of the gaseous or liquid effluent stream. Consequently, the economics of emission control can be improved if it is possible to reduce the volume of the effluent stream even if the amount of pollutant contained in the stream remains the same. It is still better if the process generating the effluent stream can be changed to eliminate completely the generation of the effluent stream: this is rarely practical. On the other hand, changing some processes and reducing the effluent stream volume is often possible. These relationships should be fully considered in the development of new manufacturing processes. Even though the value of a recovered effluent may be negligible, the other benefits associated with modifications or change of a production process may be economically very favorable.

In general, the economies that can be effected by reducing effluent stream volumes--if this is possible at all--tend to be greater than those that can be attained solely through changes in the systems treating the effluent streams.

Although the basic technology for control of pollution problems exists except in rare cases, the status of engineering development is frequently deficient. Because of a lack of basic engineering design information, design of pollution control installations is often grossly empirical--at worst, a matter of "cut-and-try." There is a need for basic data that engineers can confidently use to design installations for given service. At present, few data of this type are being developed by either private or public agencies.

The emission of nitrogen oxides from combustion processes is becoming a massive air pollution problem. It is also a problem for which no fully satisfactory basic control technology exists or appears clearly on the horizon. On the other hand, the basic technology for control of sulfur dioxide emissions does exist (though often more in theory than practice) along with the applied technology for their control in specific instances. However, the disposal of sulfur by-products remains as an economic problem and its neglect is leading to some of our most serious consequences for the environment.

Although the detection and measurement of pollutants in the environment is often difficult because of the low concentrations encountered, this problem is usually not limiting in source control, where pollutant concentrations are usually much higher. It is important, however, for assessing the ecological effects of dispersed pollutants. The effects on biological systems of such pollutants, including waste heat, are poorly defined, particularly with respect to long-term and chronic effects.

Recommendation B.1. *Research should be undertaken on a continuing basis to develop bodies of generalized design information for pollution control systems so that such systems can be designed with reasonable confidence.*

Recommendation B.2. *A strong effort should be devoted to the development of basic control technology for the nitrogen oxides emitted from fuel combustion processes.*

Recommendation B.3. *Additional research should be undertaken to develop practical uses or means of disposal for sulfur and sulfur by-products.*

Recommendation B.4. *Additional research should be devoted, on a long-term and continuing basis, to determination of the effects of pollutants on biological systems. Parallel research should be undertaken to develop the techniques and equipment for detection, measurement, and characterization of low pollutant concentrations in water, the atmosphere, and in biological materials.*

Recommendation B.5. *Research should be intensified on the effects of large heat and moisture releases on the local or regional climate in areas of high release, such as cities and large power plants.*

C. MINERAL EXPLORATION

There is need for considerably more precise information concerning mineral deposits. This observation points up the need for improved means for geological exploration. As this subject will be addressed by another COMRATE panel we shall only note here that:

Recommendation C.1. *Research should be undertaken with the aim of finding and developing better geophysical and geochemical methods for geological exploration, and better methods for interpreting the data so obtained.*

D. MINING

Future trends in conventional mining will require resort to leaner ores, and ores in less accessible places. More material will have to be moved in order to realize the necessary levels of mineral production. Unless there are some dramatic breakthroughs in techniques, mining will require increasing proportions of the nation's capital, energy, and manpower. This requirement puts premiums particularly on finding mining methods which use energy and labor more efficiently. The labor aspect may become particularly acute as mining generally does not have a good "image" as an occupation. The depressed state of U.S. mining technology is evidenced by the number of imported techniques.

Recommendation D.1. *It is recommended that increased attention be paid to rock mechanics and techniques of rock cutting and tunnel boring. In conventional approaches there is a need to determine whether harder cutting materials can be synthesized in suitable sizes and shapes (tungsten carbide and synthetic diamond have led to considerable progress recently in this respect) and to explore their capabilities and the factors affecting bitt performance in various types of rock. Novel approaches for both soft and hard rock should be fully explored and evaluated for various mining conditions. These include the use of high-power lasers, electron beam, plasma torches, thermal fragmentation, ultrasonics and shock waves, percussion and hydraulic jet techniques, and automated continuous explosive systems.*

To increase mine output without corresponding increase in manpower requires more-rapidly-operating machinery and/or a higher degree of automation. The latter will be particularly necessary for ultra-deep mines and sea-bed mining where the environment may be unsuitable for human operators.

Recommendation D.2. *It is recommended that research be accelerated on ways to automate mining operations. Such automation requires the development of suitable sensing devices (to monitor rock composition, for example), information processing equipment (mini-computers), and servo-mechanisms (robots), all of which have to be exceptionally rugged. Implicit in this recommendation is the need for cooperation between mining engineers and the electronics community familiar with signal processing.*

The ocean floor as well as what lies beneath it is beginning to offer a source for certain materials, particularly the "manganese" nodules which can be scraped off the ocean floor. Initially the needs may be met by adapting current engineering technology but new techniques need to be explored, most of which may involve development of novel forms of robotic equipment, as included in the previous recommendation, especially for mining under the sea-bed. The environmental aspects of ocean mining, while apparently benign so far, may not have received as much study as seems warranted.

Recommendation D.3. *It is recommended that research be carried out on the environmental consequences of ocean mining. The effects of disturbance of mud and associated marine life at the sea-bottom and the effects of bringing spoil up to the surface of the ocean and discharging it there need to be fully evaluated.*

A novel form of mining that is attracting much attention at present is leaching. In in situ leaching the ore deposits are soaked in acid thereby dissolving or leaching out the desired constituents which are subsequently collected from the run-off stream. For certain locations the scheme has been shown to be feasible but questions exist concerning how widely useful the technique may be, and its environmental consequences.

Recommendation D.4. *Research to determine the usefulness of in situ leaching as a mining process should be accelerated. Attention should be addressed to such questions as what ore deposits within the United States are potentially suitable for leaching operations, taking into account the geology, the possible hazards of the explosions needed to shatter the ore deposits, the hydrologic conditions and the risk of pollution of ground water, and how the surface may eventually be rehabilitated. Studies are also needed of various chemical processes that might be used for leaching various types of ores and metals, the degree of recovery obtainable, how it depends on the efficiency of the rock shattering, and the potential of leaching for recovering further values from existing mine wastes.*

E. BENEFICIATION

In most cases the crude ore produced by mining is first treated by beneficiation processes to separate the valuable mineral grains from the waste rock (gangue). Beneficiation involves crushing and grinding of the ore into its constituent mineral grains, followed by separation of the valuable minerals from the rest by flotation, magnetic separation or other means, thereby producing a mineral concentrate and a waste product called tailings. By removing the large bulk of waste rock by simple physical methods, the subsequent chemical processing of the concentrate is rendered less costly and less energy-intensive. The value of beneficiation may be judged from the fact that very few (if any) crude ores that are capable of beneficiation are not so treated.

Beneficiation processes, particularly those for fine grinding of ore, are both capital-intensive and energy-intensive. Recovery of valuable minerals by these methods generally falls between 70 percent and 90 percent, with a clustering of values near to 80 percent. Disposal of large amounts of tailings (about 160 tons of tailing per ton of recovered metal in the case of copper ores) in an environmentally acceptable manner is often costly and difficult.

The very low energy efficiency for fine grinding of ores in conventional tumbling mills is well known and is responsible for 60-75 percent of the energy used for beneficiation. The possibilities of novel grinding processes that use significantly less energy are not promising, but the potential savings from such developments are so important as to warrant thorough investigation.

Recommendation E.1. *An intensive effort is warranted to develop more energy-efficient grinding processes. Processes and phenomena requiring research and development include the study of fracture in polycrystalline rocks, the effect of chemical environment on fracture, the use of ballistic processes for shattering rock, steam shattering, and improvements in conventional grinding efficiency through engineering and developments, including more hard-wearing machinery, and computer control.*

Separation techniques in beneficiation are generally ineffective for very small mineral particles (less than 10 microns diameter). Failure to recover such small particles is frequently the main reason for low recovery of valuable minerals. Improved means to separate small mineral particles would extend the applicability of beneficiation to difficult-to-treat ores, and would improve recovery from regular ores.

Recommendation E.2. *Development of processes that make possible improved recovery of minerals, particularly in the very fine particle range (1 to 10 microns diameter), should be pursued. Promising developments of this kind include high-gradient magnetic separation, selective flocculation prior to flotation, ultraflotation, and column flotation.*

F. PRIMARY METAL PRODUCTION

Extractive metallurgy covers those steps in materials processing whereby mineral concentrates from the beneficiation step, or raw ores that are not amenable to beneficiation, are treated by chemical means to recover the metal values, to reduce these to the metallic state, and to purify or refine the metals as necessary. Extractive metallurgy is far more energy-intensive, per unit of raw material treated, than is mining or beneficiation because reduction to the metallic state involves energy expenditure for breaking chemical bonds of nearly every atom present. Extraction efficiencies are usually above 80 percent and frequently over 95 percent, so that there is room for only incremental improvements in this area. Air pollution from sulfur dioxide, particulates, and fluorides represents a serious problem for several metal industries, and disposal of wastes (for example, the "red mud" from processing of bauxite) is a nuisance problem for others.

Extractive processes show a wide range in energy utilization per unit of metal produced--from a low of about 5,600 cal. per gram (10,000 BTU per pound) of metal to a high approaching 167,000 cal. per gram (300,000 BTU per pound) of metal. A common misconception regarding energy efficiency of extractive processes is to attribute low efficiency to pyrometallurgical processes (chemical reactions carried out at high temperatures), and recommend the alternative of hydrometallurgy (chemical reactions in water near to ambient). In fact, the three processes employing the least energy per pound of metal produced are all pyrometallurgical processes (flash smelting and converting of copper, the iron blast furnace, and the lead blast furnace), and some hydrometallurgical processes are very energy-intensive (recovery of nickel from laterites).

Rather than a substitution of hydrometallurgy for pyrometallurgy, the design of more energy efficient processes must seek out the special advantages of both approaches. It must be recognized that each ore and each metal to be recovered presents special problems that require almost custom design of the process for the particular ore treated. Attempts to recover iron from high-grade iron ore by hydrometallurgy, or to recover uranium from typical low-grade ore by pyrometallurgy, would both be very wasteful of energy compared with current processes.

Hydrometallurgy has special advantages for direct treatment of low-grade ores that cannot be concentrated; it can selectively subject only the valuable minerals to chemical reaction, eliminating the requirement for energy to dissolve, heat or melt the great bulk of waste rock contained in the ore. Another advantage stems from the ease with which two or more metals in the same ore can be separately recovered. An energy disadvantage of hydrometallurgy results from the frequent need to use electrolysis, an energy-intensive process, to recover metal from aqueous solution.

Pyrometallurgy processes are best adapted to treatment of high grade ores or mineral concentrates. The processes vary in energy consumption from the lowest of all extractive processes to very high, depending on process design and the particular metal being recovered. Low energy use (high energy efficiency) generally results from continuous counter-current processes (the blast furnace), the use of natural fuel value in the raw material (burning of sulfur and iron in nonferrous metal concentrates), the use of fossil fuels rather than electric energy, the use of high-grade raw material, and the efficient recovery of waste heat.

Experience with improved process design over the past ten years indicates that important opportunities exist for reducing fuel consumption in extractive processes. Savings in fuel of 25 percent should be easy to obtain and savings as large as 50 percent to 75 percent may be possible in some cases. A side benefit of new process design in many cases is simplification and reduction in cost of pollution control for the process effluents.

> Recommendation F.1. *Intensive and extensive efforts should be directed toward design of processes with improved energy efficiency, and for which pollution control is simpler and less costly. Careful studies of energy utilization for conventional as well as new processes should be made, and the results used to encourage development of superior processes. Attention should be given to the recapture and use of waste heat.*

In many cases, ores and concentrates contain small but significant amounts of other metal values, in addition to the major metals that are recovered. These minor values are frequently lost to the waste products. In other cases, known mineral resources are not amenable to conventional extractive processes because the mixture of metals and other elements in the ore is either too complex or too "refractory" to be separated by existing processes.

Recommendation F.2. *Research and development should be directed toward new processes capable of: (a) more complete recovery of minor metal by-products now lost to waste streams; (b) treatment of complex ores containing several metal values that are difficult to separate; (c) treatment of refractory ores (frequently silicate minerals) where the metal value occurs in chemical combination that is difficult to break down. Examples of this category are: manganese in rhodonite, aluminum in clays and other silicates.*

Plants for the large scale production of the major metals (iron and steel, aluminum, copper, lead, zinc, etc.) are major users of energy resources. They also frequently generate energy (steam or electric power) for their own use or for sale. In addition they produce wastes (slag, stack gases, and warm water from various process cooling devices) that must be disposed of in an environmentally acceptable manner. In the past, metal smelters have generally been located away from centers of high population density because of problems associated with their effluents.

With the growing recognition of the need to capture effluents from both metal smelters and fuel-burning central power stations, there is good reason to consider the possible symbiotic advantages that might accrue from side-by-side location of these plants on the fringes of major population centers. Warm water produced from both plants, that now contributes to thermal pollution, could be used for space heating in the nearby community. Waste heat and by-product fuel (blast furnace gas, for example) from smelting could be used for power generation. Combination of the gaseous effluents from both plants might lead to economies in treatment for removal of particulates and sulfur dioxide. It might be possible to use high-sulfur coal as a fuel in the first stage of smelting with the production of a sulfur-free fuel gas for power generation.

Recommendation F.3. *The possible symbiotic advantages of joining the functions of metal smelting, power generation, and municipal space heating in a single plant (or side-by-side plants) should be explored.*

Another area in primary metal production in which considerable savings in overall energy expenditures might be accomplished concerns direct processing from the melt to such finished metal shapes as sheet, rod, and tube. Continuous casting processes have gained considerable acceptance for certain metals and stock shapes but the principle needs to be extended more widely. Clearly such processes avoid the energy-wasteful cooling and reheating stages associated with conventional ingot manufacture followed by deformation processing.

Recommendation F.4. *Research should be intensified on methods of direct processing from melt to finished stock shapes.*

High capital cost for extractive metallurgy plants deters adoption of already proven new processes and research and development directed toward improved processes. Recommendation 7, from Part I of this report, speaks to the need in this area.

G. RECYCLING

Efficient recycling of materials from industrial and municipal waste has become recognized as a major opportunity for conservation of materials, energy, and the environment. The greater the number of product-cycles materials can be put through the less the demand on prime resources. It is helpful to think of industrial and municipal wastes as a novel form of ore but one which is quite widely dispersed and therefore lean, especially the municipal wastes. The inefficient, energy-consuming process is the collection and concentration of the waste, a process which is more social than technical though it has to be done whether or not values are recovered from the waste. The costs of this stage may make the whole recycling operation noncompetitive with primary sources of materials unless subsidized (e.g. by calling it "garbage collection") or helped with other financial incentives. If the collection and concentration process can be carried out efficiently so that recovery percentages from wastes are very high, and if the recovery processes are flexible enough to adapt to the variable content of wastes, recycling can make a very important contribution to the conservation of materials. It must be recognized, though, that recycling cannot meet the needs of growing consumption--in full operation the recycling of material is limited to the total amount of material that was used in products at a time earlier by the average product lifetime.

Every operation in the materials cycle generates wastes but the wastes remain relatively localized up to the stage of finished, manufactured products. Thereafter they get dispersed through use and become the lean ores referred to above. The industrial wastes up to this point are, in contrast, relatively rich ores and in many cases already there is efficient recycling. For example, metal scraps resulting from cutting and forming operations on manufacturing lines are easily collected and returned to the furnaces for recasting, etc., and it is usually highly attractive, economically, to do so. For other industrial wastes, such as spent plating bath solutions and other chemicals, collection is again usually quite straightforward as are retreatment technologies, though they need to be carried out much more intensively than has been customary.

While there are countless opportunities for incremental improvements in recycling technology, the principal areas in the materials cycle where attention needs to be paid concern: (a) the enormous and growing piles of waste associated with mining and smelting operations, (b) the treatment of municipal waste, and (c) closely related to this, the designs of products so as to facilitate material recovery from municipal wastes.

Mining operations typically generate enormous piles of broken-up rock--the overburden that had to be removed to get at the mineral-bearing rocks--and finer-sized tailings from the beneficiation processes. The inevitable trend to leaner ores will cause accelerated growth of these wastes. These "wastes" can often be worked over again by leaching, concentrates can be re-ground to recover other values when economics justify it, and wastes can be used for mine fill. Nevertheless, these wastes are crying out for a major use. The big barrier is economic. The only industry that could use such material on a massive scale is perhaps the construction industry and it is the nature of things that mining operations are often in places remote from heavy construction areas, thus posing the problem and expense of transport.

> Recommendation G.1. *Efforts should be directed towards developing economic uses for mine wastes and tailings. Construction materials may offer the best prospects for such wastes. In some cases perhaps a relatively minor additional crushing and compacting will suffice to make suitable aggregates. Or new processes may have to be devised to form suitable compositions and shapes.*

The practicalities and economics of setting up collection systems are probably the main obstacles to the efficient reclamation of municipal wastes. Various approaches such as subsidies, publicity campaigns, regulation, and so on, need to be experimented with to encourage collection systems. But the main technical problems concern the sorting and segregation of wastes. Waste streams should be kept as separate as possible in order to simplify recycle, reuse, or disposal. There is much scope and need for automation in waste handling--sensors which can detect specific materials or products, and servomechanisms which can carry out their mechanical collection and sorting. The types of sensors and servomechanisms may vary according to the nature of the waste streams. Much can be done with contemporary technology, including magnetic separation, air or water separation of organics, screening, optical sorting, electrostatic separation, leaching, and hand-picking, but nevertheless it is felt that this area would benefit from more intensive attention by scientists and engineers from various disciplines.

> Recommendation G.2. *Automation methods applicable to more complete collection, sorting, and segregation of municipal waste need to be developed. Novel sensors, sensitive not only to material contents but to shapes of objects are needed as well as various forms of servomechanisms.*

A major form of municipal waste and source of useful materials, but one which has not received as much attention as it deserves, is that which results from building demolition. While reasonable effort may go into recovering the metals used in buildings--girders, plumbing, heating and air conditioning equipment, for example--relatively little heed is paid to the plastics--wall and floor coverings, plastic pipes--or to the masonry, much of which could be reused rather than end up as landfill.

Recommendation G.3. *While it is important to design buildings for longer life and lower maintenance costs, ways need to be found of improving the recovery of material values from buildings once their useful life is over.*

Certain classes of separation of municipal wastes are relatively easy to achieve. The combustible materials, for example, can be largely isolated from the noncombustible ones and used as fuels if for nothing else. Plastic, metal, and ceramic objects are also relatively easy to separate from one another. Separation becomes more difficult when these materials are joined to each other in equipment, as in household appliances, automobiles, etc. Adhesive bonding, welding, and composite structures particularly warrant attention from the disposal point of view, especially at the initial stage in product design. There is room for the development of improved "beneficiation" and "extraction" processes for treating these complex "ores," depending on the nature of the junk to be treated.

Recommendation G.4. *Intensified effort is needed to find more efficient methods of separating various classes (metals, plastics, ceramics) of industrial materials that are joined to each other in junk equipment.*

Separation of different classes of materials, awkward though it may often be, is generally more straightforward than separation within material classes--i.e. one metal or alloy from another, one plastic from another. Often, for example, the various metals get mixed up together and then it is the task in the recycling process to separate into the various constituent elemental materials the relatively mixed-up "alloys." There is considerable need for improved separation technology in these situations.

Recommendation G.5. *Research to find efficient ways of separating intimate mixtures of metals into their component elemental materials needs to be intensified. Adhesive bonds, welds, and composite structures require particular emphasis. Alternatively, for intractible mixtures efficient shredders need to be developed to facilitate more efficient extraction processes or incineration.*

In many cases it proves impractical to perform very complete separation of alloys. What results then is often a degraded alloy, i.e. admixtures of other elements or impurities to the original material resulting in an alloy of slightly different composition and, in consequence, poorer properties (such as strength, electrical resistivity). Such degraded alloys may not be suitable for reuse in their original types of application but nevertheless may still be completely suitable for less demanding applications. Further degradation can result at each recycle so that to maximize the effective use of such materials there needs to be a whole cascade of successively less demanding applications.

Recommendation G.6. *Research needs to be undertaken both on the upgrading and repurification of degraded alloys and plastics and on the properties of degraded alloys and plastics so that their prospective applications may be more readily identified.*

Composite materials can pose particularly difficult problems for recycling.

Recommendation G.7. *Research is needed on efficient ways for recycling or reclaiming material values from composite materials. Such materials include composites of dissimilar materials (e.g. glass-fiber reinforced plastic, metal-clad plastic laminates) and composites of similar classes of materials (e.g. epoxy impregnated synthetic fiber boards, copper-clad aluminum wire).*

The recycling of plastic materials is taking on increased importance because of their dependence on oil supplies. Plastics, rubber, and synthetic fibers can often be depolymerized and the various monomers recovered for reuse as feedstock (though this is expensive) or for use as fuel. The fuel value of unsegregated discarded plastic probably exceeds the replacement cost but if the plastics could be sorted they could be more valuable.

Recommendation G.8. *There is need for increasing research on methods of identifying and sorting plastics and of recovering material and monomer values from them. Standardization of plastic compositions should be encouraged to facilitate recycling.*

The recycling of glass has so far been largely inhibited by unattractive economics but as this situation changes various technical problems must be addressed. Glass is colour-mixed, and contains rocks, metal chips and ceramics no matter how well cleaned. These do not dissolve but weaken the glass. In the broader arena of ceramics there is again no recycling in general because of the problems of keeping the types separate and the chemical changes that take place on firing. Waste glass and ceramics are not very important for aggregate either because of the low cost of primary sources. However, if some of the collection costs can be internalized or covered by other incentives the technical problems would often appear tractable. Furthermore, it takes less energy to melt scrap glass (cullet) than to make new glass.

Recommendation G.9. *Research needs to be intensified on ways of sorting glass and ceramic wastes and for recovering material values from them.*

While more an institutional problem than a technical one the whole area of waste collection in order to facilitate material recycling and conservation needs special attention.

Recommendation G.10. *Federal, state and local governments are urged to take steps and develop incentives to encourage the collection and processing of wastes. Examples of possible measures include demonstration projects, development efforts, design competitions, subsidies, publicity campaigns, regulation, and so on. It is particularly desirable for waste handling and recycling to be arranged so as to capture, insofar as possible, the economies and technological advantages that accrue to large, integrated scales of operation.*

H. PRODUCT DESIGN

Product design can have a very direct bearing on material conservation. Increasingly important elements in product design are: design to minimize dispersal of scarce materials; design to facilitate product dismantlement and material reclamation; design to employ, wherever possible, more abundant materials instead of scarce ones, materials from renewable resources rather than from nonrenewable ones, and less environmentally offensive materials for those more so. In many of these areas the designer will often be responding to material costs. Opportunities for substitutes will be considered in subsequent sections.

Product design also has a direct influence on the choice of manufacturing methods to make the product. The designer should ensure, therefore, that as far as possible he develops designs which can be made using the manufacturing methods which are most economical in materials and energy, particularly those which lead directly from the fluid state to the finished shape or piece-part.

Whatever tactics the product designer uses to economize on materials, the limits to his success depend ultimately on the precision of his knowledge of the properties of the materials he is using, including their variability. Depending on the application, the designer may have to know accurately the physical, chemical, mechanical, and biological properties so that he may design to the desired levels of reliability, durability, serviceability, and safety. In addition to initial strength, electrical, optical, magnetic, or thermal properties, he must usually take into account wear, friction, fatigue, and fracture; also chemical reactivities, corrosion-resistance, flammability; and, increasingly, such properties as toxicity, rot resistance, or biodegradability. All this implies that there are greater needs than ever before for methods and techniques, especially nondestructive ones, for accurately and rapidly characterizing, analyzing, testing, and age-testing materials, and providing information pertinent to their behavior in working environments.

Recommendation H.1. *A set of principles of product design for facilitating material recovery at the end of the product life should be developed. Examples of such principles include avoidance of use of scarce materials in ways that lead to their dispersal, as in coatings for instance; and avoidance of certain juxtapositions of metals which are difficult to separate.*

While the designer must know the variability of the materials he plans to use in the properties of critical interest to him, the importance of providing him with more reproducible materials needs to be emphasized. The cost-benefit trade-off in better reproducibility should be carefully examined. Do we even know what safety factors, limiting properties, and so on, the designer uses? What does he assume about the effects of welding, plating, and other process steps on these properties? In short, how overly conservative are our designs and hence materials-wasteful?

Recommendation H.2. *Continuing research must be adequately supported on furthering basic understanding of the physical, mechanical, chemical, and biological properties of materials, the relation of these properties to manufacturing processes, particularly their relation to product performance under service conditions, and on finding thereby materials with improved performance properties. Improved techniques are needed for rapid and accurate determination of these properties as well as for determining what properties are needed to meet service conditions.*

Recommendation H.3. *It is recommended that an important contribution to materials conservation and wastage reduction can come from designing products and structures to performance and conservation specifications rather than to specified materials, constructional details, and methods. Emphasis needs to be placed on finding ways to design products for sustained performance, improved durability, maintainability, safety, and recyclability.*

I. SUBSTITUTE MATERIALS

Material substitution is not a recent concern of technology. Traditionally there has been the push-pull relationship between materials and technology: development of improved materials provides push to developing new or improved products; new, more ambitious engineering and product objectives provide demand or pull on the materials sector to come up with new or improved materials. The principal criteria for the materials in optimizing the material-product relationships are performance and cost. These will remain the principal criteria but the ground-rules are changing: performance criteria extend increasingly to include reliability, durability, safety, and impacts on man's health and environment; costs are being driven upwards by these more complex performance criteria and also by the increasing scarcities of certain materials and the energy and environmental costs associated with them.

The search for superior materials must carry on unabated. Simply to sustain man's standard of living, let alone advance it, calls for superior products to meet the pressures of growing population and concerns over environmental impact. Tougher, stronger, high-temperature corrosion resistant materials are required for higher efficiency gas turbine power generators; first-wall materials able to withstand extreme heat and radiation environments of thermo-nuclear fusion devices; superconductors with higher transition temperatures; implant materials more inert to body fluids, more transparent glass for communicating via optical waveguides; harder-wearing tool materials for drilling, mining and manufacturing machinery, and so on. The list is endless; there will always be the need to push against the constraints of contemporary property-performance levels of materials. This is the traditional aim in developing substitutes to meet specific product and performance specifications.

Sometimes substitutes appear from unexpected and radically new advances in materials science or materials technology--e.g. the prospect of catalysts based on relatively abundant oxide compounds rather than platinum; the development of fiberglass and of pyroceram; the discovery of the properties of semiconductor crystals and the method of zone refining; the discovery of superconductors capable of carrying high electrical currents. But such instances are relatively rare and usually difficult to foresee. More often, materials with superior properties come about through steady, patient research and development generating a whole series of incremental improvements; it would be unwise to assume that there will always be sudden, dramatic breakthroughs in materials science and engineering which will alleviate the shortage-induced stresses as they occur. As certain elemental materials become scarcer and more expensive we will be forced back on to a reduced range of raw ingredients out of which to concoct the industrial materials needed by our society. The process will be gradual, but in order to meet shortages that may be serious 20 years from now it is necessary to pursue continuously research programs aimed at developing industrial materials with suitable properties out of the more abundant raw materials. <u>We cannot emphasize too strongly that the discovery and development of new and improved materials as possible substitutes for existing ones takes time and that the process is generally driven by clearly-perceived functional objectives rather than by ill-placed optimism that "something will turn up when the crunch comes."</u>

A serious question is how the nation can develop substitute materials in the absence of contemporary economic incentives to do so. The development of substitute materials, including energy materials, to the point of meaningful availability, may require a great deal of capital and years of lead time; whether done by industry alone or through subsidies, or by government itself, the payoff comes only through production and consumption; it may therefore seem costly to hold substitute technologies on the shelf and it could be impossible if the development method chosen has included a guaranteed domestic market for the substitute material. The vexing questions of (1) how society should choose among alternative R & D investments designed to produce substitutes for scarce materials, (2) what are effective ways of developing substitutes to the point of significant availability without being forced to use them, and (3) how the costs of doing this should be allocated and paid, need to be given serious attention.

It is perhaps instructive to look at an example of how closely related sorts of problems are approached by the major communications company. Faced with constantly growing demands for increased and improved communications services, and with a reasonable assurance of a continuing responsibility for supplying these services on a national scale, the Bell System has continuously practiced an R & D policy of developing substitute technologies sufficiently ahead of time so that they would be adequately developed and ready for production when customer demand required them. Research programs in solid state science (which led to the transistor), in microwave transmission, in satellite communications, and in optical communications were all initiated many years, even decades, in advance of the time when it would be necessary, and economic, to have the new technologies ready to substitute for older ones. No doubt similar examples of farsighted R & D programs could be cited from other large, well-established industrial concerns. The fact that there are also many large companies which, for various reasons, have not made a practice of long-term R & D does not vitiate the argument. As Huddle has noted: (See Appendix of Source Material)

Three difficulties are encountered in substituting one material for another...: (1) that of the sheer complexity of designing around highly specialized uses of a wide variety of materials in highly sophisticated hardware and systems, like the telephone, television, computer, electric light, photocopies, gasoline pump, and many more; (2) that of achieving innumerable substitutions of single materials used in many different combinations, as for example the replacement of chromium with other materials in literally thousands of alloys employed in millions of engineering applications; and (3) that of finding a material available in the sheer volume of abundance, as well as suitable in performance, to replace such very large volume materials as steel...or aluminum...or petroleum.

The principal strategies for developing substitutes would appear to be:

1. Identify threatened materials.

The threat, short- or long-term shortages, may be through economic, military, or political actions by other nations, price mismanagement and short-sighted fiscal policies, environmental legislation, the hazards of mining, transport dislocations, labor troubles, and many more including simply a real world-wide shortage of the material.

2. Identify both critical and major uses.

For some materials, such as helium and mercury, no substitute materials may be possible. Many applications depend critically on certain materials, e.g. palladium for contacts in telephone switch-gear, platinum for catalysts and chemical processing, gold for integrated circuits, chromium for stainless steel. The major uses of certain materials are in particular areas, e.g. copper for electrical equipment, tungsten for mining, construction and metalworking machinery, gold for jewelry, tin for tinplate.

3. (i) Minimize dispersive uses of threatened materials

   (ii) Maximize recoverability

   (iii) Develop substitutes for critical and major applications.

The first two of these have been covered elsewhere in this report. The rest of this section will address the areas where substitutes appear most urgently required. Guidance is given by various official sources of information on mineral reserves, resources, and consumption rates. These data may be approximate but they are the best we have to guide scientists and engineers in choosing their research strategies for substitutes.

Unique Materials

Helium remains liquid to a lower temperature than any other gas or liquid. For the ultimate in cooling applications it has no substitute. As the essential coolant for getting most superconductors to superconduct it may prove indispensable for future energy generation and transmission technologies using superconductors to create intense, large-volume magnetic fields and for transmission lines. For those superconductors that achieve the highest transition temperatures, liquid hydrogen can be used as a substitute for helium and further research may lead to practical superconductors that can be used at temperatures which alleviate our dependence on helium. But because energy is so vital, helium should be conserved and used frugally.

> Recommendation I.1. *Dispersive uses of helium should be minimized, wasteful practices discouraged, alternatives found, as they can be, for most uses (such as in arc welding, purging rockets and spaceships), and conservation practiced, generally.*

Mercury is the only element that is a metallic liquid at room temperature. There is no direct substitute for it in mercury-wetted switches. Substitutes for these switches, for example solid state switches, would have to be developed if mercury were unavailable. Substitutes for mercury in its other applications, such as in thermometers, batteries and chemicals, need to be developed as the limited world supply of mercury and its environmental and health problems place it on the threatened list.

> Recommendation I.2. *Material or product substitutions need to be developed for those applications which use mercury.*

Asbestos has unique fibrous structure which makes it useful in various applications, in brake linings, gaskets, and particularly in construction materials. Health hazards associated with asbestos dust make it necessary to search for uses which do not create aerial dust and for acceptable substitutes.

Recommendation I.3. *Research should be undertaken to develop synthetic materials which would be suitable substitutes for asbestos where none exist.*

Specialty Metals

A number of specialty metals can be placed on the threatened list. Those for which substitutes would be highly desirable (see also Appendixes A, E) include, but are not necessarily limited to:

Chromium (HV)--for stainless steel, alloy steel, high temperature alloys;

Gold (V)--for electrical devices;

Platinum (V)--for catalysts and electrical devices;

Palladium (V)--for catalysts and electrical devices;

Tin (S)--for tinplate, solders and bronzes;

(V denotes vulnerability of U.S. to policies of other nations; HV indicates unusually-high vulnerability; S denotes a potential general world-wide shortage.)

Recommendation I.4. *Research should be undertaken to develop substitutes for materials in which the United States is deficient in resources, such as chromium, gold, platinum, palladium, tin, and their alloys in the context of their major applications. The list of "threatened" materials should be reviewed periodically. Materials for which the need for developing substitutes needs to be carefully assessed include: antimony, tungsten, vanadium, silver and zinc.*

Bulk Metals

No severe shortages in the bulk metals appear threatening yet. Limitations may eventually result from energy requirements for mining and extracting and the associated environmental problems. Copper is the nearest to being threatened though world-wide and sea-bed resources are still fairly large. It is the best material for electrical applications and if serious supply problems were to develop it may become prudent to discourage its use in other applications, e.g. plumbing, car radiators, where substitutes are relatively easy to use or develop. Even for electrical applications aluminum can often be substituted but in the long-term there is some uncertainty over the adequacy of supplies of this metal, particularly if derived from bauxite. The United States has only low-grade aluminum-bearing ores and has to depend on overseas suppliers though there are many processes whose current economic value is borderline but which are capable of extracting aluminum from nonbauxite ores. The possibility that the copper-producing countries could combine with the aluminum-producing ones to manipulate the market in the short-term should be allowed for. For the foreseeable future, however, we can expect to see a continual see-sawing back and forth between copper and aluminum as they compete with each other in the supply-demand arena. From the engineer's standpoint the primary need is to achieve and maintain flexibility of product design so that consumption patterns can be shifted correspondingly to meet changing supply situations.

Ceramics and Glass

Amidst all the concerns over impending scarcities of metals the abundance of raw materials for ceramics is in sharp contrast. The challenge is to develop a wider range of ceramic materials with a correspondingly wider range of useful properties. This challenge contains a particularly strong technological dimension. Exciting possibilities can be sensed in the progress towards building turbines for automobiles out of silicon nitride and related ceramics based on abundant materials. Examples of potential applications for new and improved ceramics and glasses, or opportunities for greater use of such materials if they can be made more cheaply by improved processes include: glass or other ceramic fibers for asbestos; bricks replacing wood and steel in building construction; ceramic (rather than asbestos) in vinyl for wall and floor tiles replacing pure plastics; concrete instead of metal for sewers and other pipes; lightweight ceramic aggregates for concrete, ceramic or glass bottles, cookware, tableware, industrial piping and plumbing, piece parts such as door knobs replacing plastics and metals, porcelain enamels for building panels; ceramic cutting tools; superstrength ceramics for turbine buckets and auto engines; glass optical waveguides for communications.

> Recommendation I.5. *Intensive research should be undertaken aimed at extending the range of useful ceramic and glass materials and at developing economic processes for producing them. Particularly desirable are ceramic and glass materials with greater ductility and fracture toughness. Engineering research should also be undertaken to improve our ability to design structures using brittle materials.*

Plastics

Petroleum is the raw material base for the major part of petrochemicals and plastics production though this represents only a few percent of total oil consumption. Though oil supplies may become increasingly critical high priority is likely to be given, as in the present oil crisis, to the petrochemical industry on account of the widespread and complete dependence that industry and society has come to place on plastics. In view of the likelihood of alternative sources of raw materials based on shale oil and coal, the prospects of shortages of plastics do not appear too threatening. In consequence, the abundance of plastics as a class may perhaps be regarded as falling between that of the scarcer metals on the one hand and the more abundant ones and ceramics on the other. Thus, it is appropriate to seek opportunities for plastics to substitute in applications where the scarcer metals are normally used while at the same time seeking ceramics to substitute for the plastics. For example, industrial piping made of chromium-containing stainless steel might be replaced by plastic pipes but in turn, these might be replaced by ceramics and glass.

No completely new major classes of plastics can be anticipated any more than new major classes of metals with novel compositions can be. But, as with metals and their alloys there are considerable opportunities for developing plastic "alloys" with various physical, chemical, and biological properties by blending different types and amounts of polymers.

Recommendation I.6. *Research should be supported aimed at extending the range of properties of plastic materials to make them even more versatile for engineering applications. Particular emphasis needs to be paid to strength, flammability, durability, and biological hazards.*

As a long-term resource base for plastics, oil seems limited and it remains to be seen what scale of supply may be feasible from shale oil and coal. One other possible source of organic raw materials is wood. The world's forests are treated rather casually at present but with careful husbandry, long-range planning, and less wasteful practices they could provide a renewable resource of both engineering materials and organic raw materials at acceptable levels of ecological impact. Enormous quantities of lignin, the binding agent in wood that holds the cellulose fibers together, are now wasted. This class of materials offers promise as an organic feedstock as well as a liquid hydrocarbon fuel material. However, to date, processes for obtaining organic feedstocks from wood have not been competitive economically, with petroleum-based processes.

Recommendation I.7. *Research should be intensified to find economic processes for converting vegetation, particularly from forests, into raw materials for use as a base for plastics. The studies should include not only the chemical processes that might be used to produce monomers and alcohols but also the range of plastics and their properties that could be obtained from a natural organic base. An important additional need is to assess the ecological and environmental consequences of using the forests as resources so that an upper limit to operations might be established.*

Wood

Besides the possible uses just referred to, wood is an important engineering material. It is used in vast quantities (greater tonnages than all metals combined) for building construction and furniture. Scraps, cuttings, trimmings, shavings and sawdust resulting from wood operations are used for making chip boards and other composites though more often these materials are wasted or uselessly burnt. Large quantities of wood are also used for making paper and cardboard. Perhaps the major problems with wood as an engineering material are its variability, though much has been done recently to reduce this by sorting using nondestructive testing techniques, and its tendency to deteriorate, rot, burn, and weather.

> Recommendation I.8. *Research needs to be extended on ways of growing more uniform and defect-free grades of wood. Efforts are needed to develop more economically viable uses for wood scraps, particularly for composite materials to be used as replacements for other constructional or structural materials in short supply. And research should be supported on ways to make wood and wooden products more durable and resistant to rot, fire, and weather.*

Composites

Structurally weak materials such as most plastics can often be strengthened by embedding in them networks of strong fibers. Fiberglass is perhaps the most familiar example. Other means of strengthening materials are based on dispersions of hard particles (as in concrete) or precipitates (as in many metal alloys). Fiber or particle strengthened materials, often called composites, can offer impressive improvements in strength-to-weight ratios over the bulk material and by the same token offer important opportunities for considerable weight and material saving. The nature of composites is that each component can be used to meet specific parts of the overall demand on the piece (such as radial-ply tires which have the different loads handled by different plies; clad metals (and plated ones); cemented-carbide tools; enamelware in general; Revere ware; etc.). Composites find application in construction materials, boat hulls (fiber reinforced plastics and ferro-concrete), car bodies, aircraft (where weight is at a premium) to name just a few.

Sandwich structures are another common form of composite materials. Plywood is perhaps the classic example, in which superior strength and rigidity is achieved over a simple sheet of wood of the same thickness by laminating together thin sheets of wood with their grains running in different directions. The outer layer which may be exposed to weather and view can be of higher quality and more pleasing appearance than the inner layers. Recent years have seen the development of plastics, many of which closely simulate the color patterns of wood, for these outer surfaces (on cheap wood) for quite durable use in interior decorating and furniture.

"Of particular significance is the new awareness that the behavior of composites derives not only from the combination of materials of different compositions, but also from structural and geometrical characteristics as well. Thus we are not limited to uniform or regular distribution of the constituents. The distribution of the constituents can be random or oriented. Also, the distribution can be nonrepetitive to form gradient composites, to provide different properties or property values in various zones of the composite. Therefore, the possible combinations are endless; and by proper choice of materials and constituents, distribution and arrangement of the structural constituents, we can design property systems to meet specific engineering needs." (Clausner, 1970)

There are some problems; composites may be troublesome to join and also, they may be more difficult to recycle. These problems have to be traded-off against their performance and initial material-saving advantages.

> Recommendation I.9. *Research is needed to extend the range and properties of composite materials available for engineering uses. Particular emphasis should be placed on developing composites based on the more abundant materials, and methods of joining and recycling them.*

Dynamic or Fundamental Materials

Materials are usually thought of as rather passive, inert parts of a product, device, or structure, that in themselves do not perform active functions. But this view is changing; increasingly materials are being used in active, dynamic roles. Again quoting Clausner (1970), "It is through this dynamic or functional approach to materials that perhaps the greatest advances in materials effectiveness will be made."

An example of what is meant by a dynamic or functional material is given by the recently developed deformable car bumpers which absorb the impact of a crash and revert to their original shape afterwards. They substitute for much more complex structures involving bumpers working on hydraulic cylinder systems. Another example is the ablating nose cone on space reentry vehicles.

Some of the most sophisticated examples of dynamic materials come from solid state electronics. The silicon integrated circuit chip is basically a single piece of material (silicon) which has been subjected to an extremely complex sequence of processing steps, numbering perhaps a hundred or more, to produce impurity, alloy, and surface oxide configurations on the silicon chip with a precision of quantitative and spatial control that is unmatched in any other area of technology and is perhaps surpassed only by nature's structures for handling signals and information in the nervous system. These minute, processed pieces of silicon, which would hardly be noticed as a speck of dust in the eye, when wires and voltages are applied to them, can be amplifiers, memory units, switches, and many other functional devices. They replace older contraptions made of vacuum tubes, sockets, capacitors, resistors, wires, support structures, transformers. The material savings are enormous (a million- to a billion-fold) as well as the savings in energy needed to operate the device.

One final example is also from communications--the prospect of optical communications in which telephone and data signals are sent along glass fiber optical waveguides instead of as electrical signals along metal wires and cables. The optical waveguide, made of ultra-pure glass and as fine as a human hair, promises much greater signal carrying capacity than its metal counterparts. To get some idea of the purity of the glass, if ocean water were as clear one could see to the bottom of the deepest oceans. The light wave carrying the signal travels along the centre of the fiber, being confined there by a precisely-controlled radial distribution of impurities which in turn cause a radial variation in the refractive index. The outer part of the fiber serves as support and protection. Thus, the properly-processed piece of glass takes the place of a wire made of metal (usually copper) and insulation (paper or polyethylene); a bundle of fibers can take the place of a cable consisting of many hundreds of wires.

The above examples give glimpses of the potential of imaginative materials and functional materials design which can have dramatic impact on materials usage patterns. It is this sort of imagination and ingenuity that will be needed increasingly to help meet future materials supply and energy constraints. But how can imagination be taught or cultivated?

SECTION II

ESTIMATION OF MINERAL RESERVES AND RESOURCES

The Fossil Fuels and Copper

CHAPTER IV
INTRODUCTION TO SECTION II

The Supply Panel of COMRATE examined methods of resource estimation and the quality of current resource estimates. At a time when the future adequacy of mineral supplies is being questioned, and the United States is becoming increasingly dependent on sources of minerals abroad, it is essential that the best possible estimates of reserves and resources be available. Such estimates are the first basis for national policy with respect to minerals.

It is true that almost every mineral commodity presents its own problems of resource estimation. A comprehensive survey of resource estimates for all minerals, however, was beyond the scope of the Panel. The Panel chose, instead, to examine estimates for two important groups of resources, fossil fuels and copper. The two groups of resources differ markedly in geological characteristics and offer very different problems of resource estimation. Studies of the two should serve, therefore, to discover major aspects of the general problem of resource estimation. Two subpanels were set up. Chapter V gives the findings of the Subpanel on Fossil Fuel Resources, and Chapter VI the findings of the Subpanel on Copper Resources.

During studies by the panels, information was accumulated not only on resources of fossil fuels and copper but also on rates at which petroleum, natural gas, and copper may be available from the resources. This information is included in the panel reports.

GENERAL CONCLUSIONS

Two general conclusions emerged:

1. Information on reserves and resources of minerals today is in many different hands. Much information is available in government files, but much is in the hands of private organizations who have acquired it through large expense and years of effort in mineral exploration and development and studies, national or worldwide in scope, of resources of commodities of interest to them. Current resource estimates could be greatly improved if these two categories of information could be brought together in such a way,

and in such timing, that a running inventory of resources could be maintained. There are, however, difficult problems involved. Among them is the problem of preserving confidentiality of data that have been acquired at private expense and constitute important assets of organizations that have obtained them.

2. As a basis for national planning, resource estimates for many minerals are seriously deficient in that they say little about the economics of mineral production and possible rates of mineral production. This is especially brought out by the study of copper. United States copper reserves are in excess of 40 times current annual production of new copper, yet there is serious question whether current rates of domestic copper production can be maintained to the end of the century. When future availability of minerals is in question, estimates of resources in the ground are a first basis of approach, but such estimates must be supplemented by engineering and economic studies of factors that determine rates of mineral production and processing.

CHAPTER V
RESOURCES OF THE FOSSIL FUELS

INTRODUCTION

The subpanel on Fossil Fuel Resources examined estimates of United States and world resources of petroleum and natural gas, tar sands, oil shales, and coal.

An attempt has been made to standardize terminology throughout this study, following the simpler of the guidelines suggested by McKelvey (1972) and Theobald, Schweinfurth, and Duncan (1972). The term reserves is restricted to materials that can be extracted economically from deposits of known extent using conventional methods. The term resources is used for materials that occur in deposits that either have not yet been discovered (undiscovered reserves) or in deposits for which technology is not yet available (water too deep, or concentrations too low for economical recovery at present).

Accompanying the report are maps designed to show annual production, proved reserves, and undiscovered resources of fossil fuels within different large regions of the world so as to contrast areas of high production and areas of high consumption. Redistribution of resources mainly by ship, is evident. Metric units are used throughout, but equivalents in English units are given in parentheses. For oil, one metric ton (tonne) is taken as equivalent to 7.5 barrels. Data are for the latest years available. For many maps separate data for 1971, 1972, and 1973 are included so as to present recent trends.

Estimation of resources of petroleum and natural gas is complicated by the fact that, while methods for the estimation of proved reserves are well established, there is no generally accepted method for estimating undiscovered resources. This is reflected in a considerable range in estimates of the latter, particularly estimates of U.S. onshore resources in the lower 48 states. Differences in estimates naturally lead to confusion over the availability of petroleum and natural gas from domestic sources. This panel has therefore paid particular attention to the methods used in arriving at various estimates. The matter is important because the size of known and discoverable resources sets limits on the ability of the U.S. to increase or even to maintain current levels of domestic petroleum production during the substantial period of time necessary for development of alternative sources of energy.

Data for reserves of petroleum and natural gas were taken as far as possible from published sources or compilations in order to avoid prejudice that might arise from use of unpublished compilations in company files. Most of the publications are articles in journals and bulletins written for industry. Government compilations might be considered unbiased by some readers, but they appear long after each year of interest. Only preliminary estimates are as yet available for 1973. When they do become available, the data are essentially the same as those of the technical journals. It is clear that an enormous quantity of public information is available on oil and gas production and reserves. It must be understood that the term _proved reserves_ does _not_ include all the hydrocarbons that are expected to be produced eventually from discovered oil and gas fields, but only that amount of hydrocarbon about which there is no question about recoverability. Published data around the world are not available on which to base an estimated amount of hydrocarbon discovered but not proved, so unpublished data have been used to estimate this category of reserves. But in the United States it is estimated that _discovered_ fields will eventually produce an additional 50 percent over and above the cited "proved reserve."

A large amount of information is also available on domestic hydrocarbon resources in oil shale, and on resources in tar sands of Alberta. Information on resources in oil shales and tar sands elsewhere in the world is incomplete and sufficient only to indicate orders of magnitude of these resources.

Methods of estimating coal resources are well developed, involving standard methods of geological mapping and correlative investigations, together with sampling by conventional methods. Estimates of the United States resources of coal presented in the report are those calculated by the United States Geological Survey on the basis of data from industry, from the geological surveys of various coal-bearing states, and from its own investigations of U.S. coalfields. The estimates are based on a large body of information, and the order of magnitude of U.S. coal reserves and resources can be considered as well established. There is need, however, for much additional information on individual coalfields, and for information that will serve as a basis for estimates of reserves of coal of various types and qualities. Information on the sulfur content of coals in various fields is especially needed in view of the problems of sulfur pollution examined in Section III by the environmental panel.

Estimates of world resources of coal are likewise those calculated by the U.S. Geological Survey. Data for various continents are somewhat uneven in quality, but the orders of magnitude of resources in the world's major coal producing regions are probably well established at this time. It should be noted that data are for total resources of coal in the ground, roughly one-half of which is estimated to be recoverable.

Conclusions

Petroleum and Natural Gas Resources

1. World resources of petroleum and natural gas, discovered reserves and undiscovered recoverable resources, will be seriously depleted by the end of the century if present trends of world production and consumption continue.

2. An estimate of 15 billion tonnes (113 billion bbl) for United States undiscovered, recoverable resources of petroleum appears realistic.

3. An estimate of 15 trillion cu. meters (530 trillion cu. ft.) for U.S. undiscovered, recoverable resources of natural gas appears realistic.

4. A large increase in U.S. annual production of petroleum and natural gas is very unlikely.

5. The largest untapped oil resource, probably in excess of 27×10^9 tonnes (202×10^9 bbl), is in the oil in known oil fields of the United States that is unrecoverable with present technology.

6. The second largest, but probably most accessible domestic oil and gas resource is under the continental shelves.

7. U.S. resources of petroleum in oil shale are extremely large, but future rates of production from this source are speculative.

Coal Resources

8. World and U.S. coal reserves plus resources are adequate for hundreds of years at current or even doubled rates of consumption.

Recommendations

We recommend:

1. That in view of the long lead time required for development of alternative sources of energy, energy policy place a strong emphasis on conservation.

2. That research and development aimed at increasing recovery of petroleum from known oil fields be actively encouraged.

3. That there be speedy investigation of the continental shelves for oil and gas resources.

OIL

Production of Crude Oil and Natural Gas Liquids

Data on annual production of crude oil in the United States and Canada are provided by the American Petroleum Institute in cooperation with the American Gas Association and the Canadian Petroleum Association (Anonymous, 1972a, 1973a). However, production data for the other nations of the world are assembled by the Oil and Gas Journal (Anonymous, 1973b, 1973d, 1974a) and they do not appear in government bureau reports until a year or more later, when they are published by the Bureau of Mines' International Petroleum Annual (Southard, 1973, 1974). When published, these data for foreign production are essentially the same as those given earlier by the Oil and Gas Journal. Two secondary sources are the International Petroleum Encyclopedia (McCaslin, 1973a, 1974a) that is edited and published by the Oil and Gas Journal, and the Minerals Yearbook that is published by the Bureau of Mines: the statistics in both compilations are essentially the same as the earlier published ones by the same organizations. Data on natural gas liquids (NGL) are more difficult to obtain, but the total for crude oil plus NGL production during 1971 was 553 million tonnes (41.5 x 10^9 bbl) (Albers et al., 1973, p. 1, 126, 127) versus 464 million tonnes (3.5 x 10^9 bbl) for crude oil alone in the United States. The tonnages of crude oil plus NGL for South America during 1971 were 240 versus 226 millions (1.8 v. 1.7 x 10^9 bbl). Information on NGL for most of the rest of the world is not reported, is included with the crude oil, or does not exist because the NGL was flared off with the natural gas. The world distribution of reported crude oil production alone is given by Figure 1 for 1971, 1972, and 1973; that for crude oil plus NGL is given by Figure 2. Annual production of crude oil alone (Figure 1) is concentrated in the Middle East, which during 1973 produced about 38 percent of the world total, followed by Asia (mainly U.S.S.R.) with 17.1 percent, followed by the United States with 16.7 percent. This was the first year that oil production in the United States has not exceeded that in Asia. In fact, Figure 1 depicts a steady decline in oil production during the three year period by the United States and a steady increase by both the Middle East and Asia.

Annual Production of Crude Oil and Natural Gas Liquids from Offshore Fields

Production of crude oil plus NGL from the continental shelf or other underwater regions during 1971 was tabulated by Albers, et al. (1973) who considered Lake Maracaibo (Venezuela) production to be from land rather than from underwater. Data for later production of crude plus NGL are lacking, so Figure 3 is based upon production of crude oil alone taken from the Oil and Gas Journal (McCaslin, 1972, 1973b). Unfortunately, the Oil and Gas Journal listings for crude oil production in underwater parts of other countries during 1972 and 1973 are only for giant offshore oil fields and not for all fields (the giant offshore fields contribute about 95 percent of the total offshore crude oil production). Thus the data for 1971 are not quite comparable with those for the world during 1972 and 1973. The main trend observed in Figure 3 is a steady decrease in offshore production of crude oil from the continental shelves of the United States, in contrast with a slight increase in production from other underwater areas of the world, averaging about 18 percent of total production for the three years. Concentrations are in the Persian Gulf, Lake Maracaibo, Gulf of Mexico, Gulf of Guinea, southern California, Bass Strait (Australia), South China and Java seas, and the Caspian Sea in decreasing order of annual production.

Annual Consumption

Data on annual consumption of refined petroleum products were compiled and published for the various countries of the world by the World Petroleum Encyclopedia (McCaslin, 1973a, 1974a) and a year later by the Bureau of Mines' International Petroleum Annual (Southard, 1973, 1974). Data are not available from the Bureau of Mines for 1973. Data for individual Communist countries are not available for any years. Results from both sources are similar but not identical; therefore, in order to present comparable data for 1971, 1972 and 1973, Figure 4 is based entirely upon the World Petroleum Encyclopedia (McCaslin, 1973a, 1974a) that also contains the latest revised figures for consumption during the earlier two years. The regions of consumption of oil products (demand for refined oil) differ markedly from the regions of production (Figure 4 versus Figures 1 and 2). For example, the United States consumed about one-third more than it produced in terms of crude oil (increasing from 31 percent more in 1971 to 41 percent in 1972 to 47 percent in 1973), but Europe consumed about 21 times more than it produced. When production is expressed in terms of crude oil plus NGL the consumption for the United States during 1971 exceeded production by about 14 percent. Shipments to the United States and Europe came from most of the other regions, with the Middle East shipping 93 percent of its production. Consumption per capita ranged from 3.5 tonnes (26 bbl) for the United States to 0.1 tonne (0.75 bbl), for Africa, with the average for the world minus the United States and Europe at 0.3 tonne (2.25 bbl). Comparison of the data within Figure 4 shows that consumption increased 19 to 25 percent between 1971 and

1973 for Asia, South America, and Africa, but only 13 to 14 percent for the other regions including the United States and Europe.

Cumulative Production

Cumulative production of crude oil plus NGL was tabulated for 1971 by Albers, et al. (1973), but as no later data on NGL are available, the only way to have more up-to-date cumulative production figures is to base them upon crude oil alone. Accordingly, cumulative production in the different regions is taken from the only public source, the Oil and Gas Journal (Anonymous, 1973d), which as for previous years, cumulates to July 1 (Figure 5). As this compilation gives no data for cumulative crude oil production for the United States or for communist nations, data for these areas had to come from the American Gas Association (Anonymous, 1973a, p.10) through 1972 for the United States and, from Albers, et al. (1973) for communist nations through 1971 (assuming that the communist nations produced little NGL). Both cumulations were updated to July 1, 1973 using annual production data from the International Petroleum Encyclopedia (McCaslin, 1974a). Noteworthy is the fact that the United States has produced more than one-third of the world's total crude oil produced to date. Comparison of the data in Figure 5 with those of Figure 1 shows little relationship between cumulative crude oil production and annual crude oil production. For example, the cumulative crude oil production for North America, Europe, and South America is 27 to 22 times 1973 production, whereas for Asia, Oceania, the Middle East and Africa it is only 13 to 7 times 1973 production. This lack of relationship is due to differences in the dates when initial large production began.

Cumulative offshore production of crude oil from the continental shelves and other underwater areas was tabulated by McCaslin (1974b) to the end of 1973. His tabulations to the end of 1971 were for total offshore crude oil (McCaslin, 1972), but those to the ends of 1972 and 1973 were only for giant oil fields. The best approximation seemed to be that of adding to the cumulative offshore inside-oil production through the end of 1971, the annual offshore productions for 1972 and 1973. The minor fields appear to add only about 5 percent to the world total for giant fields during 1972 and 1973, so the presentation in Figure 6 is nearly correct. The ratio of cumulative total crude oil production to cumulative offshore crude oil production at the end of 1973 (Figures 5 and 6) ranges from 2.0 for South America (where production from Lake Maracaibo dominates) to 49 and 32 for Asia and Europe where offshore oil has been minor through 1973.

Proved Reserves

Total proved reserves of crude oil as of January 1, 1974 (Figure 7) are entirely from a compilation of data by the Oil and Gas Journal (Anonymous, 1973c). An earlier compilation, at the end of 1971, by Albers, et al. (1973) of the United States Geological Survey, includes NGL for the United States and a few other nations, but omits it for most nations because they do not report NGL. Most of those that produce NGL include it with the crude oil. Comparison shows the results given on Figure 7 to be within 25 percent of the estimates by Albers, et al. and of two oil companies that provided records from their files; differences between these various sources are erratic as though due to unsystematic differences in methods of estimation from drillhole data in various regions. It is reiterated that estimates of proved reserves are smaller than the amount of oil expected to be produced ultimately from presently discovered fields. Comparison of Figures 7 and 1 shows little relationship of proved reserves to 1973 production of crude oil. The ratio ranges from 10 for the United States to 45 and 68 for the Middle East and Europe, with 31 as the world average. This ratio is not nearly as significant as popularly considered, because it is not a simple measure of how much oil remains to be produced. Instead, it is influenced considerably by the intensity of exploration, the dominant kinds of oil field traps and the price of the oil. Another ratio, that of crude oil reserves to 1973 annual consumption, also is of interest. This one ranges from only 3.2 for Europe (destined to increase with expanded exploration in the North Sea) and 5.6 for the United States to 183 for Africa and 650 for the Middle East. The ratio essentially is a measure of the number of years that proved reserves in a region can supply crude oil for the consumption in the same region, in the unlikely event that no oil is shipped, that no new reserves are found, and that consumption remains constant.

Offshore reserves of crude oil were tabulated by the _Oil and Gas Journal_ (McCaslin, 1974b) to January 1, 1974, and these data are the basis for Figure 8. Comparison of Figures 7 and 8 shows that proved reserves of offshore crude oil in different regions range from 1.4 to 36 percent of total proved reserves in the same regions. The lowest percentage is for Asia (1.4) and Africa (7.6) and the highest is for Oceania (36.4). For the United States it is 19.6, which is slightly greater than the ratio of annual offshore to total annual crude oil production (Figures 1 and 3).

Weeks (1973) estimated the proved world offshore reserves of oil plus oil equivalent of gas (1000 m^3 of gas = 0.78 tonnes of oil or 6,040 ft^3 = 1 bbl) to be 19,000 million tonnes (143 x 10^9 bbl). This figure differs considerably from the total of 13,000 million tonnes (98 x 10^9 bbl) of crude oil from Figure 8 plus an estimated 700 million tonnes (5,250 x 10^9 bbl) of oil equivalent from Figure 13, or a total of 13,700 million tonnes (103 x 10^9 bbl) of oil plus oil equivalent.

Total Discovered Reserves

In addition to the proved reserves discussed above, there is a sizeable increment of reserves that have been discovered but are not considered to be "proved." This increment, estimated to average 50 percent, is the difference between ultimate production actually obtainable from any given reservoir, and the proven reserves assigned to that reservoir at any given time.

The reserves shown on Figure 9 represent best estimates of the total amount of recoverable oil actually discovered, including both proved reserves and the increment discussed above. Since there are no published estimates of the discovered but unproven increment, estimates of Company D were used in compiling Figure 9. There is a sizeable amount of published information on the techniques of such reserve estimation; Company D's estimates were made using well known and widely accepted techniques.

GAS

Annual production of natural gas was compiled from data in the Oil and Gas Journal (Anonymous, 1972b, 1973b, 1974a) and shown in Figure 10 for the same regions as for crude oil production. The measurements are within about 25 percent of those compiled by Albers, et al. (1973) and those in unpublished files of Company D. Little change in gas production is exhibited for the three years 1971, 1972, and 1973 by the United States, North America, and South America. Gas production decreased for Africa, but it increased 50 percent for the Middle East, 37 percent for Europe, 33 percent for Oceania, and 17 percent for Asia. The distribution of the 1973 annual production of gas (Figure 10) is very different from that of oil (Figure 1). The ratio of gas to crude oil m^3/tonne ranges from 5,000 m^3/tonne (23,500 ft^3/bbl) for Europe and 1,450 m^3/tonne (6,800 ft^3/bbl) for the United States (and North America) to less than 100 m^3/tonne (471 ft^3/bbl) for the Middle East, Africa, and Oceania. The ratio for the entire world is 500 m^3/tonne (2,350 ft^3/bbl). The high ratios for the United States and Europe are due to the nearness of the gas-producing fields to industries that can use the gas. Low ratios elsewhere are due to the high cost of liquefying and shipping the gas to industrial centers. If the amount of gas brought to the surface has the same ratio throughout the world as the average for the United States actual production (1.45), then about 2,560 billion cubic meters (90×10^{12} ft^3) of gas is brought to the surface but not used. This is nearly twice the amount that is produced and used in the world. Some of this gas is pumped back underground in order to improve recovery of oil, but most of it is flared and wasted. On the other hand, about two-thirds of the gas produced in the United States is not associated with oil. Presumably, little such unassociated gas is produced elsewhere except in Europe or Japan, and so the amount of natural gas flared to waste is correspondingly reduced below the estimate above.

Partial confirmation of the waste of natural gas in undeveloped nations is provided by incomplete statistics on gas production, gas that is marketed, gas used in repressuring, and gas that is vented or flared. These data were assembled and published by the Bureau of Mines (Koelling, 1974). The statistics for 1972 were plotted in Figure 11 to permit comparison with data presented by the other figures. The totals are similar to those of Figure 10, but the similarity is taken as an indication that losses due to flaring are grossly underreported. Nevertheless, Figure 11 clearly shows a greater utilization of natural gas (or lesser wasting of it) in industrialized regions than in undeveloped ones. Expressed in another way, the United States, all of Europe, Canada, Australia, Japan, and New Zealand market 94 percent of their gas, whereas all other nations together market only 31 percent. In the United States and Canada 84 percent of the rest of the gas is used for re-pressuring, but in the rest of the world 80 percent of it is listed as flared, but probably this is a minimum percentage.

Offshore gas production was mapped (Figure 12) from data of Albers, et al. (1973) for 1971. No data for 1972 or 1973 were found in the literature. For 1971 the United States produced 62 percent of the total offshore natural gas while producing only 19 percent of the total offshore crude oil. This has resulted in a much greater cumulative production of natural gas in the United States as compared with other oil producing regions of the world (Figure 13). The principal areas where large amounts of gas are still being flared and wasted are the Middle East, North Africa and the Gulf of Guinea. But if plans presently in process in these areas are consummated, much of this waste will be stopped in the reasonably near future. Considerable but unknown quantities of NGL are flared off with the natural gas.

Proved Reserves

Proved reserves for natural gas (Figure 14) are only 11 times 1973 production (Figure 10) for the United States, in comparison with 42 times for the entire world. The largest ratios of proved reserves to 1973 production are for Africa (530), Oceania (310), Middle East (220), and Asia (80). However, the high ratios in industrially undeveloped regions are simply the result of low annual reported production of gas (instead, pumping it back underground or flaring) where its use is impractical or where liquefying and shipping is too costly.

UNDISCOVERED RECOVERABLE RESOURCES OF OIL AND GAS

United States

Many estimates of undiscovered resources of oil or gas in the United States have been made during the past three decades by men associated either with oil companies or with the U.S. Geological Survey (McCulloh, 1973). Best known are those of Weeks (1948, 1958 and 1960) formerly of Standard Oil Company of New Jersey (now Exxon Company) and now a consultant, and Hubbert (1956, 1959, 1962, 1966, 1967, 1969, 1971 and 1974) formerly of Shell Oil Company and now of the U.S. Geological Survey. Essentially, Weeks estimated the areas and volumes of sediment in major basins of the world and multiplied these by the concentrations of oil in similar basins arranged in three groups according to degrees of favorability for petroleum. His estimate of undiscovered crude oil plus natural gas liquids was 22,300 million tonnes (167 x 10^9 bbl) for land and ocean floor of the United States in 1960. In contrast, Hubbert (1967) based his estimates of undiscovered reserves of oil and gas of the United States upon statistical projections of past oil or gas production and of drilling experience. Using this method, Hubbert estimated, as of January 1, 1967, 3,200-8,500 million tonnes (24-64 x 10^9 bbl) of undiscovered oil plus NGL and 5,000-14,000 billion m^3 (180-480 x 10^{12} ft^3) of natural gas for land and adjacent continental shelf exclusive of Alaska.

Many large oil companies that have active research programs continuously compile information upon petroleum potential in new and old areas. Mostly, this information is considered proprietary, but some of it was provided by Company C (did not wish to be identified here), whose estimate in 1973 for for recoverable crude oil plus NGL likely to be discovered in the United States between 1973 and 1985 is 7,300 million tonnes (55 x 10^9 bbl.) Far more complete information was provided by Company D, which was highly cooperative in this study. Its estimates for the United States updated to April 15, 1974 are 11,900 million tonnes (89 x 10^9 bbl) of crude oil plus NGL, including 7,200 million (54 x 10^9) in offshore areas. For natural gas its estimates are 12,600 billion m^3, including 4,800 billion m^3 (170 x 10^{12} ft^3) in offshore areas. Another independent estimate of 12,000 million tonnes (90 x 10^9 bbl) was formulated by Company E as of 1974. Larger estimates of undiscovered recoverable natural gas for the United States were made by Rossinier (1973): 33,000 billion m^3 (1,165 x 10^{12} ft^3) including 7,000 billion m^3 (250 x 10^{12} ft^3) from offshore fields.

The other group of estimates was made by men of the U.S. Geological Survey. First was Hendricks (1965) who extrapolated the production per unit area of drilled basins throughout the rest of the basins or to similar basins, and thereby developed an estimate of undiscovered resources of crude oil amounting to 40,000 million tonnes (301 x 10^9 bbl) and of natural gas amounting to 37,000 billion m^3 (1,300 x 10^{12} ft^3) in the United States as of January 1, 1962. Recently, Theobald, Schweinfurth, and Duncan (1972) of the U.S. Geological Survey made new estimates of undiscovered producible resources of the United States using an extension of the method of Hendricks (1965). Their estimate for crude oil plus NGL was 61,000 million tonnes (458 x 10^9 bbl) and for natural gas it was 56,000 billion m^3 (1,980 x 10^{12} ft^3). Separate computations indicated that 26,000 million tonnes (195 x 10^9 bbl) of the crude oil plus NGL and 24,000 billion m^3 (850 x 10^{12} ft^3) of the natural gas was from offshore areas. Several times as much oil and gas was reported present but in concentrations too small for recovery by present methods. Publication of these estimates aroused considerable opposition from men who had studied the question of petroleum resources because the Geological Survey figures are so much larger than other ones. In a Senate hearing, Hubbert (1974) pointed out that the estimates made by a Survey Geologist, A. D. Zapp, in 1962 considered that the richest parts of the basins that were selected for drilling by oil companies were typical of the entire basins. In other words, the amount of oil to be produced, according to the U.S. Geological Survey, would be proportional only to the number of wells drilled, with no importance attached to differences in the geology within different parts of the basins. Actual drilling experience, however, shows that the oil produced per well in a given field or region decreases with the number of wells drilled, sometimes expressed also as barrels of oil discovered per foot of exploratory drilling.

A new set of estimates then was prepared by the U.S. Geological Survey (McKelvey, 1974) with lower figures attributed to consideration of newly available geophysical data. These estimates are 27 to 54 x 10^9 tonnes (200-400 x 10^9 bbl) of crude oil plus NGL, including 8 to 17 x 10^9 tonnes (64-128 x 10^9 bbl) in offshore areas. For natural gas the new results are 28-57 x 10^{12} m^3 (990-2,000 x 10^{12} ft^3) including 11 to 23 x 10^{12} m^3 (390-810 x 10^{12} ft^3) in offshore areas. The revised estimates are much lower than the ones of 1972, but they still are considerably higher than those made by other men and organizations.

Table 1 compares the estimates of undiscovered recoverable resources of crude oil plus NGL and of natural gas in the United States.

TABLE 1 Estimates of Undiscovered Recoverable Oil Resources of the United States

	Oil & NGL		Gas	
	(10^9 tonnes)	(10^9 bbl)	(10^{12} m^3)	(10^{12} ft^3)
Oil Companies				
1. Company A (Weeks, 1960)	22.3	(168)	----	
2. Company C (1973)	7.3**	(55)	----	
3. Company D (1974)	11.9	(89)	12.6	(450)
4. Company E	12	(90)	----	
U.S. Geological Survey				
5. Hendricks (1965)	46	(346)	37	(1,300)
6. Hubbert (1967)	3.2-8.5*	(24-64)*	5-14*	(180-500)*
7. Theobald, et. al. (1972)	61	(458)	56	(1,980)
8. McKelvey (1974)	27-54	(200-400)	28-57	(990-2,000)
9. Hubbert (1974)	9.6	(72)	15.3	(540)

* Exclusive of Alaska
** Estimated discoverable between 1973 and 1985

The Panel's review of various estimates and consultations with various men and organizations involved indicate that five different methods of estimating undiscovered hydrocarbon resources have been employed: (a) straight volumetric, (b) geologic basin analysis, (c) probabilistic exploration/ engineering analysis, (d) analysis of historical production and discovery data, and (e) analysis of discovery index. Method (a) was used in estimate 6. A combination of methods (a) and (b) was used in estimates 7 and 8. Method (b) was used in estimate 5. Methods (c), (d), and (e) were used in estimate 4. Methods (d) and (e) were used in estimates 2 and 9. Method (a) and the combination of methods (a) and (b) used by various members of the U.S. Geological Survey yielded results which appear high, whereas the estimates 2, 3, 4, 5, and 9 above, using methods (b), (c), (d), and (e) yielded reasonably consistent results.

A breakdown of estimate 4 furnished by Company D, and a comparison with a breakdown of the 1974 Geological Survey estimate (see Appendix) indicates that most of the differences between estimate 8 and estimate 4 and 9 lie in resources estimated for the conterminous 48 states.

The estimates were reviewed by the panel with various men involved. In attempting to reconcile the differences between estimate 8 and estimates 2, 4, 5, and 9 it became evident that certain factors used in estimate 8 could have been more rigorously derived. Particularly critical is the discovery ratio assumed for unexplored parts of basins in making estimate 8 (see Appendix). The low figures for undiscovered resources were calculated on the basis of a discovery ratio of 0.5, the high figures on the basis of a ratio 1.0. Both ratios appear to be too high to be used in calculating undiscovered resources of the conterminous 48 states in which exploration has been carried on for more than 100 years. Hubbert (1974b) has rigorously appraised the value of the ratio based on drilling, discovery, and production data covering all explored basins in the conterminous United States. He found the value of the ratio, with a high degree of certainty, to be very near 0.1. When this ratio is applied to the portion of estimate 8 representing undiscovered resources in the conterminous 48 states, estimate 8 is reduced to 16 billion tonnes (approximately 120×10^9 bbl).

Upon review of these several estimates and the methodologies upon which they are based, it is the judgment of the panel that the undiscovered hydrocarbon resource base of the United States including Alaska onshore and offshore, approximates 15 billion tonnes (113 billion bbl) of crude oil and NGL, and 15 trillion cubic meters (530×10^{12} cubic feet) of gas. Although there is unavoidable uncertainty in these figures, the uncertainty is insignificant when viewed in the context of the enormous difference between the size of these resources and those of coal and of shale oil (see Table 3).

All estimates are in agreement that the bulk of undiscovered oil resources will be found either offshore or in Alaska. In both areas development will be slower and more costly than on-land development. Also, both terrains present added problems stemming from the need to consider the effects of oil production on relatively unknown or extremely delicate ecosystems.

FUTURE RATES OF U.S. PETROLEUM PRODUCTION

The analyses by Hubbert (1956, 1957, 1959, 1962, 1966 and 1967, and Appendix) indicate there are definite mathematical relationships between ultimate reserves, changes in rates of discovery with time, changes in proved reserves with time, and rates of petroleum production. Given an ultimate reserve (production to date + proved resources + unknown recoverable resources) of around 33 x 10^9 tonnes (247 x 10^9 bbl), a substantial increase in U.S. annual production of crude petroleum, even for a short period, is very unlikely. Given the long lead time necessary for development of alternate sources of energy, it seems evident that conservation of petroleum should receive a strong emphasis in U.S. mineral policy if dependence upon imports of petroleum is to be reduced.

UNRECOVERED RESOURCES OF PETROLEUM IN KNOWN OIL FIELDS OF THE UNITED STATES

Oil produced in the United States plus proved reserves of oil are approximately 18 x 10^9 tonnes (136 x 10^9 bbl). It is well known, however, that by use of present methods only a part of the total oil in place in known oil fields is recoverable by use of present technology. Percent of recovery varies widely from field to field, depending on the characteristics of the contained oil and the characteristics of the reserves. In a recent symposium on tertiary recovery methods (Snyder, 1974) a range from 13.5 to 46 percent was cited. A firm figure for the average recovery percentage for all oil fields of the United States is not available, but 30 percent appears reasonable, whereas 40 percent is probably too high. Even if the latter figure is accepted, however, it means that in known oil fields some 27 x 10^9 tonnes (202 x 10^9 bbl) remains unrecoverable, roughly twice the estimated unknown recoverable resources. Known unrecovered oil thus appears to constitute the largest single untapped oil resource of the United States.

Research aimed at improving recovery percentages has been carried on for many years, and substantial improvements have been achieved since the earlier days of the petroleum industry. Primary recovery from ordinary wells has been supplemented with marked success by secondary methods of gas reprocessing and water flooding. It is generally agreed that further improvement by developing tertiary recovery methods will not be easy, but in view of the energy resources at stake, research and development of improved methods of recovery should be actively encouraged as a part of national mineral policy.

One means of increasing recovery from known fields is the mining of oil-rich sands from reservoir beds of oil fields where wells have reached the point of uneconomic production. Where the sands are shallow enough, they can be mined by stripping, like shallow coal beds. According to Herkenhoff (1972) and Anonymous (1974b), there are 383 known shallow oil fields (overburden less than 150 meters [500 feet]) in the United States. If these were mined, the oil recovery might be increased from the usual 25 to 40 percent attained by wells to perhaps 90 percent. Similarly, if new techniques of tertiary recovery are developed, about twice as much additional oil might be produced as has come from past cumulative production, perhaps yielding 14,000 million tonnes (105 x 10^9 bbl) for the United States and 39,000 million tonnes (290 x 10^9 bbl) for the world (Figure 4).

WORLD OIL RESOURCES

Estimates of undiscovered resources of oil and gas for the world have been compiled only by oil companies; material is available from Weeks (1960), from Company C for 1973, and from Company D for 1974. The estimates for crude oil plus NGL have been plotted together on Figure 15. The wide range of the estimates is expectable in view of differences in information available to each organization. Undiscovered resources of natural gas were estimated only by Company D, with results presented in Figure 16. Comparison in Figures 15, 7 and 1 shows that undiscovered resources of crude oil in the world exceed proved reserves and they are 25 to 75 times the 1973 production of crude oil. Similarly, Figures 16, 14 and 10 show that undiscovered resources of natural gas are about 100 times 1973 world production.

Published estimates of offshore undiscovered resources of the world have been made by Weeks (1973, 1974), who combined crude oil plus NGL with natural gas (using a ratio of 1000 m^3 of gas equals 0.78 tonne of oil (6,040 ft^3 gas = 1 bbl oil)). His results for undiscovered total petroleum resources amount to 183,000 million tonnes (1,370 billion bbl) for the continental shelves, small basin shelves, and shallow seas; 61,000 x 10^6 tonnes (460 billion bbl) for the continental slopes; 12,000 x 10^6 tonnes (90 billion bbl) for the continental rises; and 3,500 x 10^6 tonnes (26 billion bbl) for deep-sea trenches and associated ridges. This total of 260,000 million tonnes (1,950 x 10^9 bbl) for undiscovered resources on the ocean floor approaches the 320,000 million tonnes (2,400 x 10^9 bbl) for Weeks' (1960) estimate for the crude oil plus NGL of the world plus the Company D's estimate of undiscovered natural gas of the world (Figure 16) converted to oil equivalent.

Soviet interest in undiscovered petroleum resources of the ocean floor is illustrated by publications on general geological factors (Fedynskiy and Levin 1970) as well as by quantitative estimates (Kalinko, 1969). The latter estimated 34,200 million tonnes (257 billion bbl) of oil and 13,444 billion m^3 (475 trillion ft^3) of gas beneath water-covered regions of the world; Soviet estimates are thus much lower than those of Weeks (Figures 15 and 16).

SPECULATIVE RESOURCES OF OIL AND GAS

The most spectacular petroleum accumulations are those in the giant oil and gas fields of the world. In fact, 70 percent of the past cumulative production plus proved reserves of oil and 50 percent of the same for gas is in the giant fields (Halbouty, et al., 1970). Probably an even higher percentage of offshore oil and gas is from giant fields, as the high costs there preclude development of small fields on the ocean floor. As shown by Figure 17, most of the giant fields occur in two broad curved belts, one in northern South America and western North America and the other in northern Africa, the Middle East, and the boundary between Europe and Asia. There and elsewhere the fields occur in clusters except in mainland China, where scattered single fields attest to incomplete exploration and an expectation of future substantial addition to production and reserves.

Oil and gas fields are widespread along many continental shelves (Figure 18). Noteworthy is their absence or rarity off eastern Asia, (except Indonesia) southern Asia, eastern Africa, northwestern Africa, eastern United States, eastern South America, western South America, northern North America and Asia, and off Antarctica even though many of these shelves appear to have high potential (Figure 19). Many of the gaps can be ascribed to climatically inhospitable regions; others are due to politically inhospitable host nations. As politics change, considerable filling of gaps in the distribution pattern of offshore oil fields may occur. Particularly promising are the ancient deltas of large rivers of the world (Figure 20). Many of these deltas are major producing areas of oil and gas. Most others are inadequately explored owing to difficulties of terrain or politics. When explored, these deltas should materially increase oil production and reserves.

Belts of thick marine sediments of Mesozoic and Tertiary age (Figure 21) contain fields that produce about 60 percent of the world's oil and gas. Most of these belts underlie coastal regions, where they have been localized by marginal troughs bounded on their oceanward sides by dams of tectonic, diapiric, or reefal origin (Emery, 1970). Because the sediments in these troughs are thick and contain much organic matter produced from nutrients in continental runoff, the quantity of oil and gas in them may well exceed the average for continental areas that are underlain by sediments. Again, most of the continental shelves of the world are less well explored than the land, so the concentrations of oil and gas beneath the shelves probably are greater than expected and listed among undiscovered reserves in Figure 15.

Lastly, nothing really is known about the oil and gas potential of the continental rises (Figure 22). The volume of sediments beneath these rises probably exceeds the total beneath the shelves. Much of the sediment is fine grained and some of it probably is rich in organic matter, having slid oceanward from positions of accumulation on the continental slope within the depth range of low oxygen content in the ocean water (Emery, 1969; and in press). Seismic-reflection records also show the presence of many velocity discontinuities, probably most of which are layers of sand distributed by turbidity currents, thus being potential reservoir beds. The same seismic records reveal the presence of numerous folds, faults, and stratigraphic traps, all of which could be sites for concentrations of oil and gas. In spite of the promise presented by continental rises, no exploratory drilling has occurred on them, largely because of the difficulty in controlling flows that might result from the drilling. Probably new methods of well completion on the deep-ocean floor will be developed during the next decade or two, and these may be followed by testing of the oil and gas potential of the continental rises of the world.

SHALE OIL

Oil has been produced from oil shales in Scotland, China, Queensland, the East Baltic (Estonia-Leningrad), and South Africa, and production from the East Baltic area furnishes about 0.8 percent of the oil production of the U.S.S.R. Production in recent decades has never been, however, more than a small fraction of annual world oil production. In the United States oil has been produced only in experimental runs and one large pilot operation.

Data for the present report are drawn from reports by Duncan and Swanson (1965), Duncan (1967), Padula (1969), the National Petroleum Council (1972), and Culberson and Pitman (1973), and are summarized in Table 2. It must be stressed, however, that data even for the Green River oil shales, which have been more extensively explored and sampled than any other major shale oil deposits, are still incompletely tested. The figures in Table 2 serve only to indicate that United States and world resources of shale oil are very large, far larger than estimated total United States and world resources of conventional petroleum.

The sharp increase in prices of crude petroleum by the Organization of Petroleum Exporting Countries during 1973-74 has placed shale oil resources in an entirely new economic context. Marginal at best in 1972, some of the richer shale-oil deposits may now be economic. Tracts in the Green River oil shale basins have been leased, and mining and processing projects have been undertaken. Estimates of oil resources in the Green River formation differ considerably.

Duncan and Swanson estimated 21.3 x 10^9 tonnes (160 x 10^9 bbl) of oil in shales of the Green River formation, averaging 10.5-12 wt. % of oil (30-35 gal./ton) of which half was considered recoverable under conditions of 1965. The National Petroleum Council (1972) estimated 12 x 10^9 tonnes (90 x 10^9 bbl) to be worth present consideration, the remainder of the total of 240 x 10^9 tonnes (1,800 x 10^9 bbl) in the formation being deeply buried, too low in grade, or insufficiently explored. Even this amount, however, is nearly equal to the estimated total of proved and undiscovered recoverable resources of conventional petroleum of the United States. Tracts leased by the Department

TABLE 2 Shale-Oil Resources

		Resources (Billions)				Grade	
		Identified		Hypothetical		Wt. Percent	
	Area	Tonnes	(Bbl)	Tonnes	(Bbl)	Oil	Gal./Ton
United States	Green River basins, Wyo., Colo., Utah	56	(418)	3.3	(25)	9-35	(25-100)
	Green River basins, Wyo., Colo., Utah	190	(1,400)	80	(600)	3.5-8.8	(10-25)
	Chattanooga shale, Mid-continent	27	(200)	107	(800)	3.5-8.8	(10-25)
	Alaska			33	(250)	3.5-8.8	(10-25)
	S.W. Montana			0.7-1.3	(5-10)	3.5-5.2	(10-25)
Brazil	S.E. Brazil (Irati Sh.)	40-110	(300-800)	430	(3,200)	3.5-8.8	(10-25)
	S.E. Brazil (Tertiary)	0.3	(2.0)			4-13	(11-37)
Scotland		0.04	(0.3-0.5)	Possibly >0.13	(>1.0)	5-14	(16-40)
Estonia-Russia	S.E. Baltic area	1.3	(10)			18	(50)
Russia	North Siberia	10	(78)			8.8	(25)
				470	(3,500)	3.5-8.8	(10-25)
Yugoslavia	Morava Valley	0.03	(21)		?	8.8-14	(25-40)
	Kolubara Valley				Large ?		
China	Various areas	1.9	(14)	19	(140)	5.2	(15)
Zaire	Stanleyville Basin	13.3	(100)			8.8	(25)
Zaire	Mayumbe	?			Large	?	
South Africa	Karroo	0.02	0.130		Large	7-8.8	(20-25)
Australia	Port Curtis, Qsld.	0.03	(200)			5.2	(15)
	Various, N.S.W.	0.03	(200)			?	

of the Interior during the past year are considered to cover resources of not less than 340 million tonnes (2.6 x 10^9 bbl).*

The size of the resources of shale oil, both in the United States and in the world as a whole can easily arouse false hopes of their rapid development as an alternative to conventional petroleum as an energy source. There is little prospect, however, that shale-oil deposits can provide such an alternative. The problems involved in the shale-oil development are formidable, ranging from problems of mining and processing technology to environmental problems of disposal of waste and availability of water for processing. Capital investment required for production at a level of a billion barrels a year, roughly 15 percent of current U.S. annual consumption of petroleum, is enormous. At best, shale oil can be expected to serve only as a supplement to other sources of energy within the next 10 to 15 years.

TAR SANDS

The tar sands resources of the world are incompletely known, but it is already clear that they are major world resources of petroleum. The best known deposits, and by far the most productive, are the tar sands of Alberta, with a current production of about 2,200,000 tonnes (165 x 10^6 bbl) of crude oil per year. Total resources of oil in tar sands of three areas in Alberta have been estimated at around 80 x 10^9 tonnes (600 x 10^9 bbl). Pow and others (1963) estimated 40 x 10^9 tonnes (300 x 10^9 bbl) recoverable oil, whereas Humphrey (1973) estimates 47 x 10^9 tonnes (350 x 10^9 bbl). New plants planned or proposed will greatly increase the scale of production. Large deposits of tar sand are also reported to occur on Melville Island in Arctic Canada.

United States resources of oil in tar sands are estimated at 3,900 million tonnes (29 x 10^9 bbl.). Tar sand deposits in eastern Venezuela (*Oil and Gas Journal*, 1973) are reported to contain about 93 x 10^9 tonnes (700 x 10^9 bbl) of oil, of which about one-tenth is considered recoverable with present technology.

Total world resources of tar sands are at present unknown.

COAL

The annual production of coal (Figure 25) is rather different from that of oil (Figure 1) and gas (Figure 10), although rather similar tonnages of coal and oil were produced in the United States as well as the entire world.

---

*During the past year, four tracts were leased, overlying a total of 3.6 billion tonnes (4 billion tons) of shale containing not less than 10.8% oil (30 gallons per ton), with a mean oil content of about 12.5% (35 gallons per ton). (Source: L. Schramm, USBM, by phone.)

Asia and Europe were the dominant producing regions during 1971, with 40 to 37 percent of the total respectively, as compared with their 17 and 1.3 percent of the total world oil production. Only 0.3 percent of the world total of coal was produced in the Middle East as compared with its 38 percent of the world oil. About one-quarter of the coal was lignite and the rest was bituminous and anthracite.

Proved reserves and undiscovered resources of coal in the ground (Figure 25) are even larger than those for shale oil (Figures 23 and 24). Owing to losses in coal mining and processing, however, it is generally estimated that only one-half of the coal will be recoverable. Rather close confirmation of the government estimates for coal is provided by independent estimates from the files of Company D. Reserves and resources in the United States are about 5,800 times the annual production, and for the entire world they are about 5,000 times. There is, therefore, no cause for alarm about future needs for coal during the next hundred years. Moreover it, like oil shale, can augment the supplies of natural oil and gas, both as a fuel substitute and as material for distillation of oil and gas. About 110,000 million tonnes (121,000 x 10^6 short tons) of reserves in the United States (200 times the annual production) are at depths shallow enough to be strippable, although one-third of this tonnage may require advanced machinery. Essentially 100 percent of the coal in place is obtained by stripping, but only about 50 percent is recovered by present underground mining methods. If only half of the coal reserves and resources of Figure 26 is recovered, the amount of energy available from coal is still enormously greater than the total available from oil and gas. Reserves and resources are thus very large relative to United States needs; however, serious environmental problems must be resolved before these reserves can all become available. For the present, environmental problems set limits on the scale of using coal for energy.

COMPARISON OF ENERGY FROM FOSSIL FUELS WITH EARTH'S ENERGY FROM CERTAIN OTHER SOURCES

For ease in making comparisons, some of the more pertinent statistics were drawn from the preceding figures, rounded off, and compiled in Table 3. It is evident that the proved reserves of shale oil and coal are many times larger than those for petroleum, but this does not tell the whole story because of differences in heats of combustion.

The average heats of combustion of oil (plus NGL), gas, and coal were taken as 11 Kcal/g (5.8 x 10^6 BTU/bbl) 9 Kcal/liter (1,260 BTU/ft^3), and 7 Kcal/g (12,600 BTU/lb), respectively. Multiplying these numbers by the latest data on world production (Figures 2, 10 and 25), we find that the heats produced by combustion are within a factor of 3 for these materials (Table 4). They would be more nearly equal were much of the gas used rather than being returned underground or flared, as its heat value is about 0.23 x 10^{20} cal/year (0.09 x 10^{18} BTU/year). The sum of the heat energy from the fossil fuels actually used is 0.6 x 10^{20} cal/year (6.25 x 10^{18} BTU/year); this is three times the energy of the tides, and about one-fifth the energy of the

TABLE 3 Comparisons of Production and Recoverable Reserves*

	United States		World	
1973 production of oil	450×10^6 Tonnes	(3.4×10^9 bbl)	$2,700 \times 10^6$ Tonnes	(20×10^9 bbl)
1971 production of NGL	80×10^6 "	(0.6×10^9 ")	130×10^6 "	(10^9 bbl)
1973 production of gas	0.65×10^{12} m^3	(23×10^{12} Ft^3)	1.36×10^{12} m^3	(48×10^{12} Ft^3)
1971 production of coal	0.51×10^9 Tonnes	(0.56×10^9 Tons)	3×10^9 Tonnes	(3.3×10^9 Tons)
Proved reserves of crude oil	5×10^9 "	(37.5×10^9 bbl)	80×10^9 "	(600×10^9 bbl)
Undiscovered resources of crude oil	15×10^9 "	(113×10^9 ")	150×10^9 "	($1,130 \times 10^9$ ")
Proved reserves of natural gas	7×10^{12} m^3	(250×10^{12} Ft^3)	60×10^{12} m^3	($2,100 \times 10^{12}$ Ft^3)
Undiscovered resources of natural gas	15×10^{12} m^3	(530×10^{12} Ft^3)	140×10^{12} m^3	($4,900 \times 10^{12}$ Ft^3)
Proved reserves of shale oil	0.5×10^{12} Tonnes	(3.75×10^{12} bbl)	1.6×10^{12} Tonnes	(12×10^{12} bbl)
Undiscovered resources of shale oil	3×10^{12} "	(22.5×10^{12} ")	44×10^{12} "	(330×10^{12} ")
Incomplete reserves of tar sand oil	4×10^9 "	(30×10^9 ")	100×10^9 "	(750×10^9 ")
Proved reserves and undiscovered resources of coal	3×10^{12} "	(3.3×10^{12} Tons)	10×10^{12} "	(11×10^{12} Tons)

* Numbers in this table are estimates of recoverable reserves and resources and are rounded off from the numbers given by the accompanying figures which have an unrealistic number of significant figures owing to their origin by addition of estimates.

TABLE 4 Energy Produced from Fossil Fuels Compared with Earth's Energy from Certain Other Sources*

	10^{20} Cal/Year	(10^{12} Watts)	10^{20} Cal/Year	(10^{12} Watts)
Combustion by Man				
Crude Oil and NGL	0.31	(4.1)		
Natural Gas	0.12	(1.6)		
Coal	0.21	(2.8)		
Total			0.6	(8.0)
Dissipated by Tides (Earth, water, air)			0.2	(2.7)
Radioactivity of Earth (if like chondrite)			2.5	(33.0)
Geothermal losses of Earth			3.2	(42.0)
Solar Energy at Earth's Surface			6,500	(86,000)

* Partly from Williams and Von Herzen (1974).

Earth's interior produced by radioactive decay and manifested by geothermal gradients. However, it is only 0.01 percent of the solar energy; in fact, it equals only 48 minutes of solar radiation striking the entire Earth. Even the total reserves (including undiscovered ones) have energy equal only to two days of solar radiation on the Earth. The unwary reader might conclude that solar energy offers a free ride with respect to supplies of oil, gas, and coal. However, in order to match the rate of fossil energy use, all of the sun's energy that reaches the Earth's surface within an area of 100 by 100 km near the equator would have to be captured. Moreover, at present solar energy is much less efficiently converted to electricity than is fossil fuel energy.

All in all, the low concentration of energy from the sun, the tides, and from the Earth's interior makes them unattractive at present as large scale energy sources compared with fossil fuels. Although natural oil and gas have limited lives at the presently increasing rates of use, the reserves of shale oil, tar sands, and coal are so great that fossil energy is likely to be available for several centuries to come. Their use, however, involves environmental costs: consumption of water, pollution of streams, and use of land areas for dumping of slag. These costs cannot be precisely evaluated against the value of the energy that is produced, owing to changing standards of public concern for environment versus energy.

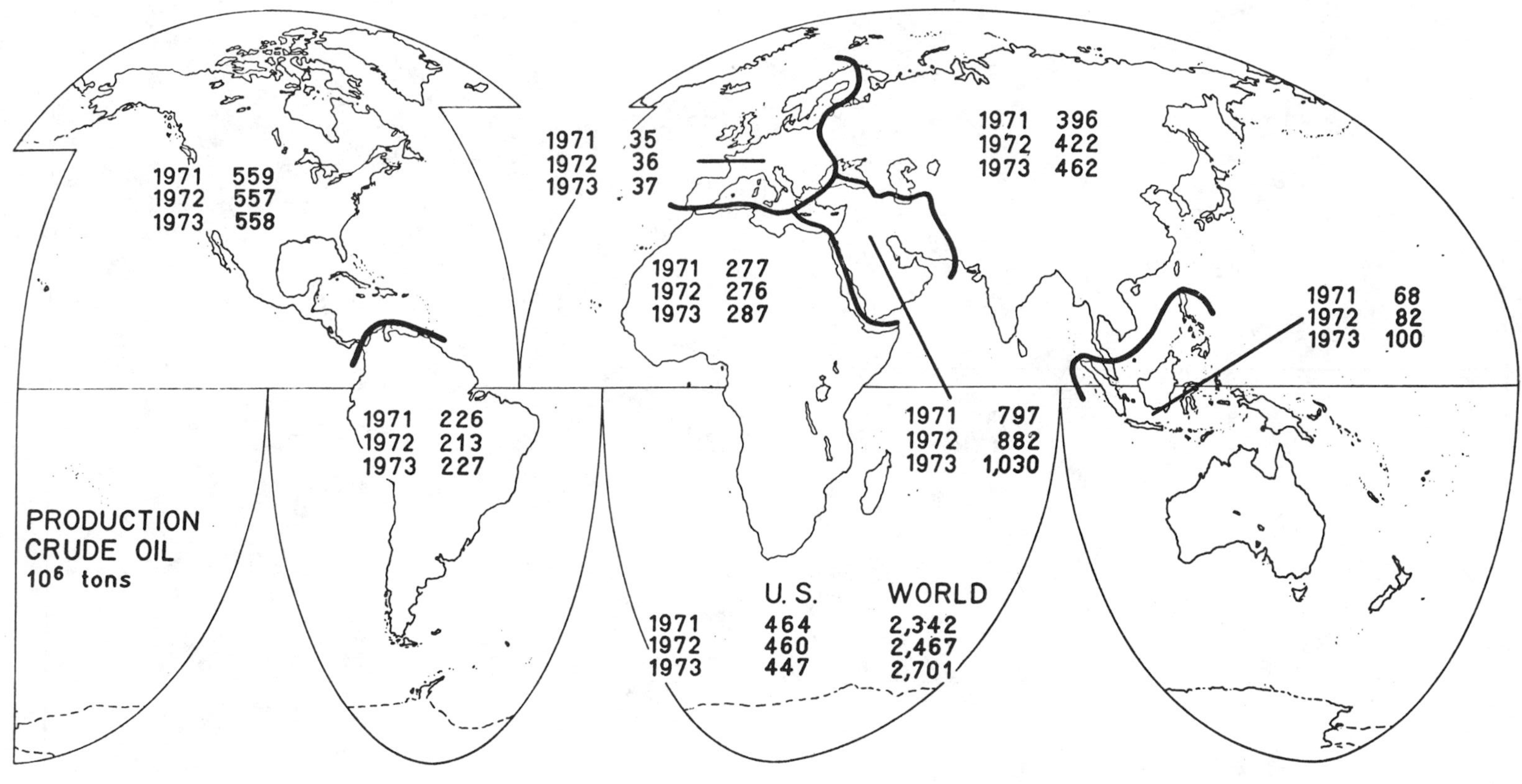

FIGURE 1: Annual Production of Crude Oil Alone (in millions of metric tons). The boundary lines separate the seven compilation regions: North America, South America, Africa, Europe, Asia, Middle East, and Oceania. Data are from Anonymous (1972a, 1973a, 1974a).

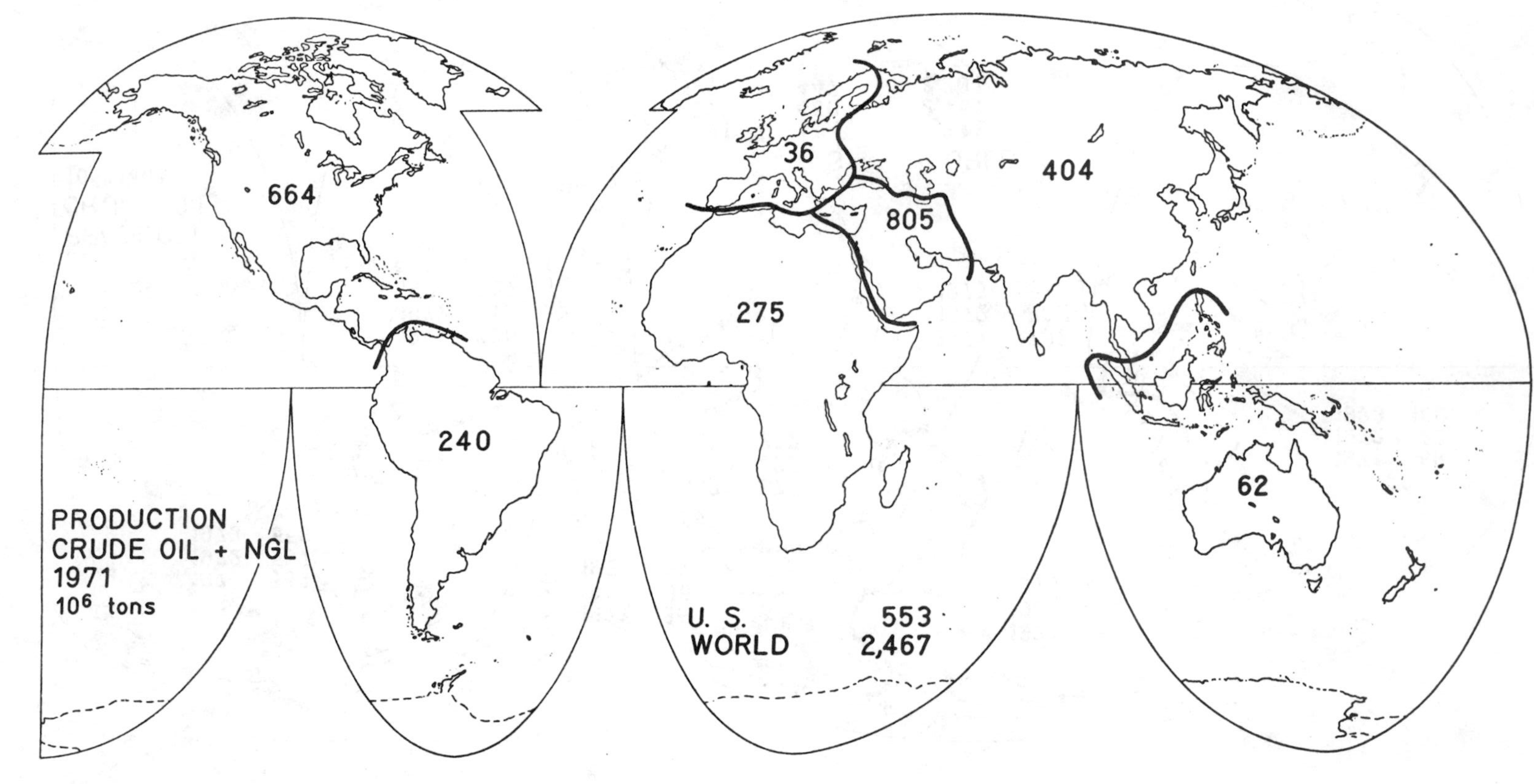

FIGURE 2: Annual Production of Crude Oil Plus Natural Gas Liquids (NGL) For 1971. Data are from Albers et al. (1973, p. 126, 127).

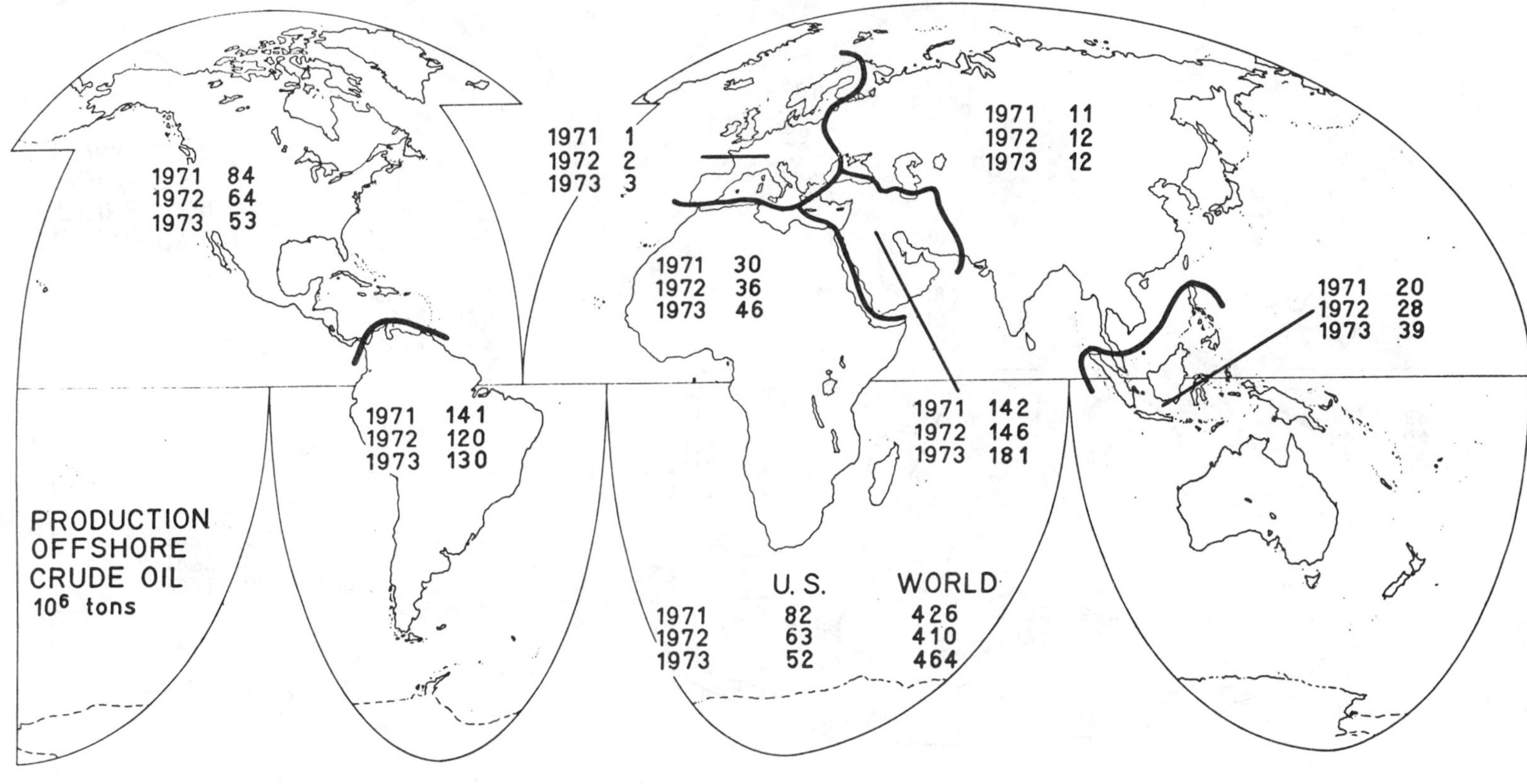

FIGURE 3: Annual Production of Crude Oil From The Continental Shelf and Other Underwater Areas. Data are from McCaslin (1972, 1973b, and 1974b).

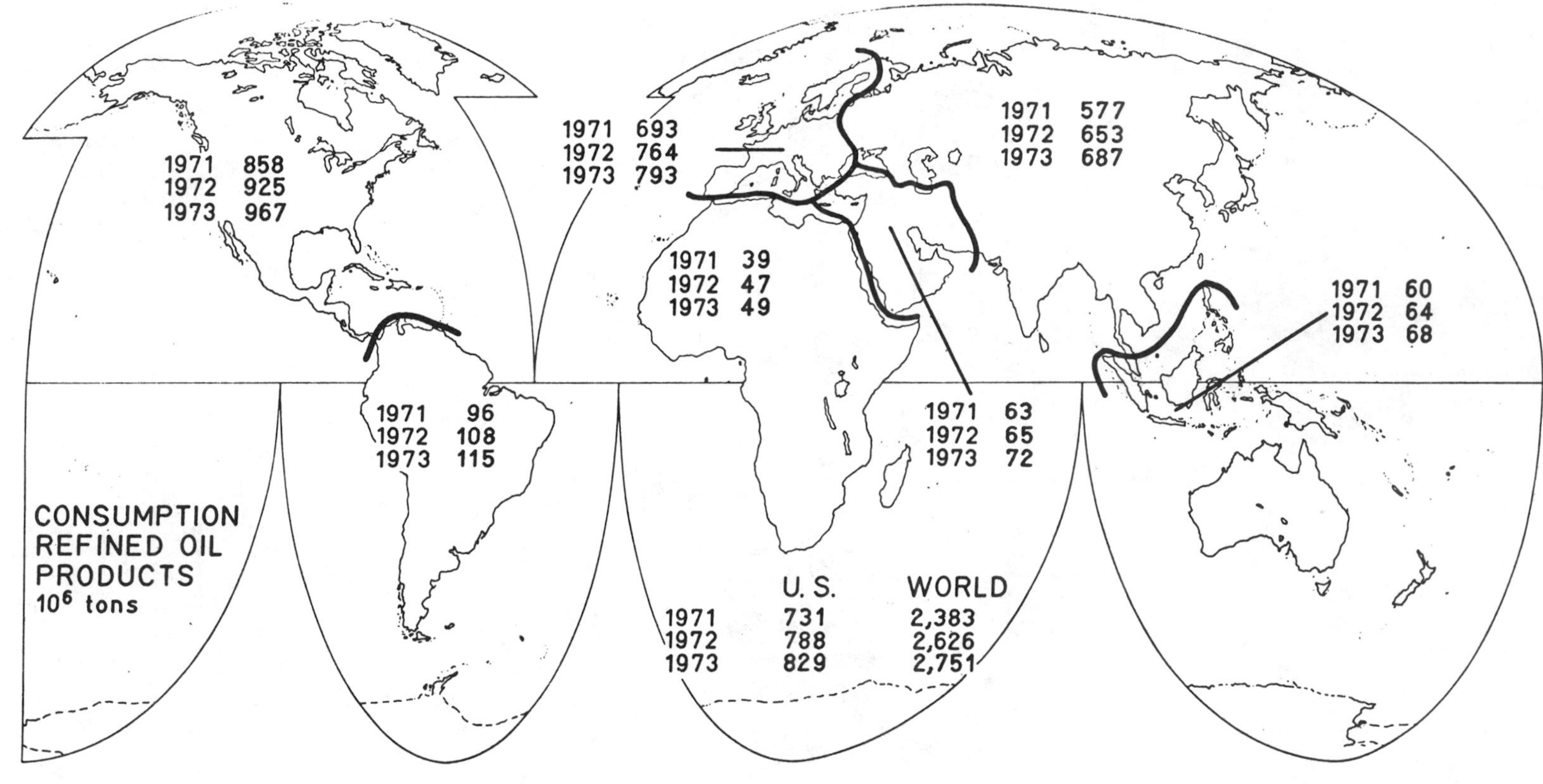

FIGURE 4: Annual Consumption of Refined Oil Products (Demand). Data are from the World Petroleum Encyclopedia (McCaslin, 1973a, p. 266-272, 1974a).

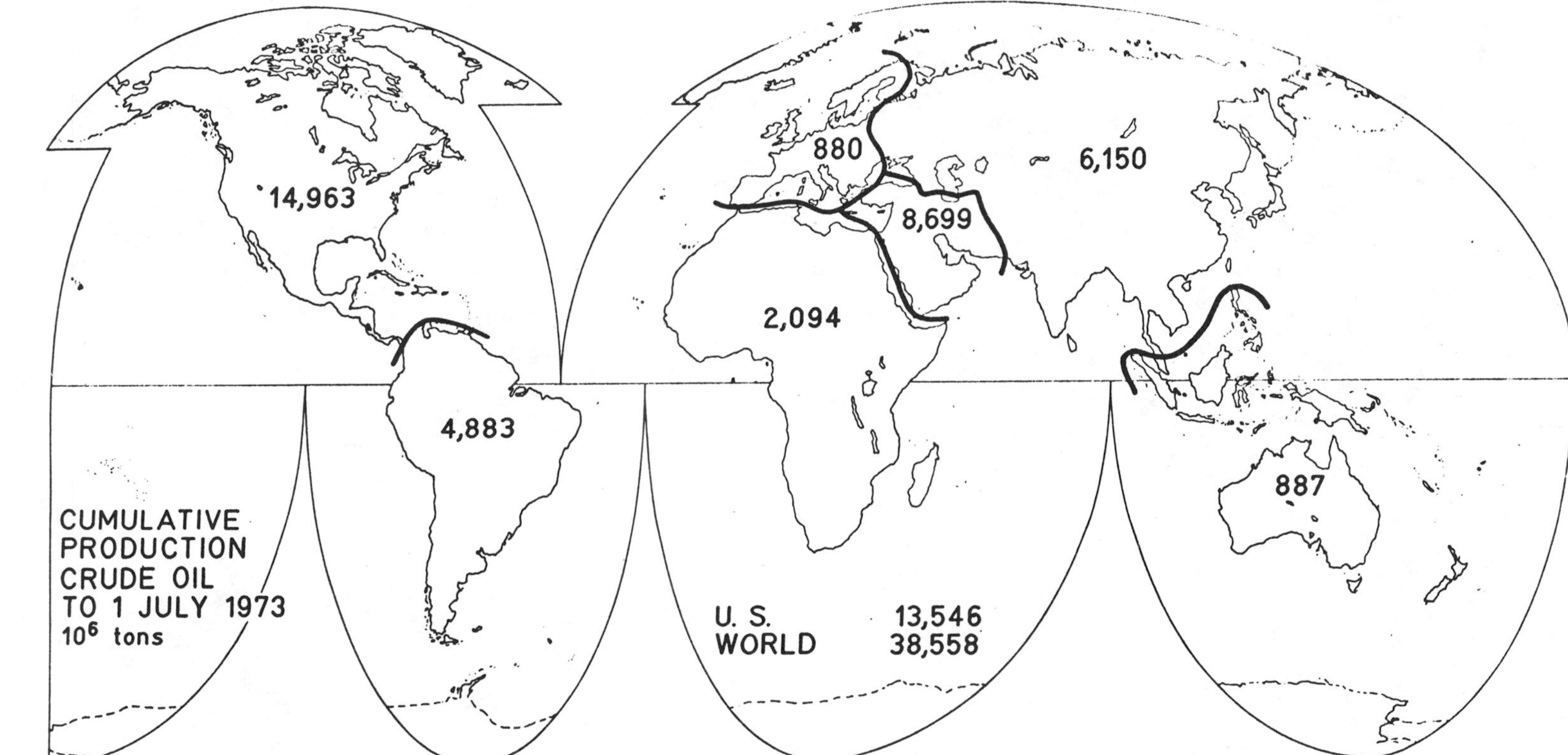

FIGURE 5: Cumulative Total Crude Oil Production. Data are from Oil and Gas Journal (Anonymous, 1973d) and Albers et al. (1973), updated by annual production data from the American Gas Association (Anonymous, 1973a) and the International Petroleum Encyclopedia (McCaslin, 1974a). Note that the compilation is to the middle of 1973, not through the end of the calendar year, as are statistics given in other maps.

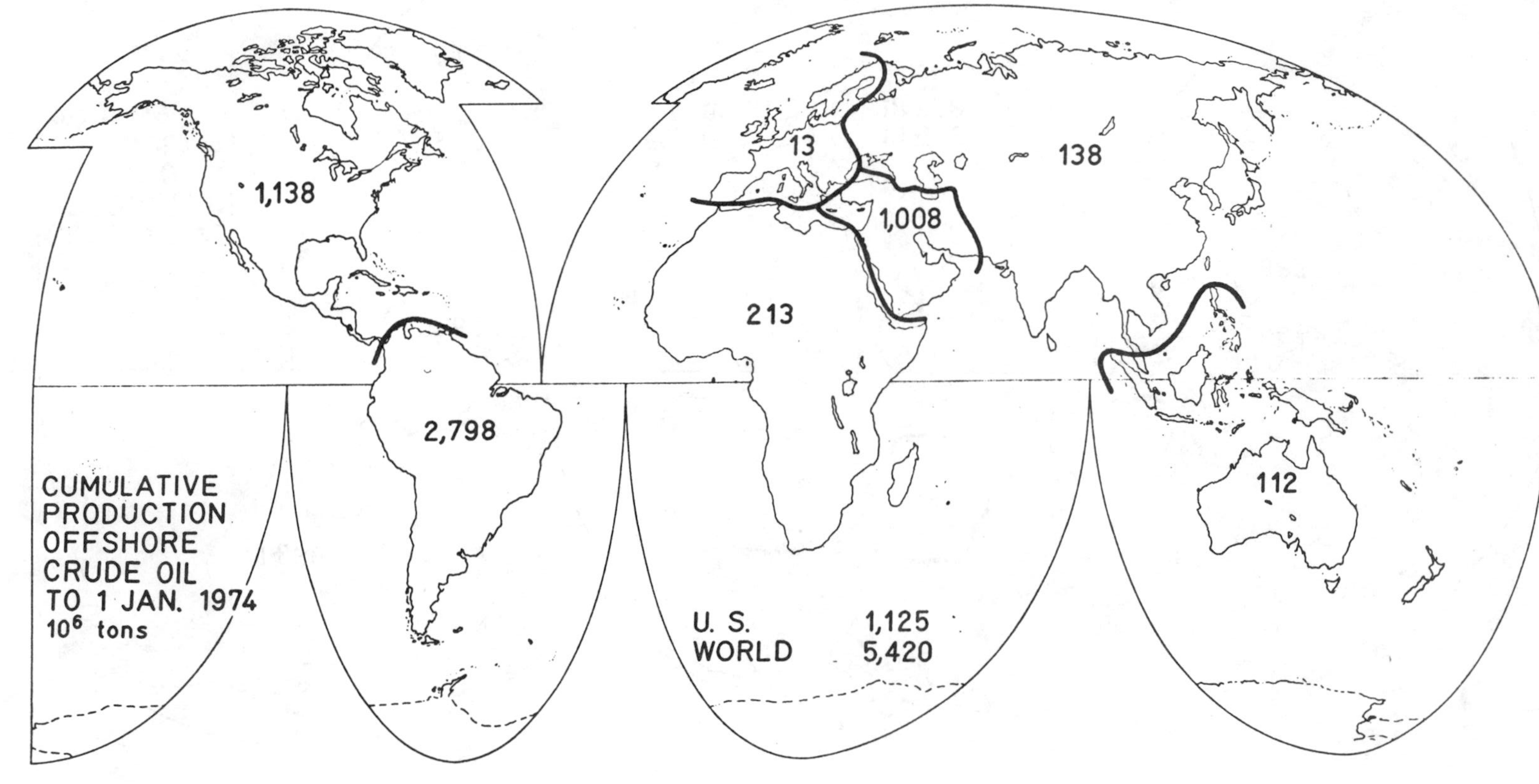

FIGURE 6: Cumulative Offshore Crude Oil Production. Data are from McCaslin (1972, 1973b, 1974b).

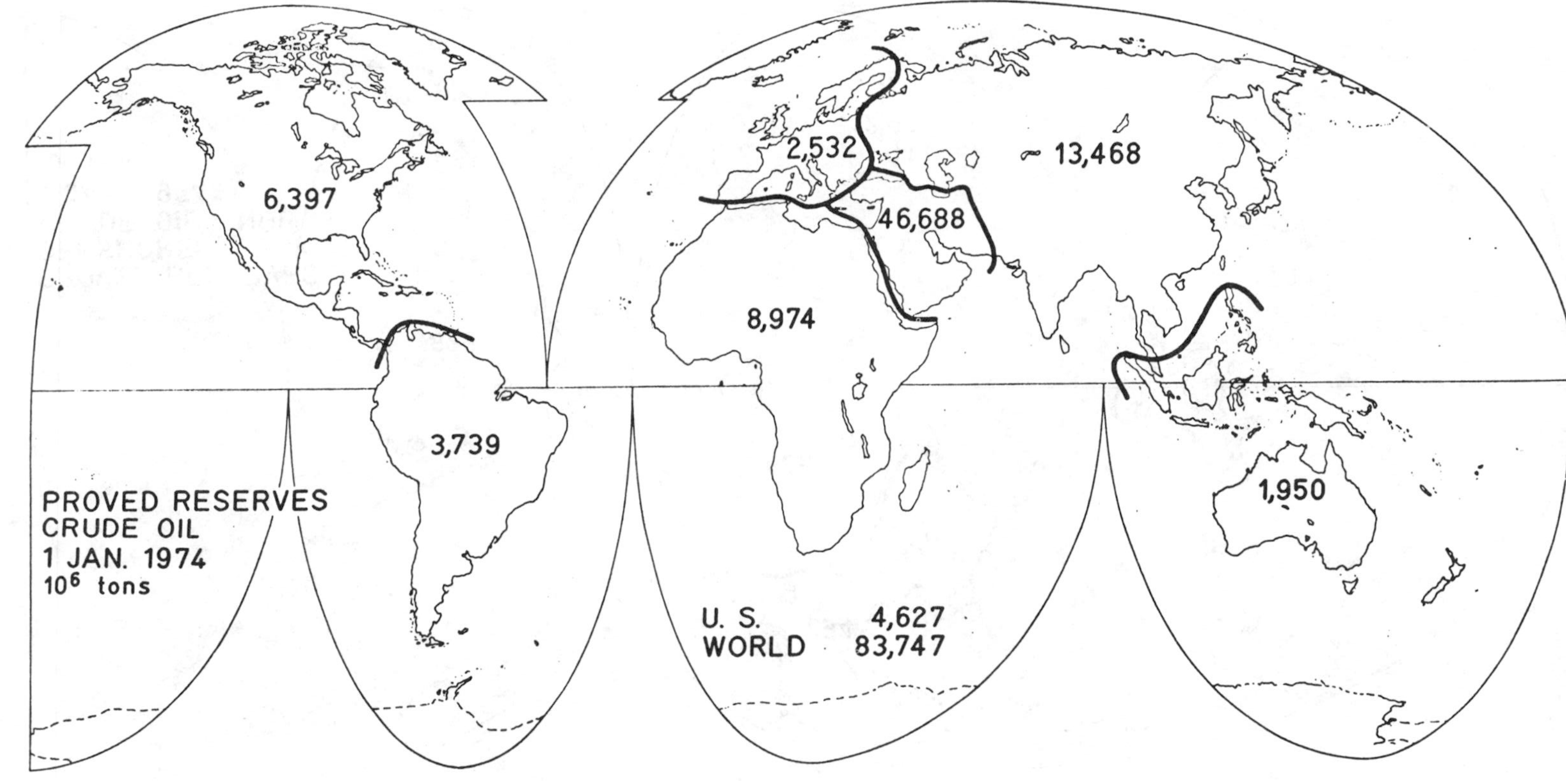

FIGURE 7: Total Proved Crude Oil Reserves. Data are from Anonymous (1973c).

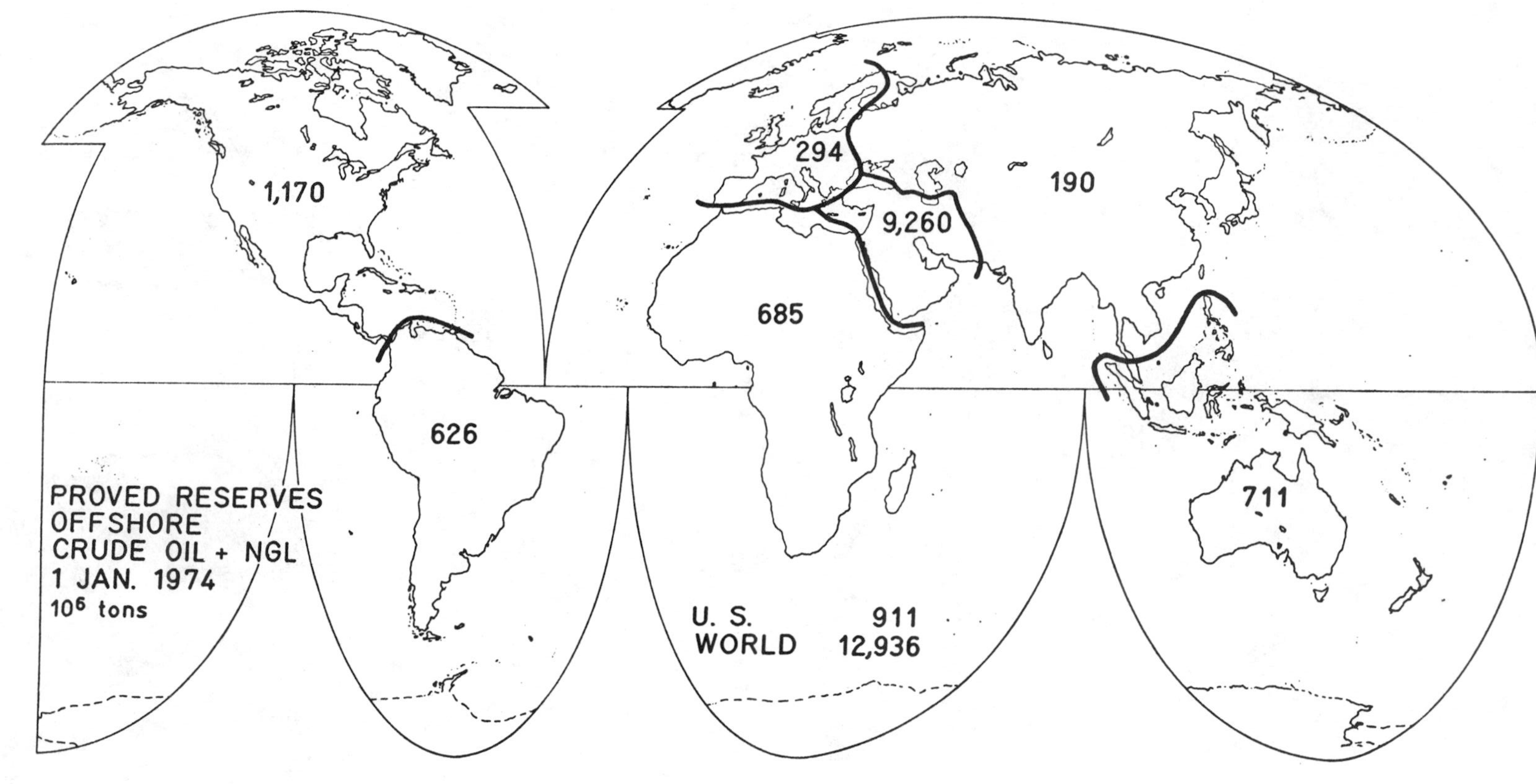

FIGURE 8: Proved Crude Oil Reserves from the Continental Shelves and Other Underwater Areas. Data are from McCaslin (1974b).

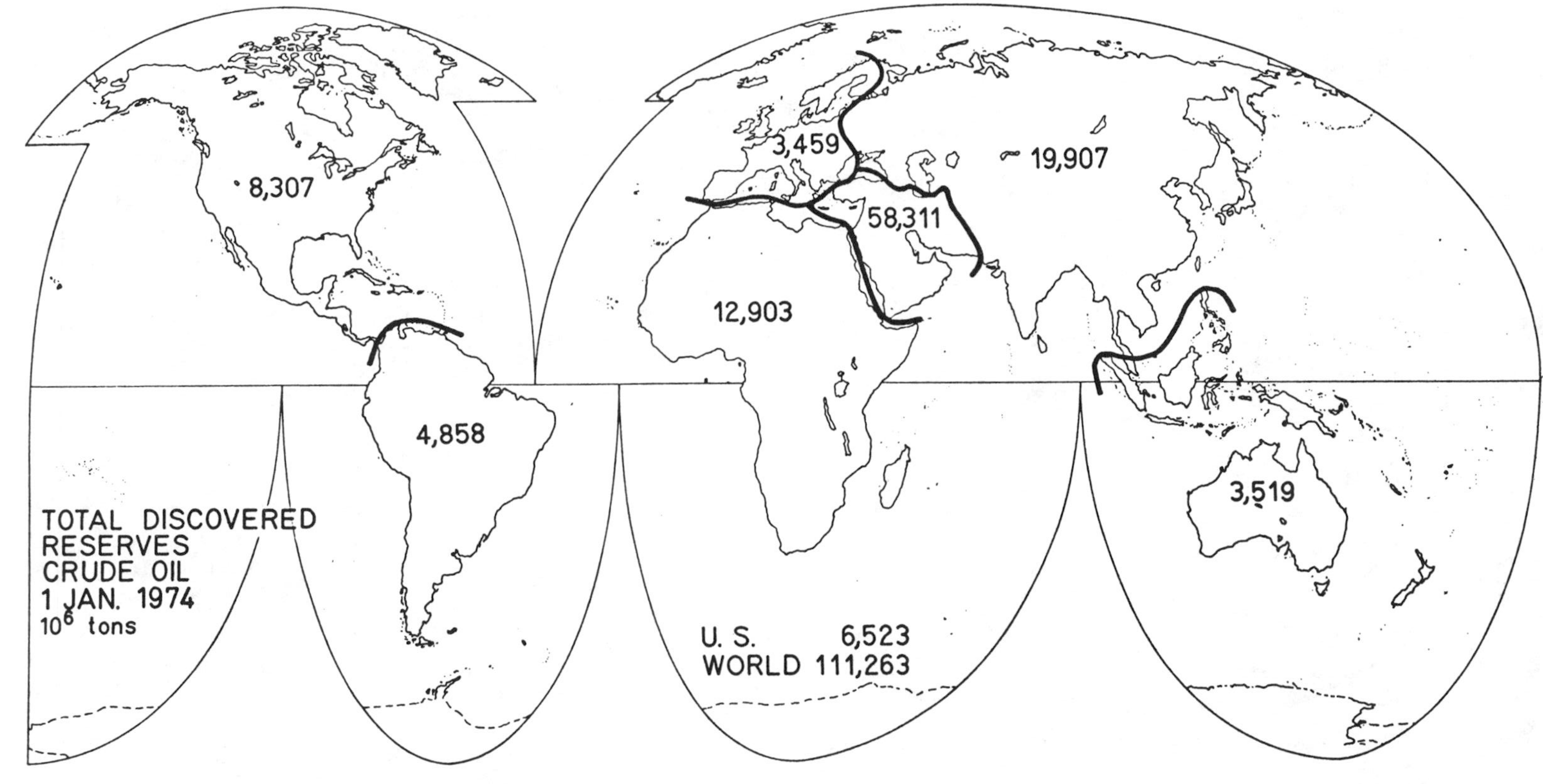

FIGURE 9: Total Discovered Reserves of Crude Oil. The increment beyond proved reserves of Figure 8 are from unpublished files of Company D.

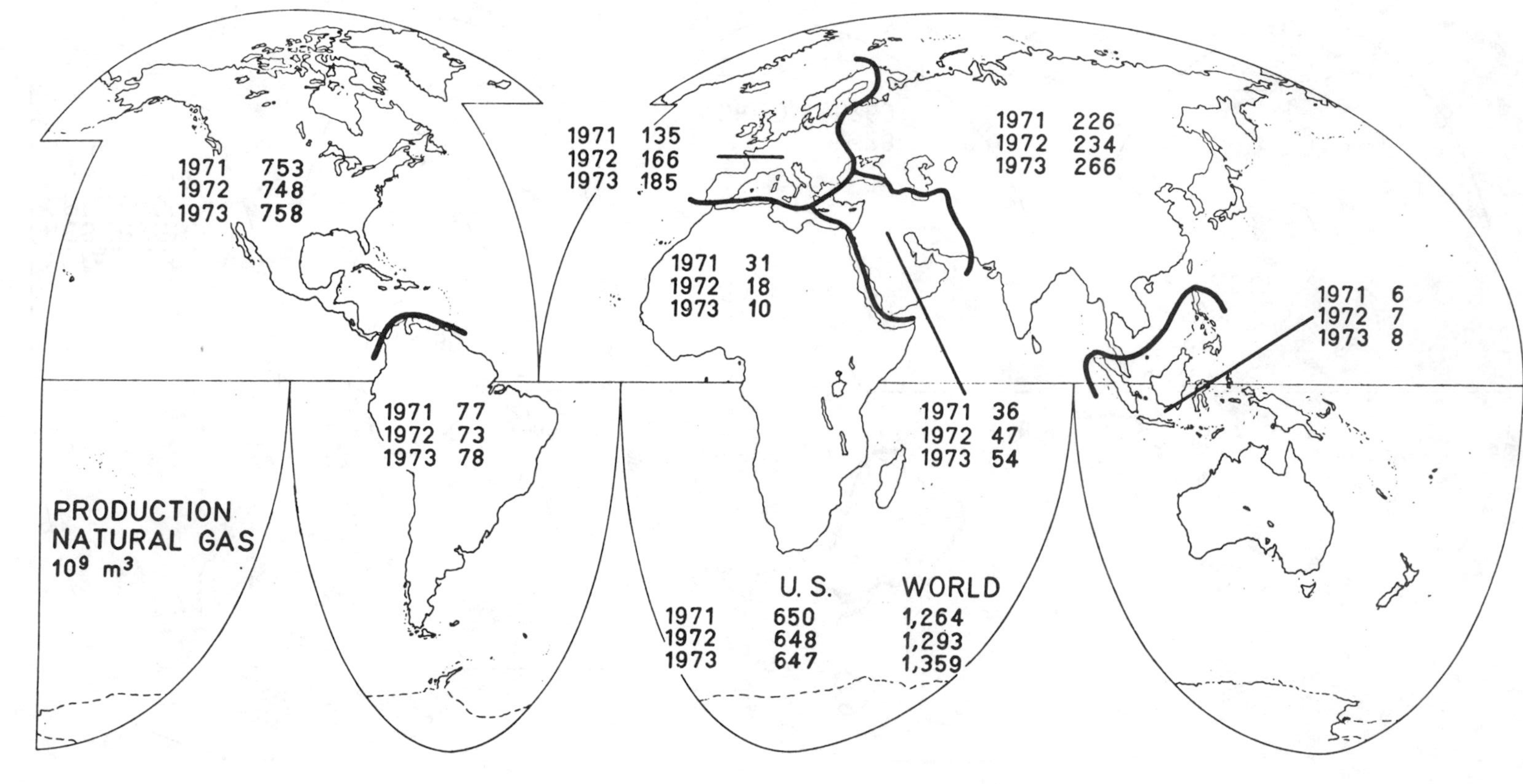

FIGURE 10: Annual Production of Natural Gas (in billions of cubic meters). Data are from Anonymous (1972b, 1973b, 1974a).

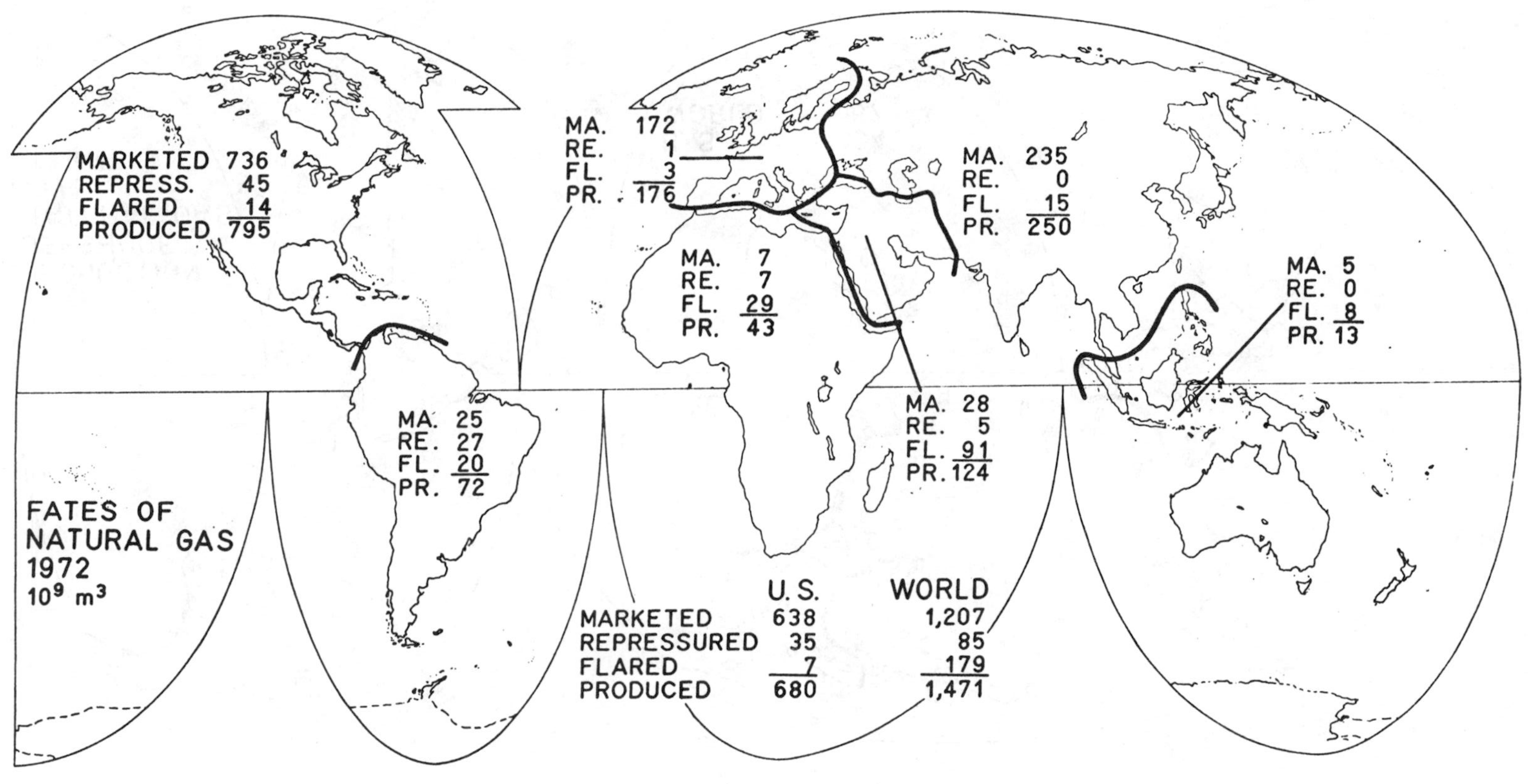

FIGURE 11: Fates of Natural Gas That is Brought to the Surface; Quantities That are Marketed, Used for Repressuring, and Vented or Flared to Waste. Data are from Koelling (1974).

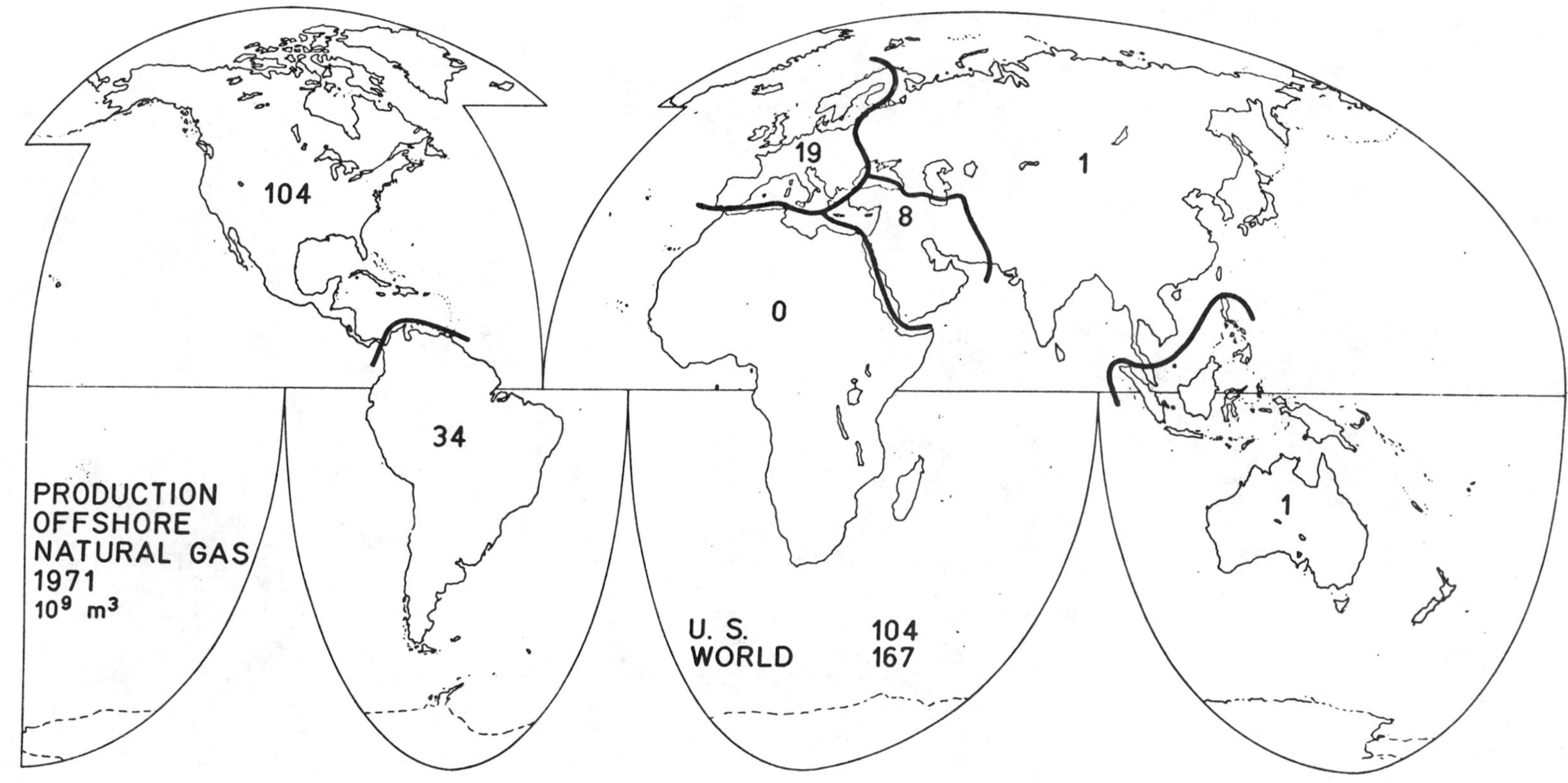

FIGURE 12: Annual Production of Natural Gas from the Continental Shelf and Other Underwater Areas. Data are from Albers et al. (1973) with estimated adjustment for production from Lake Maracaibo.

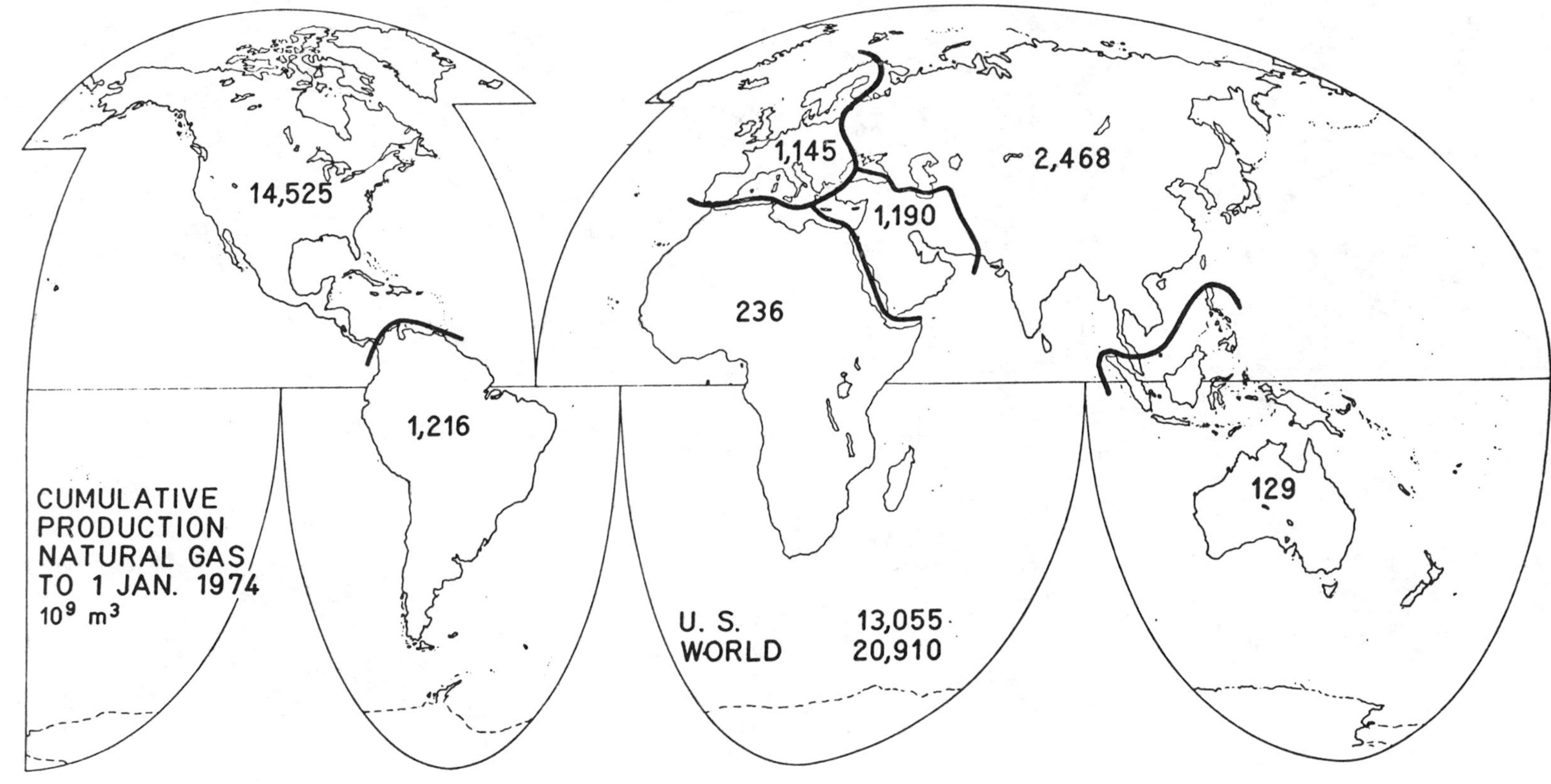

FIGURE 13: Cumulative Production of Natural Gas. Data through 1971 (Albers et al., 1973) were supplemented by 1972 and 1973 gas production from Figure 10. The total for the United States is confirmed by the estimate of Rossinier (1973).

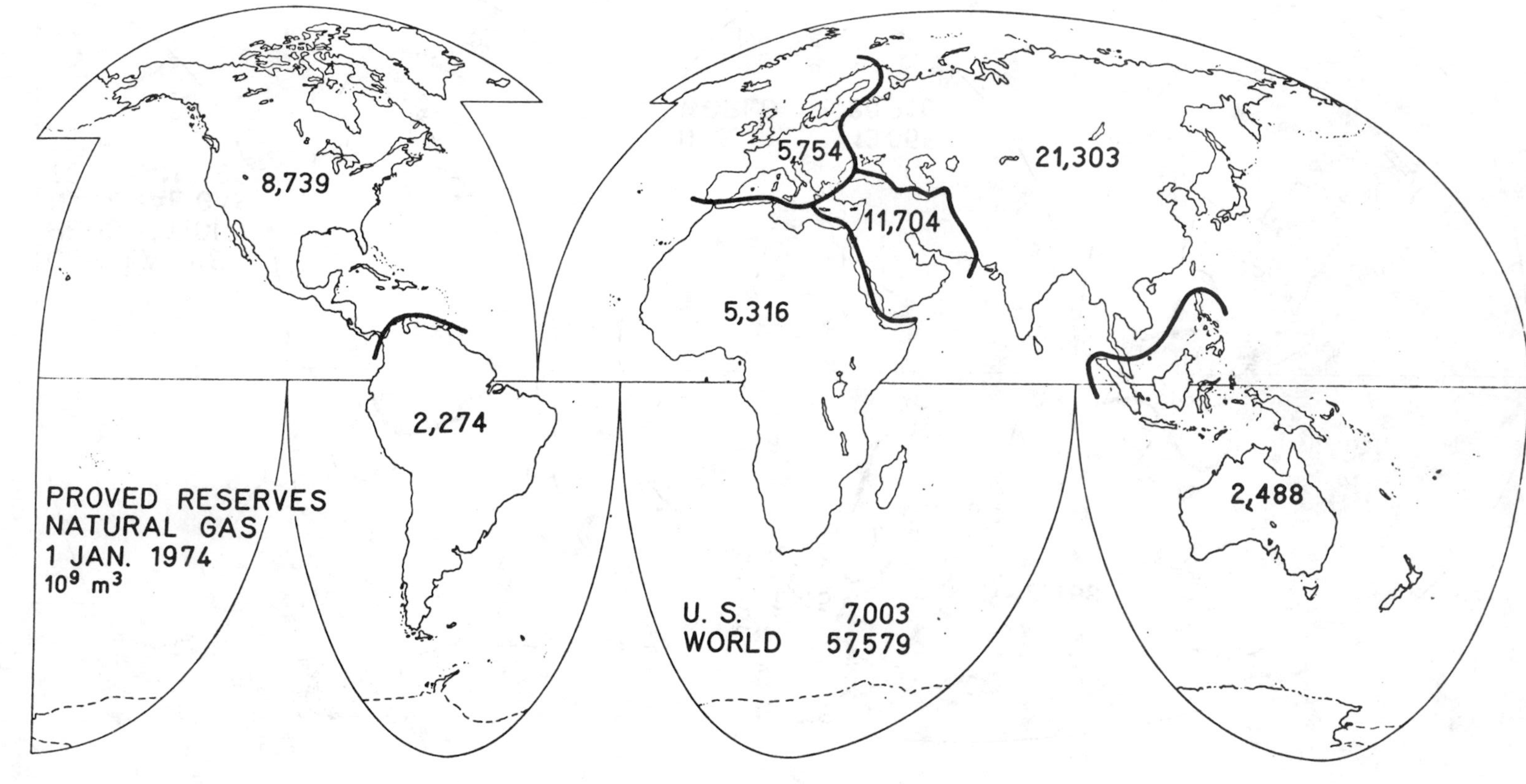

FIGURE 14: Proved Reserves of Natural Gas. Data are from Anonymous (1973c) and Rossinier (1973).

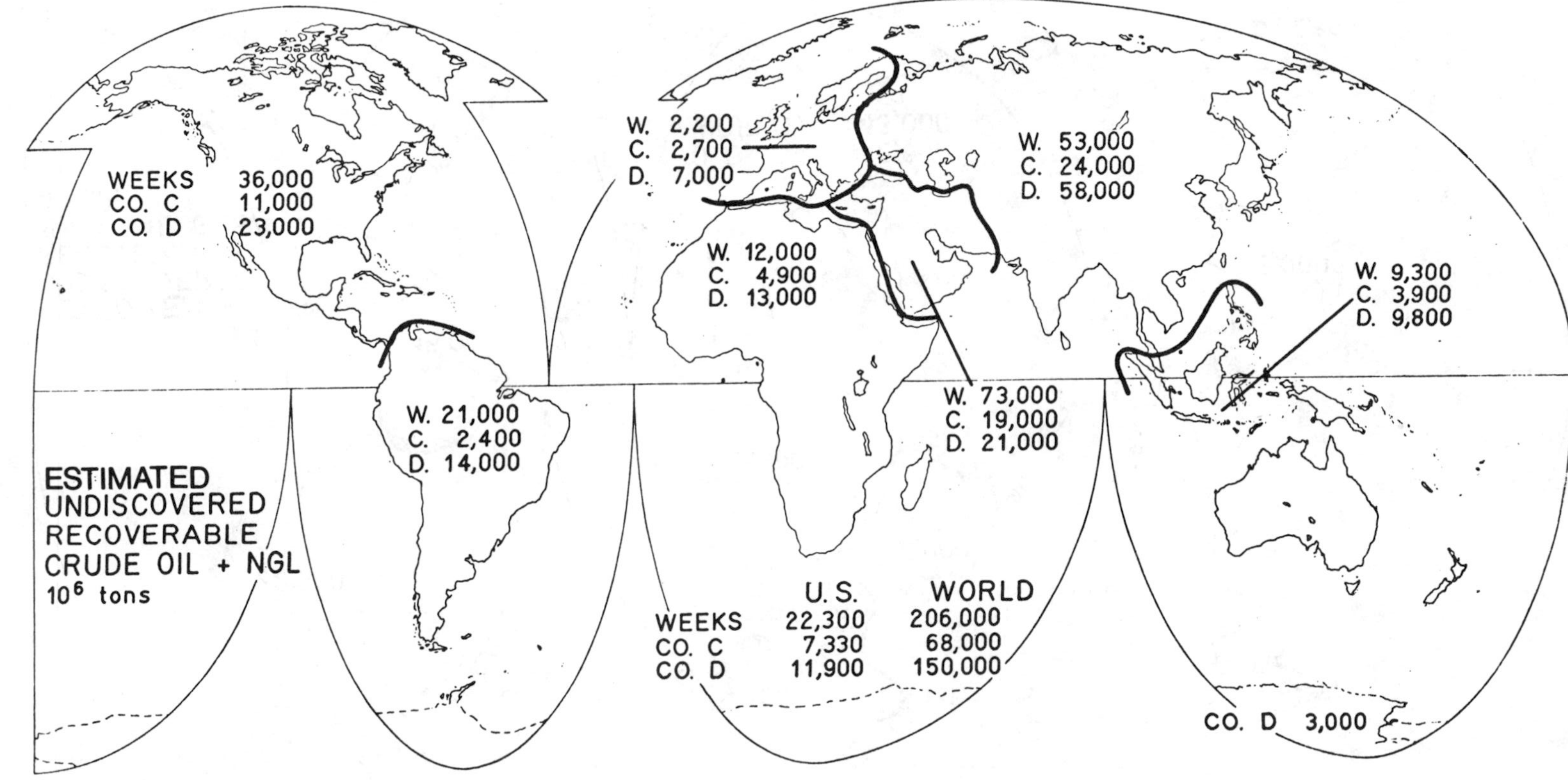

FIGURE 15: Comparison of Different Estimates of Undiscovered Resources of Recoverable Crude Oil Plus NGL in the Various Regions of the World. Data are from Weeks (1960) and from unpublished reports of two oil companies, one of which (D) listed resources to be discovered between 1973 and 1985.

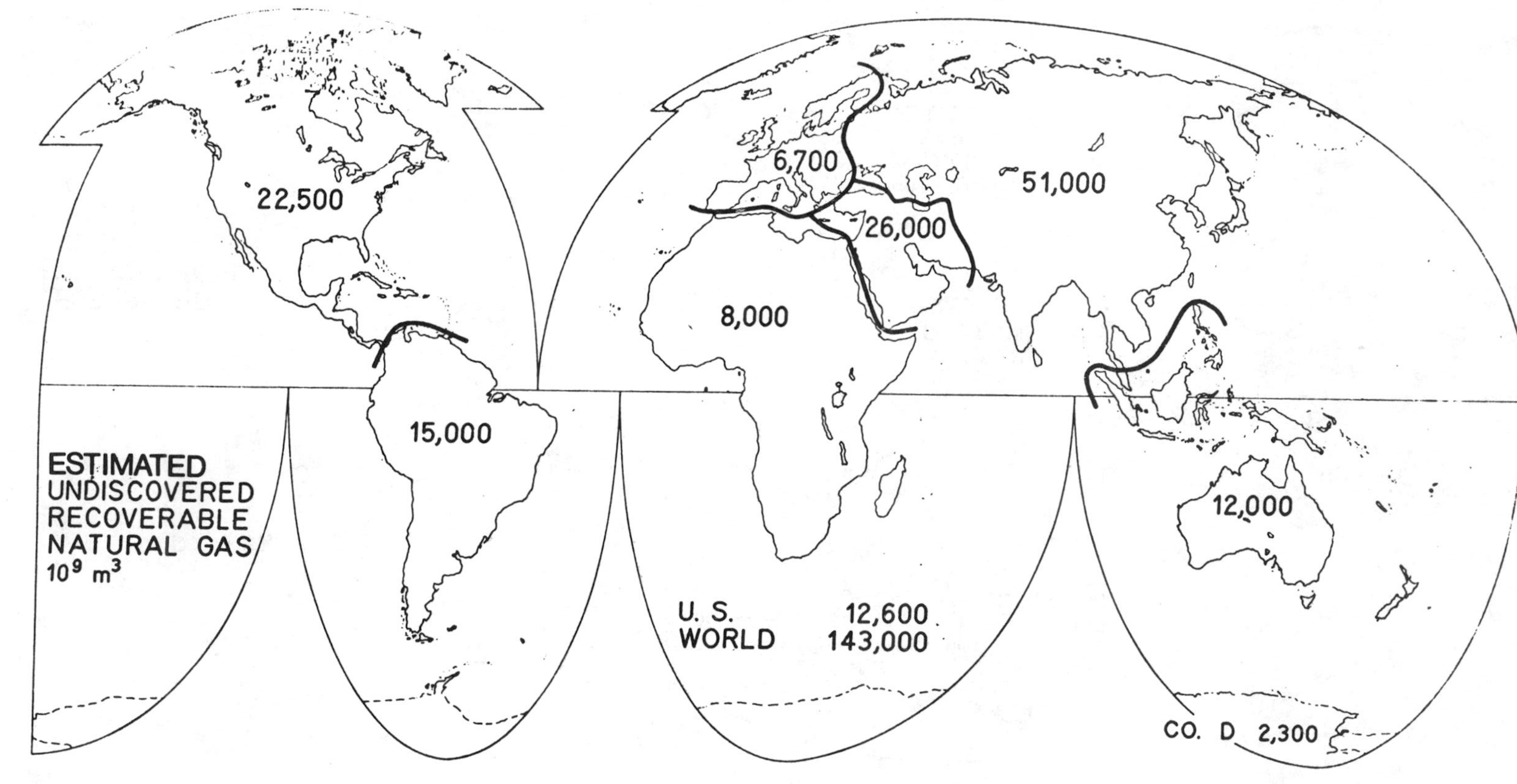

FIGURE 16: Estimates of Undiscovered Natural Gas Resources. From unpublished reports of Company D.

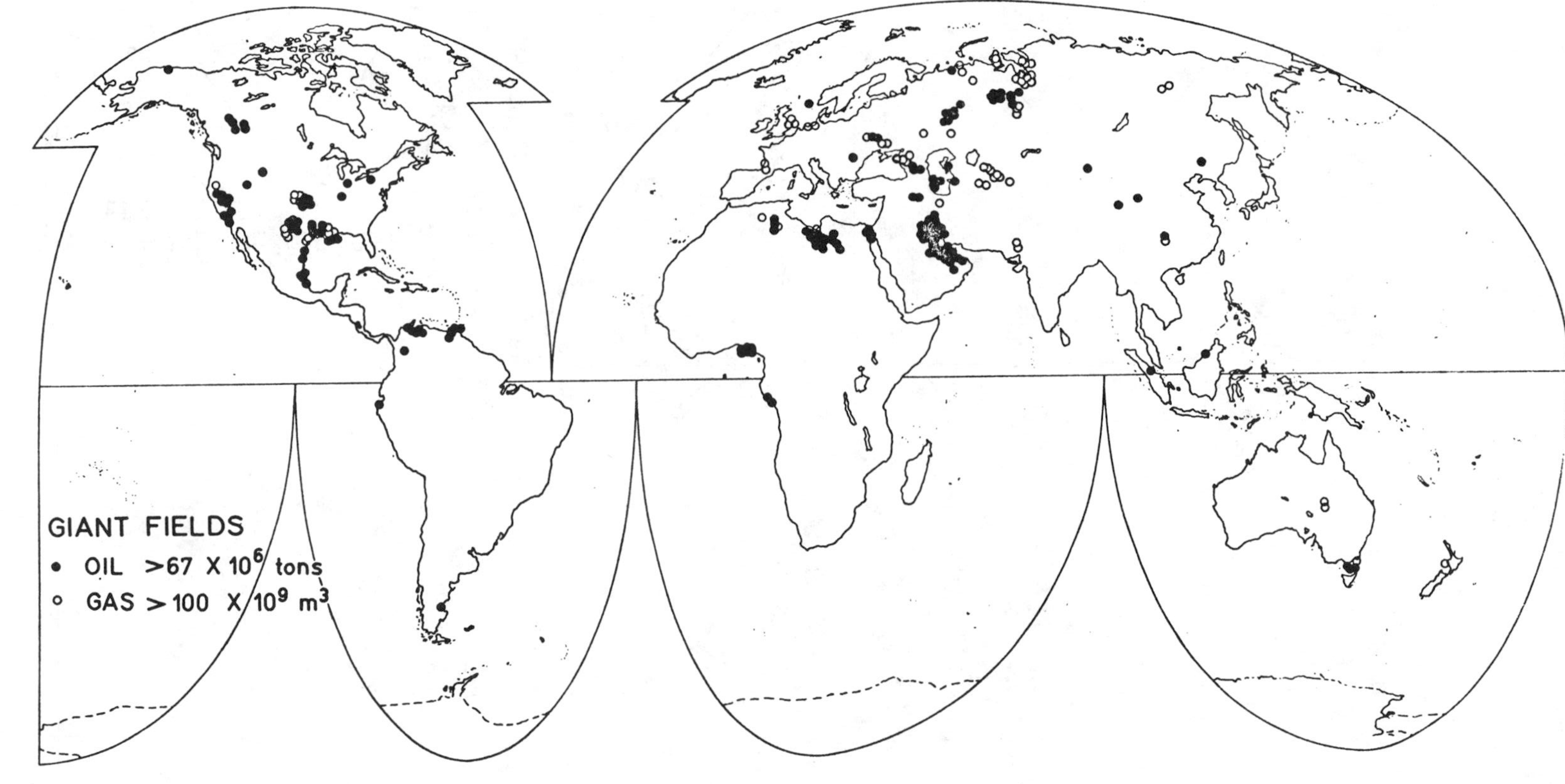

FIGURE 17: Distribution of the World's Giant Oil and Gas Fields. From Halbouty et al. (1970).

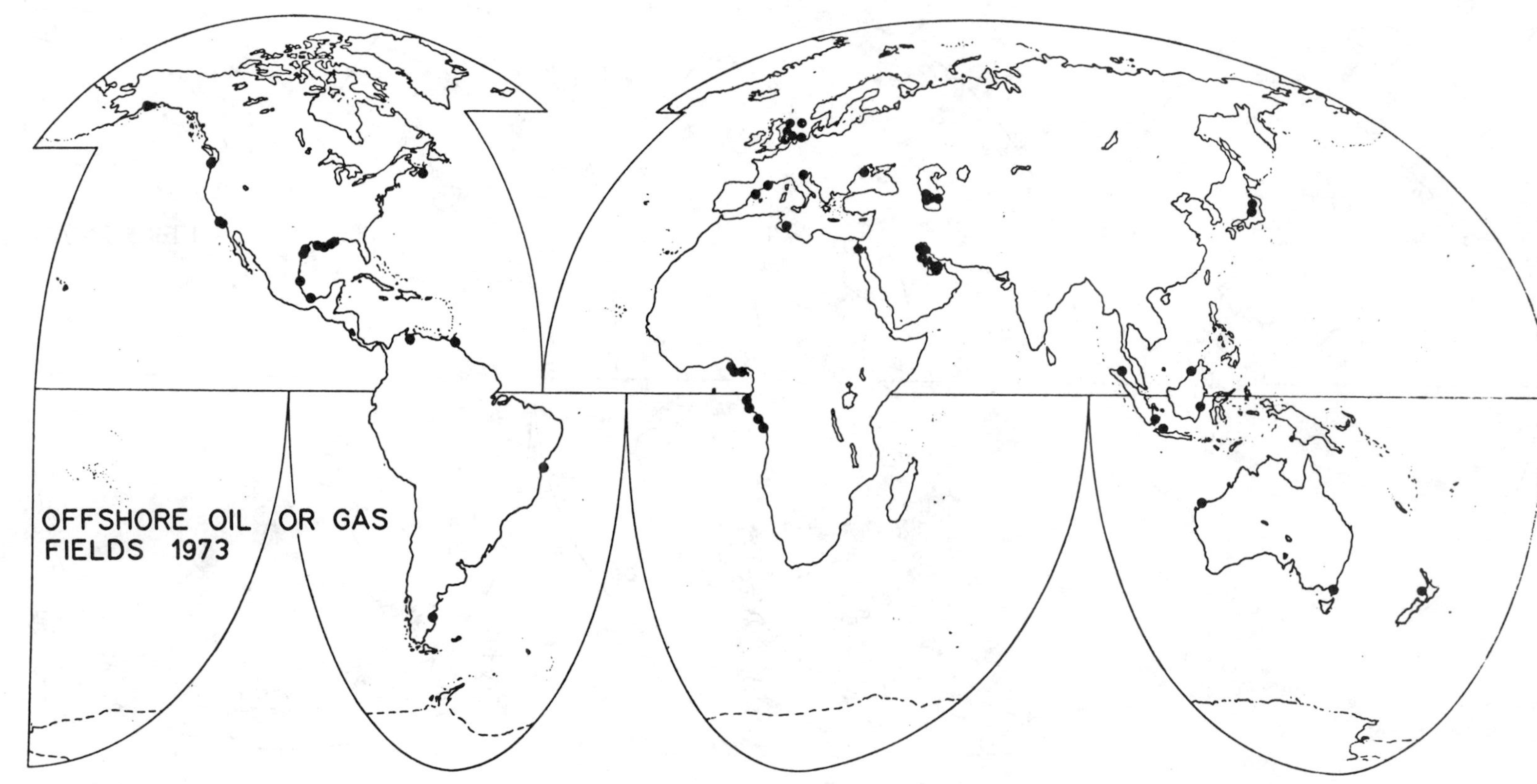

FIGURE 18: Distribution of Offshore Oil and Gas Fields. From McKelvey and Wang (1970).

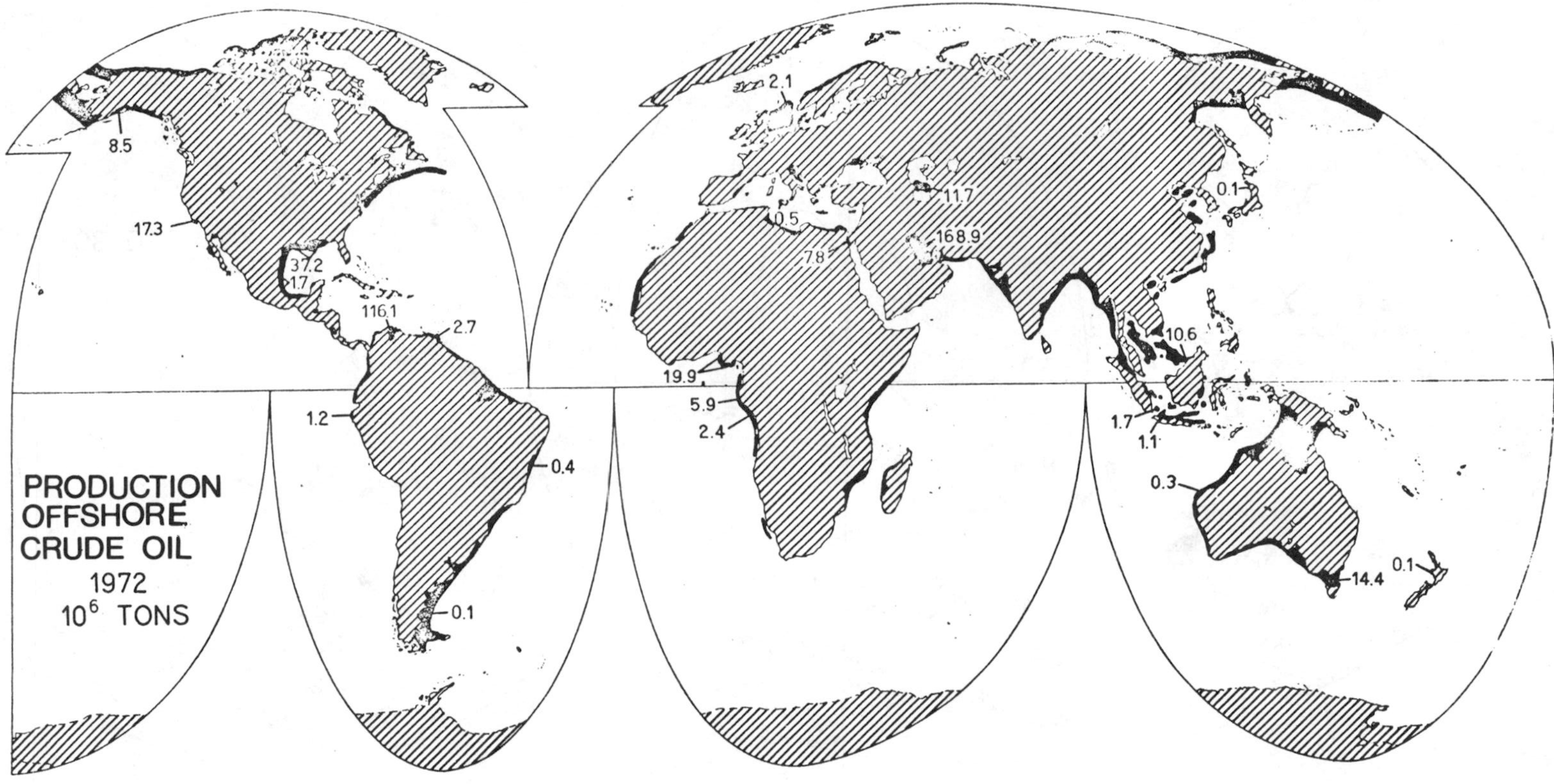

FIGURE 19: Offshore Production of Crude Oil for 1972 from the Most Favorable Parts of the Continental Shelves of the World.

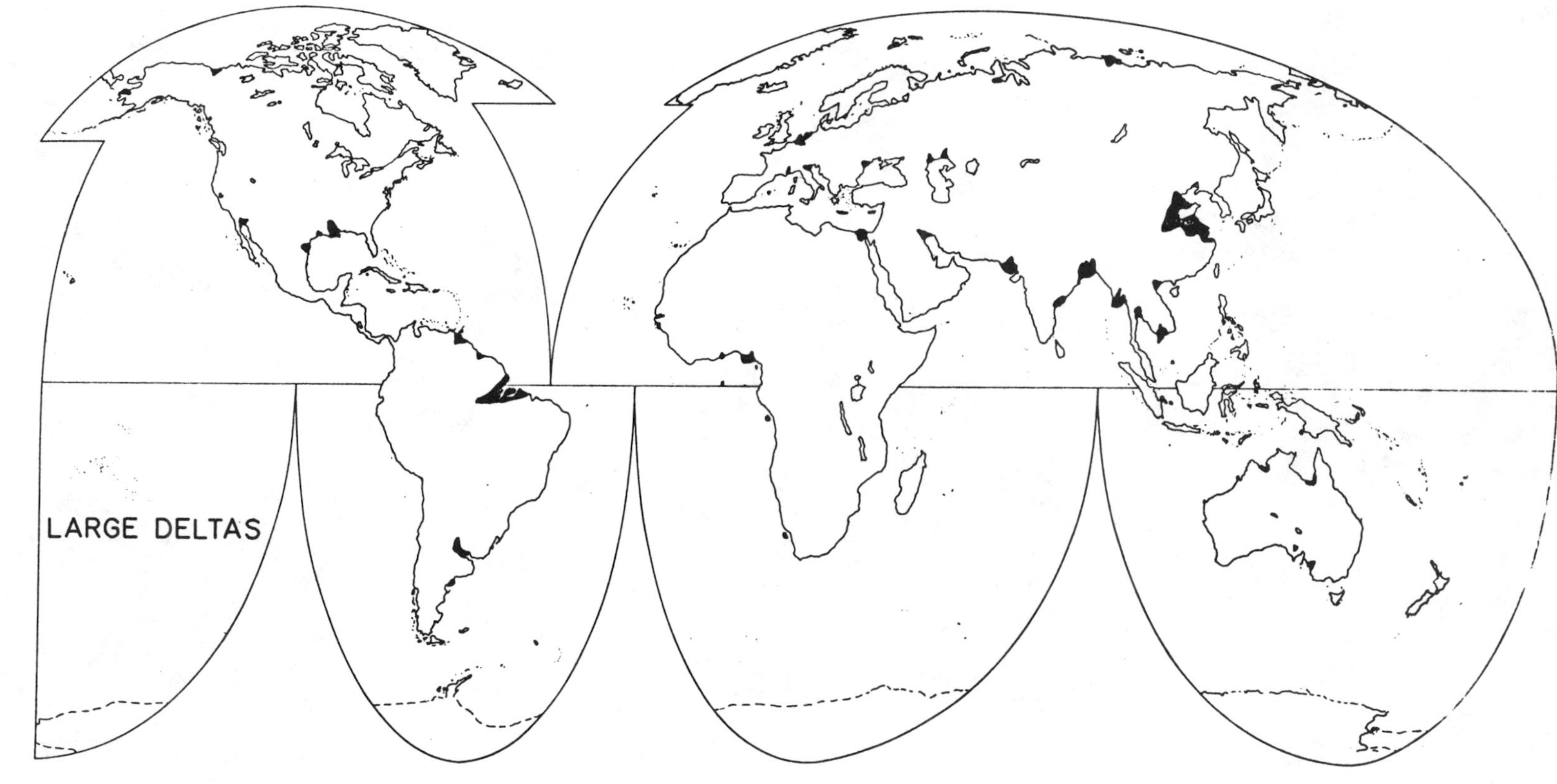

FIGURE 20: Distribution of Large Deltas.

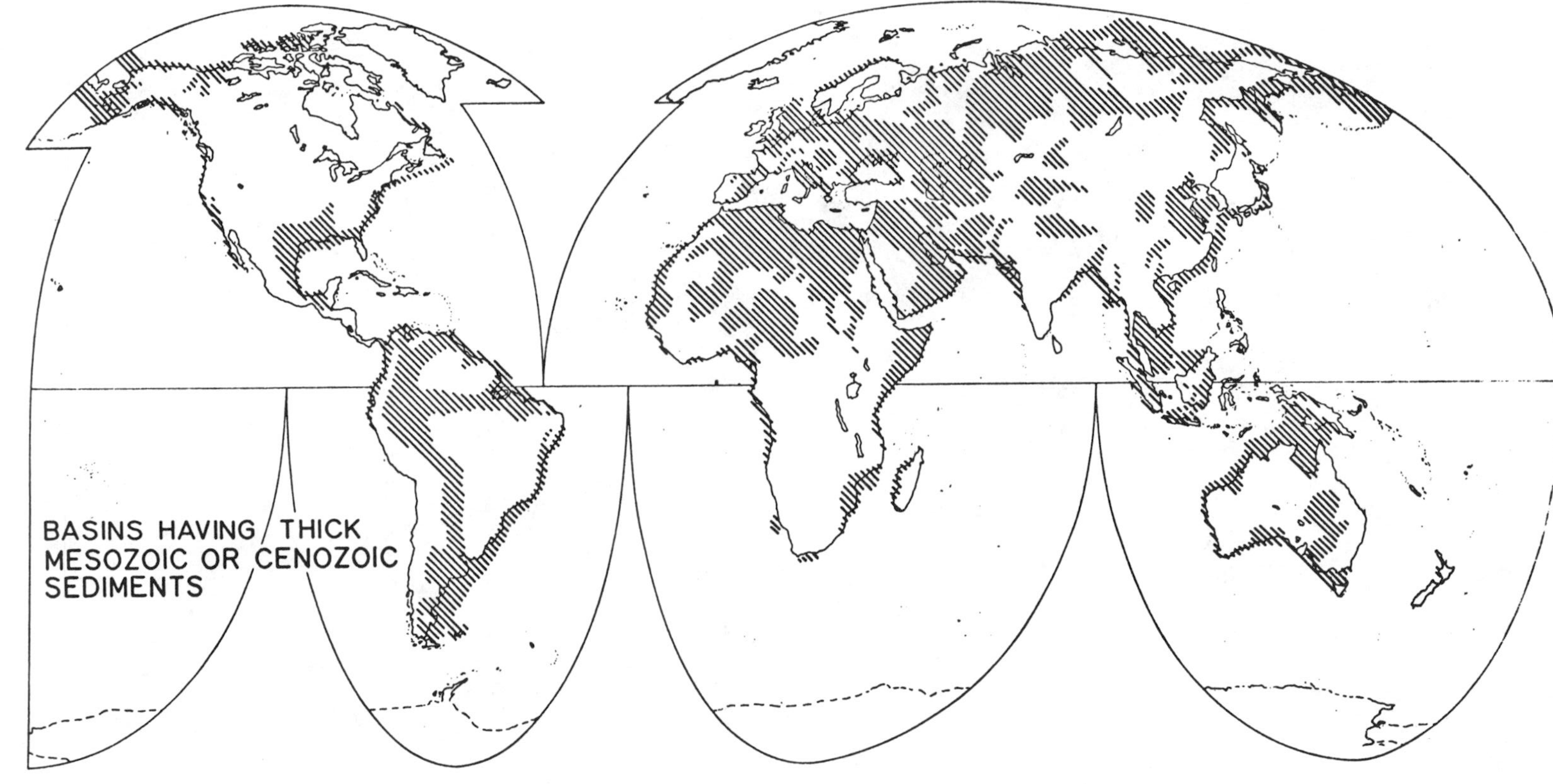

FIGURE 21: Areas of Thick Marine Mesozoic or Cenozoic Sediments. Note their frequent occurrence along coastal zones. Compiled from many geological and structural maps of continents and of the world.

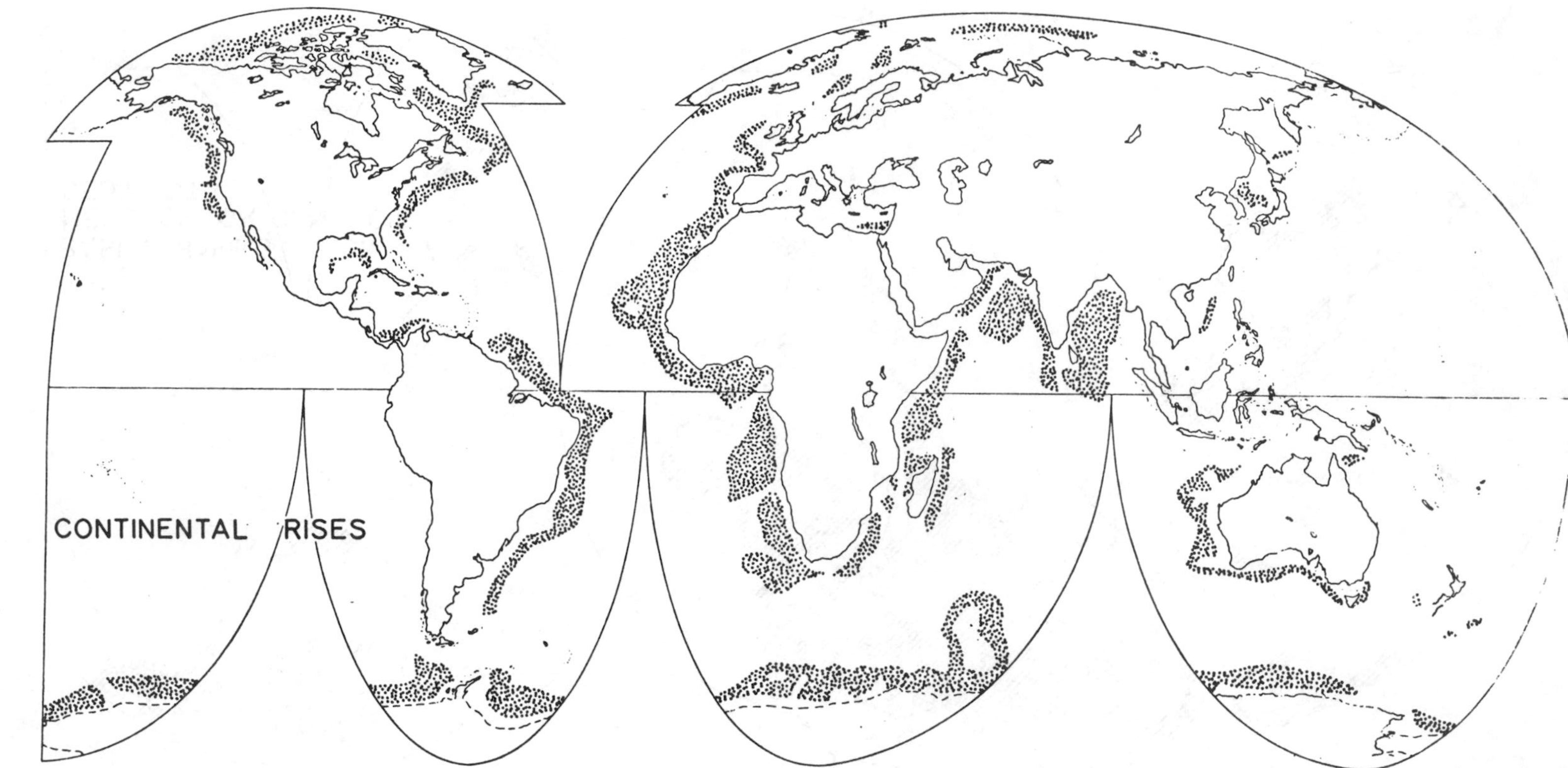

FIGURE 22: Distribution of Continental Rises. Note their presence along both sides of the Atlantic and Indian Oceans, along the northern coasts of North America, Asia, and Antarctica, as compared with their rarity along both sides of the Pacific Ocean. Thus the rises are common within crustal plates and absent at plate convergences. From Emery (1969).

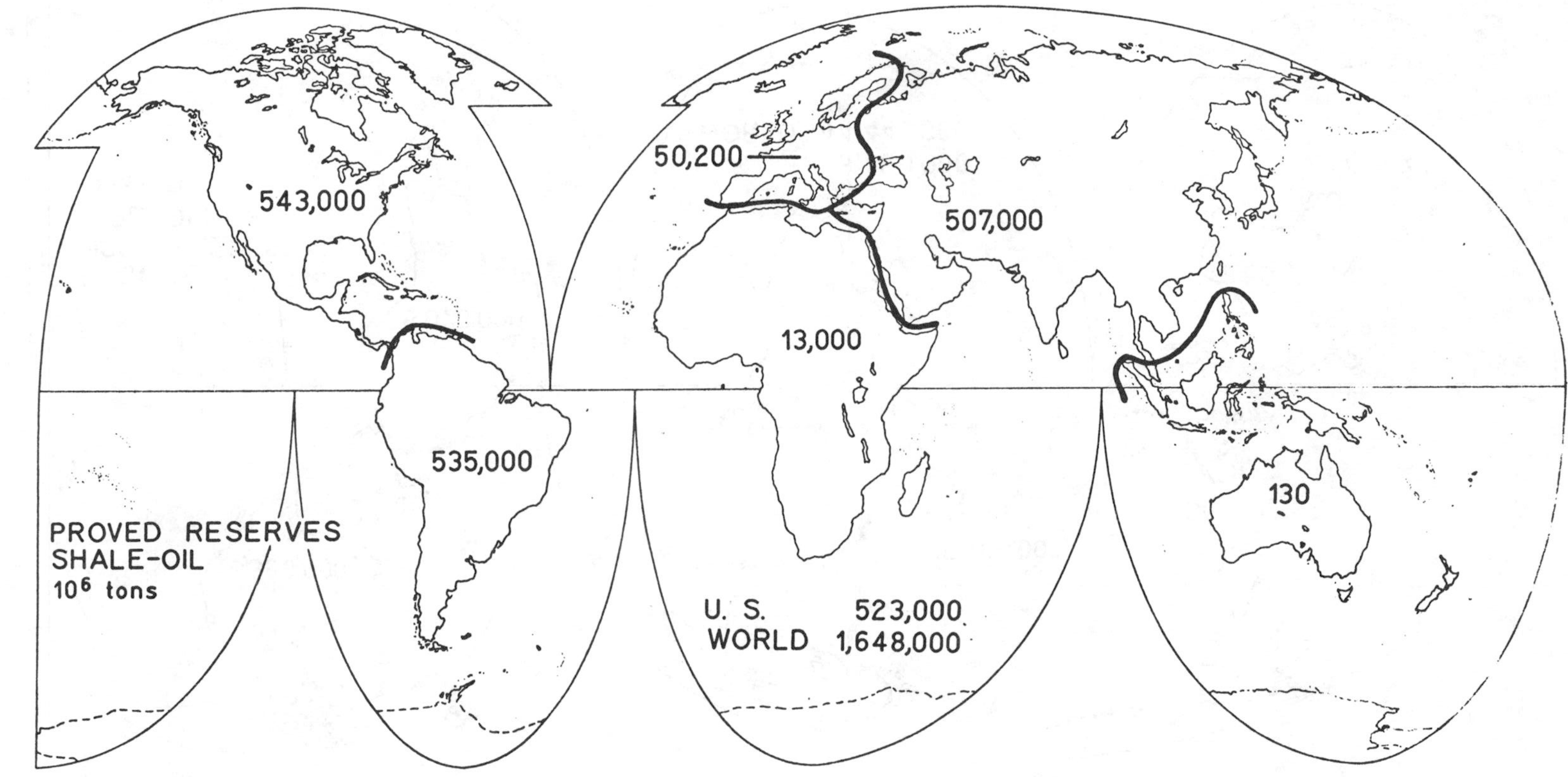

FIGURE 23: Proved (Plus Hypothetical--Extrapolated from Outcrops or Well Borings) Reserves of Shale Oil. Note that the regions of Asia and the Middle East are merged only for Figures 23, 24, and 26. Data are from Culberson and Pitman (1973).

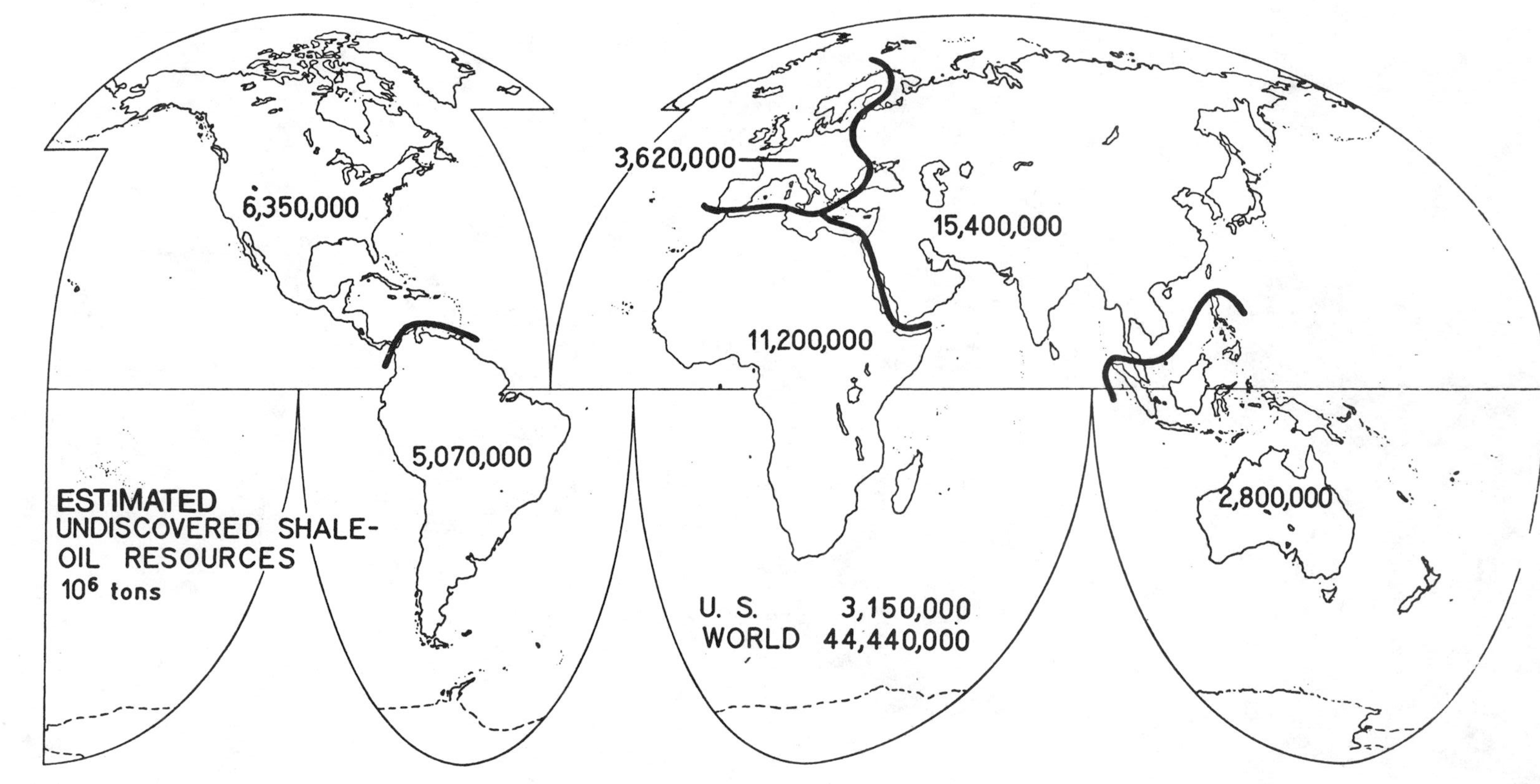

FIGURE 24: Undiscovered Resources of Recoverable Shale Oil. Data are from Culberson and Pitman (1973).

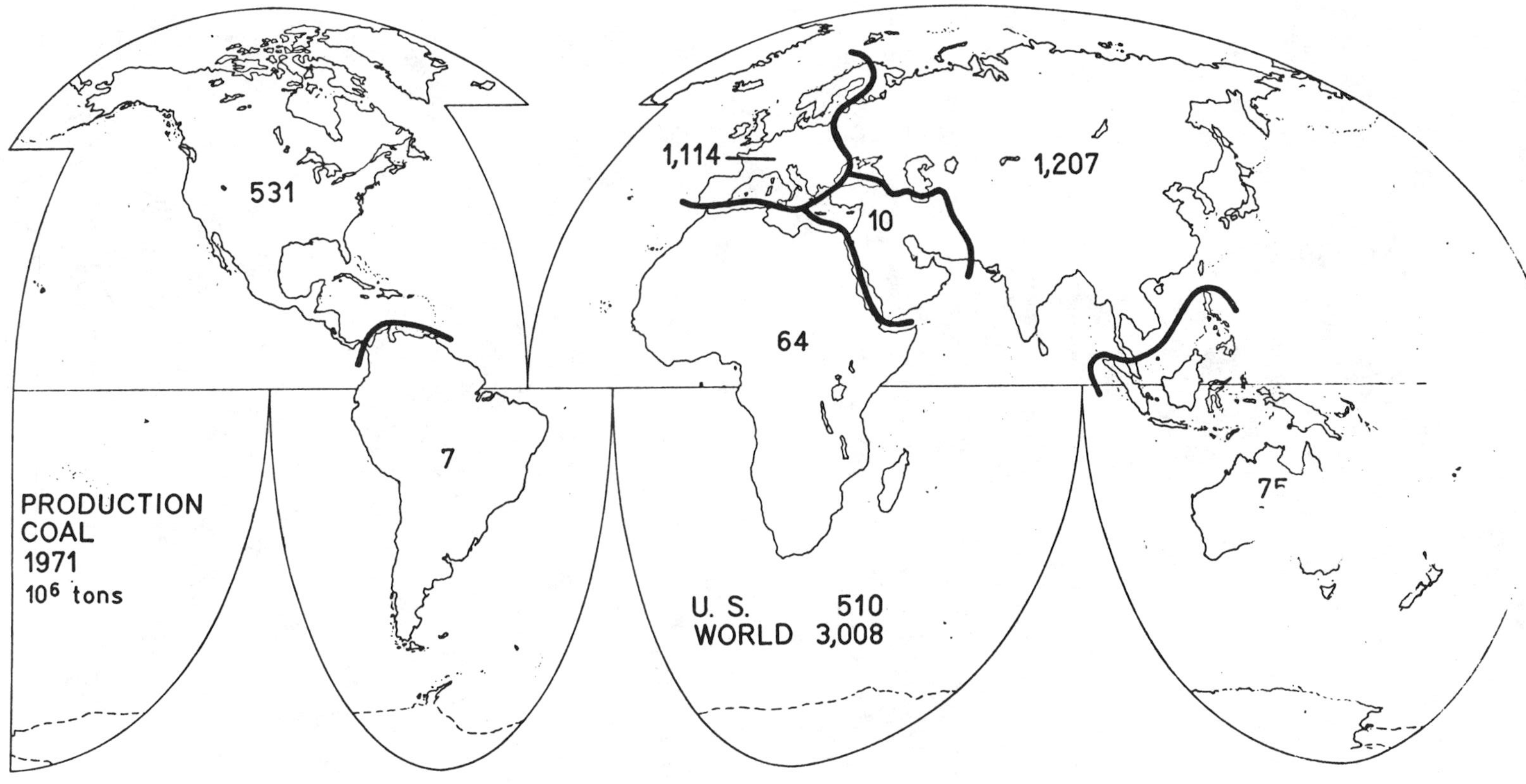

FIGURE 25: Annual Production of Coal. Data are from Federoff (1973) and Westerstrom (1973).

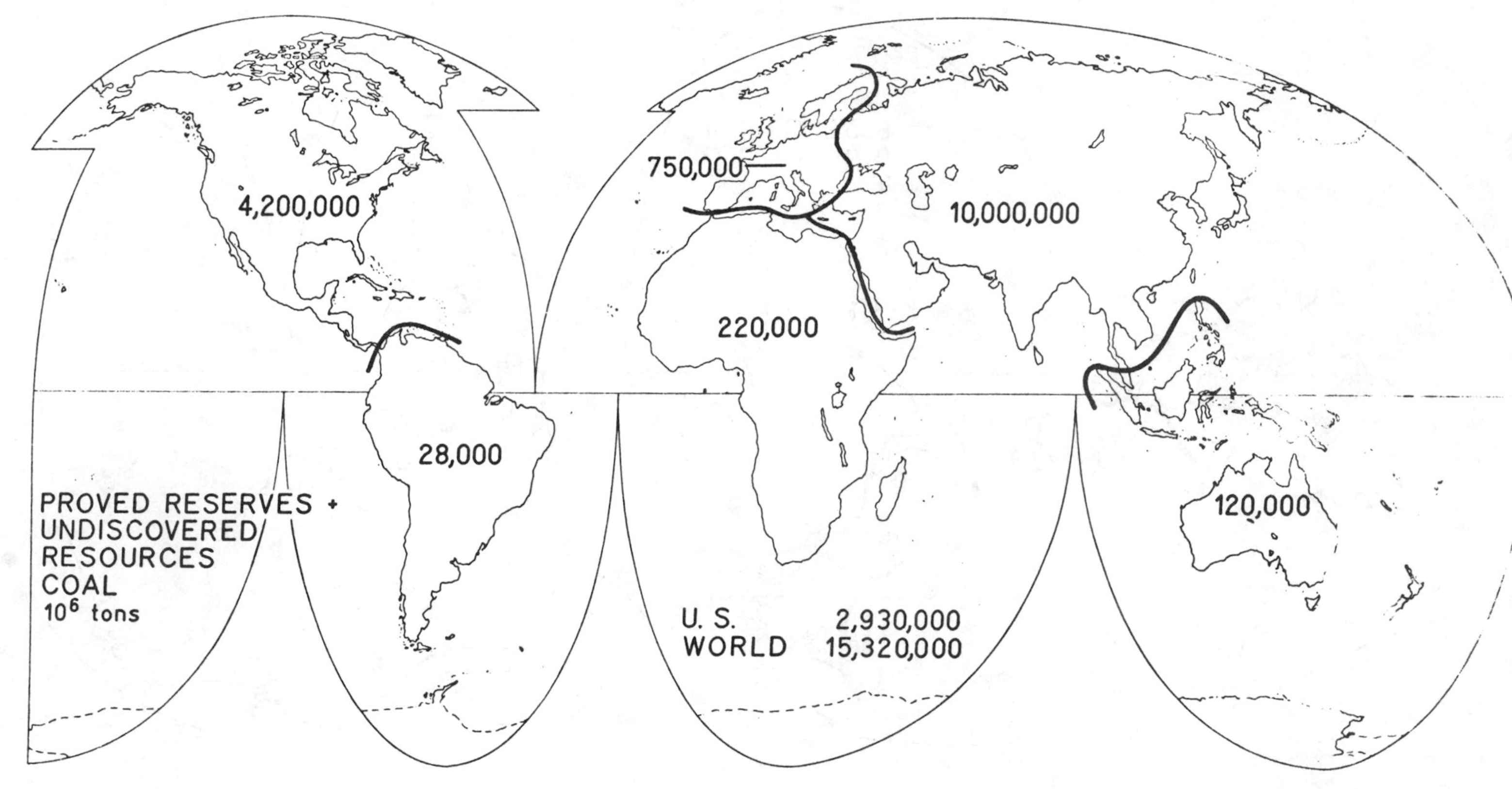

FIGURE 26: Proved Reserves and Undiscovered Resources of Coal in Place as of January 1972. Data are from Averitt (1973).

CHAPTER VI
RESOURCES OF COPPER

INTRODUCTION

The subpanel on Copper Resources examined estimates for United States and world resources of copper, the way estimates can be made, the uncertainties inherent in the different ways of making estimates, and the problems of maintaining self-sufficiency of primary copper production in the United States.

Copper was chosen from among possible metals as a topic of study for many reasons, but two were of paramount importance in the choice. Firstly, it is a commodity that interacts closely with fossil fuels; about half of the primary production is used in the production and transmission of electricity, and as use of energy in the form of electricity continues to rise, the use of copper may also be expected to rise. Secondly, copper is a commodity produced in sufficiently large volume and from such a diversity of sources that it is reasonable to expect that a useful body of data on its occurence and production is available.

Although the initial objective of the study of copper was evaluation of current estimates of copper resources, the panel accumulated, in the course of its study, a substantial amount of information on factors governing the availability of copper in the United States between now and the end of the century. This information is included in the report. Copper reserves and resources, both in the U.S. and the rest of the world, are large relative both to current production and consumption and to production and consumption anticipated to the end of the century, regardless of the anticipated consumption curve. Despite the large size of resources, however, it is not at all certain that rates of production can satisfy demands to the end of the century. Rates of production will depend on such factors as the price of copper, the rates at which new copper deposits can be discovered, the rates at which resources can be transferred to reserves, and rates of construction of smelting and refining capacity. These in turn lead to two problems which affect copper availability, but have not yet been fully evaluated. They are environmental considerations of both mine production and smelter capacity and the question of access to potential copper-bearing areas for purposes of prospecting, exploration, and mining. The panel suggests that both questions deserve further study.

Reliable information concerning the annual production and consumption of most mineral resources is available from several publications, notably those from the U.S. Bureau of Mines. Some information concerning unmined reserves is available from the same sources, but for most commodities the magnitude of undeveloped and undiscovered resources is speculative, and estimates are generally of low reliability.* COMRATE set itself the task of evaluating approaches to resource estimates and selected copper as a metal for which sufficient information might be available to make a useful analysis.

Shortly after beginning the study, Professional Paper 820 was published by the U.S. Geological Survey and its chapter on copper addressed itself to, and answered, many of the initial questions raised by COMRATE. The Committee has found the information from the U.S.G.S. and the U.S.B.M. exceedingly valuable in carrying out its study and in reaching its conclusions.

Conclusions

1. Availability of copper from domestic copper deposits during the remainder of this century depends both on success in discovery of new deposits and on resolution of the problem of inadequate smelter capacity. We do not foresee continued United States self-sufficiency in production of primary copper and therefore recommend a strong policy of conservation.

2. Means of improving the effectiveness of exploration should be sought by industry and encouraged by the government.

3. The only large, identified class of copper resources not presently being exploited appears to be manganese nodules on the deep sea floor. The magnitude of the recoverable nodule resources is apparently as large as the identified copper reserve on land.

THE AMOUNT OF RECOVERABLE COPPER IN THE EARTH'S CRUST

One can make loosely theoretical, but not entirely satisfactory, estimates of the amount of recoverable copper. The average content of copper in Earth's crust is 58 ppm so that the copper content of the outer 45 km is

---

*In this report, the use of the words "reserves" and "resources" and their classification conform with the November 1973 joint U.S.B.M.-U.S.G.S. resource classification. When contemplating any numerical expression of resources, one should keep in mind that "measured reserves," for which the uncertainty is at a minimum, are usually computed with a margin of error of 20 percent or less. Any other resource category involves a much greater margin of error; the estimation of undiscovered resources is an elusive task for which there is no accurate procedure.

approximately 4.6 x 10^{15} tonnes, (5 x 10^{15} tons). Assuming that the deeper mantle is inaccessible to mining, this is the total potential supply of copper regardless of what economic and technologic changes lie ahead. But most of Earth's crust is not really accessible to mining, so 4.6 x 10^{15} tonnes is an unrealistic figure. The 70 percent of the crust that lies beneath the sea can be excluded as an unlikely site for large scale, low-cost mining. Of the remainder, which is the exposed continental crust, only the upper 10 percent will probably even be accessible because of problems such as rock strength and rising temperatures. Thus, 1.4 x 10^{14} tonnes (1.5 x 10^{14}tons) is the amount of "accessible" copper, and to recover it we would have to mine and process the entire land mass exposed above sea level to a depth of 4.5 km (2.8 miles). 1.4 x 10^{14} tonnes (1.5 x 10^{14} tons) is clearly an academic number. We might predict a technological society that exploited its metallic mineral resources by mining and processing 1 percent of its exposed land mass, but the processing of 100 percent or even 10 percent seems unreasonable.

A more realistic evaluation of the amount of copper potentially available can be obtained by considering how copper occurs in the crust. Most of the copper present is locked in solid solution in common silicate minerals, and in this form it cannot be concentrated by physical beneficiation to produce a copper-rich concentrate ready for the smelter. Recovery of copper, or any other trace metal, from silicate solid solution generally involves disruption of the silicate mineral structure. This is an energy-intensive process. Where geological circumstances raise local copper contents to levels of 0.1 percent or more (that is, local enrichments of 18 times or more above geochemical background), we observe that copper may either form individual minerals which can be physically concentrated, or it may sometimes occur in a state amenable to selective chemical extraction without disruption of silicate minerals. Physical beneficiation and most chemical leaching processes are greatly less energy-intensive than extraction of metals from silicates. A content of about 0.1 percent copper seems, therefore, to represent a sort of geochemical barrier to a regular reduction in grade of minable copper ores. Therefore we consider a grade of 0.1 percent copper to be a lower cut-off value below which copper resources may never be recovered.

The amount of copper present in local concentration and amenable to recovery must be vastly less than 1.4 x 10^{14} tonnes (1.5 x 10^{14} tons). Even in heavily mineralized areas such as the southwestern portion of the U.S., the area of the crust known to be underlain by deposits containing copper at grades above 0.1 percent Cu is no more than 0.01 percent. If it is assumed that the same "mineralization" factor could be applied around the world, and retaining the figure of 10 percent as the accessible portion of the crust, the copper tonnage in deposits down to a grade of 0.1 percent Cu would only be 1.4 x 10^{10} tonnes (1.5 x 10^{10} tons). Inasmuch as the southwestern U.S. is one of the most richly mineralized copper regions in the world, 1.4 x 10^{10} tonnes can be postulated as an upper limit for the amount of copper that can be produced by present or future mining methods without overcoming the geochemical barrier presented by metals trapped in silicate solid solutions.

Using the kind of analysis just given for the United States, which accounts for approximately 6.2 percent of the world's land area, the upper limit for U.S. copper production is suggested as 0.9 x 10^9 tonnes (10^9 tons).

Reserves of Copper

The U.S. Bureau of Mines published a Commodity Data Summary in January 1974 listing the U.S. reserves of contained metallic copper, estimated to be producible at a price of 50¢/pound, to be 75 x 10^6 tonnes (83 x 10^6 tons). Information on world reserves is less satisfactory because of inadequate reporting in some countries, but the U.S.G.S. figure of 313 x 10^6 tonnes (344 x 10^6 tons) given in P.P. 820 is probably close to correct. The present annual world production of copper is about 7.3 x 10^6 tonnes (8 x 10^6 tons) and past production can be estimated between 90 and 140 x 10^6 tonnes (100 and 150 x 10^6 tons). Reserves plus past production already amount, therefore, to approximately 3 percent of the 1.4 x 10^{10} tonnes (1.5 x 10^{10} tons) suggested as the world's ultimate yield. By the same kind of reasoning, the U.S. reserves, plus past production, are about 16 percent of the U.S. ultimate yield.

Reserves, however, imply a high degree of certainty both as to tonnage and grade, and it is a common practice for mine operators to report as a reserve only that material specifically tested and developed for mining in the immediate future. A mine may therefore have a large annual production but report constant or even increased reserves from year to year. What is happening is that resources are being transferred by development to reserves. Estimation of ultimate yields by means other than the simple approach already used, therefore, would involve indirect and consequently less certain means of evaluating size, grade, and distribution of ore bodies.

SIZE OF IDENTIFIED COPPER DEPOSITS OF THE WORLD

During the last six years, three studies have provided extremely valuable information about the size of known copper deposits. Such information is, of course, the cornerstone of any consideration of undiscovered resources. These studies evaluated copper deposits as follows (see the Bibliography at end of report for references to these studies) and found the average size of all deposits to be slightly larger than 900,000 tonnes (1 million tons) of contained copper, but the range in size to be very large.

Investigator	No. of Deposits Investigated	Total Copper Metal Contained (Past Production plus "Reserves")
Pelissonnier, 1968	374 (USSR included)	356 million tonnes (392 million short tons)
Pelissonnier, 1972	522 (USSR included)	396 million tonnes (436 million short tons)
Amax, 1970	315 (USSR excluded)	390 million tonnes (429 million short tons)

The three studies only include material that is minable by present technology and economic standards. The low grade ore, between grades of 0.5 and 0.1 percent Cu, therefore, is inadequately represented in the estimates. Nevertheless, some very significant observations can be drawn concerning the size, type, and distribution of copper deposits.

Concentration of Copper

The Amax study is particularly valuable in that it indicates that the bulk of the copper in the kind of deposits so far discovered is contained in a small percentage of the deposits--28.9 percent of the deposits contain 84.8 percent of the copper. This is summarized in Table 1, where the deposits are grouped by size classes. For each class the upper tonnage limit is equal to the lower tonnage limit multiplied by the square root of 10. The information contained in Table 1 is graphically presented in Figure 1 as a frequency distribution by class, and in Figure 2 with a cumulative distribution. Figure 2 shows that the first three classes of deposits (more than 900,000 tonnes, or 1 million tons of metal contained) account for the major part of the copper tonnage in known deposits.

Deposits in Class #1, with from 9 million to 28.7 million tonnes (10 million to 31.6 million tons) of metal contained, can be described as "super-giant" deposits, whereas those in Class #2, with 2.9 to 9 million tonnes (3.1 million to 10 million tons) of copper contained, may be labeled as "giant" copper deposits. These two classes provide the bulk of the current production, and it is reasonable to suspect that their undeveloped resources are sufficiently large to ensure that the same will be true in the future. The concept of giant and super-giant deposits is apparently just as important in mineral deposits as it is in petroleum and gas fields. Giant and super-giant deposits with grades between 0.1 and 0.5 percent copper will presumably account for the bulk of low grade copper resources, when they are identified. We conclude, therefore, that the low grade copper resources will be found within geological features that are relatively large and that deposits formed in completely unsuspected geological environments are unlikely.

Distribution by Geologic Type of Deposit

Any resource appraisal or resource development program must pay special attention to an analysis of geological environments most favorable to deposits found in Classes #1 to #3, and especially to those of Classes #1 and #2. Many different geologic types of copper deposits have been recognized and mined. For definitions, the reader is referred to Professional Paper 820 and to Pelissonnier (1972).

Table 2 indicates that 89 percent of the copper mined and developed as reserves occurs in three kinds of deposits and emphasizes the critical importance of porphyry copper and strata-bound copper deposits, which together contain 79 percent of the known copper reserves. It is probable, but not proven, that undiscovered resources will show the same distribution. The geological age distribution of the three main types has also been shown in Table 2. It appears there may be a definite age range for each type. This is another geologic factor which should be used in resource appraisal of geological belts of known age.

It is interesting to note that of six deposits in tonnage Class #1, four are porphyry copper and two are strata-bound; in 40 deposits in Class #2, 28 are porphyry copper, nine strata-bound, two massive sulfides and one associated with basic and ultrabasic rocks. Thus, the two main geologic types of deposits provide the largest deposits.

The U.S.G.S. has initiated a study of size and grade characteristics of selected copper deposits by geologic type and location. Results published in Professional Paper 820 are shown in Figure 3; they confirm Pelissonnier's data that porphyry and strata-bound copper account for the largest deposits. The U.S.G.S. data also suggest that marked differences in size may be found within a given geologic type in different parts of the world.

After assembling data for the copper chapter in P.P. 820, Cox, Singer, and Drew of the U.S.G.S. continued upgrading their inventory of known copper deposits and released the data in Figure 4 to COMRATE. Although it does not refine the indication in P.P. 820 that there are geographic size distributions within classes, it does indicate that exploration and new development can significantly change the reported world reserve figure. Their new estimate is 340 million tonnes (370 million tons), significantly higher than the earlier number of 310 million tonnes (344 million tons).

The porphyry population, which accounts for 50 percent of the copper reserves, has also been the object of a recent special analysis by the U.S.G.S. group. Their conclusion, that both tonnage and grade have a geologic upper limit but an economic lower limit that changes with time, can be deduced from Figures 3 and 4. The wide scatter of tonnage at each grade level can be expected for lower grade levels. This means, however, that total metal content of lower grade deposits will also be lower than for deposits known to date.

Changing Size and Grade of Deposits with Time

One important result of recent U.S.G.S. studies is documented by Figure 5 and Table 3. The data show that recent discoveries are, on the average, lower in grade than deposits brought into production prior to 1960. For instance, the median dimensions for 19 pre-1960 porphyry coppers are 450 million tonnes (495 million tons) of 0.84 percent copper ore containing 3.8 million tonnes (4.2 million tons) copper metal. This can be compared to median values for all of the 78 selected porphyry coppers known to date of 224 million tonnes (246 million tons) of 0.73 percent copper ore with a metal content of 1.7 million tonnes (1.8 million tons). This illustrates that, in most areas, the first discoveries were the largest and it may be explained by the fact that, in comparison with the smaller deposits found later, the large dimensions of the first discovery allow greater chances of surface expression of mineralization and, therefore, easier discovery. Table 3 illustrates again what Figure 3 pointed to: there are geographic variations in grade and tonnage of porphyry coppers.

The bimodal grade distribution observed in Figure 5 is explained by the process of secondary surficial enrichment superimposed on primary copper deposition processes and/or high cut-off grades in politically sensitive areas. The tonnage displays a unimodal distribution.

With a view to exploration planning, Amax analyzed the size distribution of porphyry copper deposits in two parts of the North American Cordillera. Figure 6 shows that, as of 1968, giant or super-giant deposits were not known in British Columbia and the Yukon, whereas, in Arizona and southwestern New Mexico, many giant deposits had been discovered. The situation changed after 1968 when the first British Columbian giant copper deposit, Valley Copper, was discovered. This illustrates one of the pitfalls in forecasting undiscovered resources and demonstrates the need for a continually upgraded data base.

Another demonstration of the need for frequent updating of reserves and revisions of resource forecasts is clearly pointed out by the changes over time in the appraisal of the tonnage and grade of the Bingham porphyry copper deposit. This is illustrated in Table 4. The remaining proven ore reserves, as of 1970, are five times larger as to ore tonnage and 2.5 times larger as to metal content than the reserves as of 1915--and this after extracting, between 1915 and 1970, 3.4 times the ore tonnage and 2.2 times the metal tonnage of the 1915 reserves. Bingham is a super-giant, Class #1 deposit which was explored in steps. By comparison, many modern discoveries are evaluated thoroughly by drilling the entire deposit, including its submarginal mineralization, before an optimum production plan is decided upon. This modern approach allows improved estimates of "subeconomic" identified resources; however, such information is rarely available from mining companies. Nonetheless, the information is critically needed for resource evaluation prior to mineral policy formulation.

Discussion of Lasky's Rule

Studying domestic porphyry copper mines operating before 1950, S. G. Lasky of the U.S.G.S. discovered that within the narrow range of then economically minable grades, an arithmetic decline in grade was associated with a geometric increase of the tonnage of reserves, which he expressed as:

$$\text{Grade} = K_1 - K_2 \text{ (log tonnage)}$$

in which K_1, K_2 are constants.

For the average U.S. porphyry copper deposit, he found that a decrease in grade of 0.1 percent Cu was associated with an increase in tonnage of about 18 percent.

In his paper, Lasky stated that similar tonnage-grade relationships seemed to apply to such widely-divergent deposits as the low-grade syngenetic manganese deposits in the Artillery Mountains of Arizona, the vanadium and phosphate deposits of Idaho and Wyoming, the Falconbridge nickel deposit, and the Alaska-Juneau gold deposit.

In the erroneous belief that the Lasky principle could be extended to all kinds of deposits and over virtually all ranges of ore grades, various writers have arrived at grossly optimistic and completely unrealistic estimates of undiscovered reserves of mineral resources without sufficient regard for the geological evidence.

The French school of geostatisticians headed by Matheron has shown that, in a general manner, ore deposits over large areas tend to have grade distribution, which, when plotted as logarithms on the abscissae against the tonnage plotted on the ordinate, yield a normal Gaussian bell-shaped curve. They have also recognized, as a sequel of this more general principle, that Lasky's rule will only hold true for a restricted range of values in economic deposits above some lower grade limit. Outside of this range, however, Lasky's rule will lead to absurd predictions and unrealistic expectations on tonnage-grade relationships. It will manifestly fail, for example, with high-grade iron and aluminum ores.

T. S. Lovering, in an excellent study published in the volume "Resources and Man," pointed out some of the limitations of Lasky's rule, on both the high and low tonnage sides, both of which lead to obviously meaningless conclusions. He argued that such processes as fracturing of rocks, weathering, and sedimentation often result in higher grade minable ore bodies, but with a more or less pronounced break in the continuity of ore grades when approaching the surrounding protore or barren rock. He warned that Lasky's A/G (arithmetic-geometric) ratio should be found notoriously ineffective for the majority of mercury, gold, silver, tungsten, lead, zinc, antimony, beryllium, tantalum, niobium, and rare earth deposits.

It is evident that Lasky's principle, while useful in a restricted, practical range of values for many individual mines and mining districts, the geological evidence permitting, must be rejected as a universal tool for the statistical evaluation of mineral distribution over larger areas. In particular, the use of Lasky's rule to predict future giant-sized, still lower-grade, resources is erroneous, overly optimistic, and unrealistic. The number of giant and super-giant ore bodies with grades between 0.1 and 0.5 percent is probably very small if, in fact, any exist at all.

Geographical Distribution: Copper Provinces

Pelissonnier demonstrated that the concept of metallogenic provinces is definitely meaningful with regard to the distribution of copper deposits around the world. 98 percent of the copper inventoried by Pelissonnier is contained in 11 copper provinces, which cover only a small part of the continental crust. These 11 provinces are listed in Table 5 and shown in Figure 7. Thus, the distributions of copper deposits, with regard to metal content, geologic type, location, and geologic environment (i.e., geologic province) demonstrate patterns which might be used in resource appraisal. This knowledge allows extrapolation as to what may be found in unexplored areas and under cover of barren rocks.

Following a workshop held by COMRATE, one of the attendees, Paul Eimon, served as coordinator in assembling and preparing a map on which all known copper deposits, whether under production or not, are plotted. From this valuable map (Figure 19) the location of several metallogenic provinces is readily apparent; for example, the Urals, a belt along western South America coinciding with the Andes, and a remarkable linear belt in Zambia and Zaire. We consider it possible that within these metallogenic provinces, as much as 0.01 percent of the surface may be underlain by deposits with grades of 0.1 percent or better. Outside of the provinces, the frequency of deposits is certainly much lower.

COPPER RESOURCES OF THE WORLD

Just as we seek to learn from the past in order to plan better for the future, so can we use our knowledge of discovered copper deposits (including past production and "reserves") to advantage in considering undiscovered copper resources of the world. This knowledge can be employed either in subjective appraisal by geologists using their experienced judgment, or as the foundation for a quantitative appraisal model taking into consideration cost and price factors.

"Experienced Judgment" Geological Appraisal (U.S.G.S. Approach)

Depending upon the degree of geological knowledge in the areas involved, such appraisal can be made as follows:

1. In known copper provinces, the potential for discovery of new copper resources can be assessed for each geologic type of deposit known, or expectable, in the province:

(a) as to the number of discoveries to be made, according to the degree of exploration of the geologically favorable environments and to the density of mineralization in known mineral districts and belts, and

(b) as to size of new deposits according to tonnage and grade distribution of known deposits in the province.

2. In new copper provinces, newly recognized because of a recent discovery, e.g., Australasian Island Arcs, Central America and Central Iran porphyry copper provinces, the potential can be assessed by comparison with geologically similar known provinces where the largest known deposits of the same geologic type are similar in size and geology to the recent discovery in the new province.

3. In potential future copper provinces, which are regions considered geologically favorable for the occurrence of copper deposits, because of geological similarity with known copper provinces. The provinces can be assessed simply by comparison with an average of geologically similar copper provinces, using statistics of the deposit population in similar areas. The Pacific Belt of Far Eastern Siberia, the eastern side of the Ural Mountains, and parts of North Africa may be considered as potential future provinces.

The U.S.G.S. copper resource appraisal reported in Professional Paper 820 employs the judgment method. We have removed from their estimates such resources as copper-bearing nodules on the bottom of the deep oceans, and have increased world reserves to take account of new data and summarized their results in Table 6. The indicated total of 0.13×10^{10} tonnes (0.142×10^{10} tons) of contained copper is, thus, 10 percent of the 1.3×10^{10} tonnes (1.4×10^{10} tons) considered a recoverable maximum. The "experienced judgment" method cannot reasonably take into account the potential of regions of unknown geology, such as parts of Antarctica or the sand and laterite covered regions of Central Australia. We do not believe it likely, however, that the "experienced judgment" method is low by more than a factor of two or three. The "experienced judgment" method cannot account for completely unconventional resources, such as manganese nodules, for which we lack experience. What the "experienced judgment" estimate suggests, therefore, is that a very significant fraction of the world's recoverable copper resources will be found in deposits of the kind with which we are already familiar.

Quantitative Geologic-Economic Resource Appraisal Model

This approach, pioneered by the European Economic Communities, was applied to copper resources by J. Brinck. Figure 8 (reproduced from Brinck's 1971 article) is based on a study by Brinck, using 1963 copper data. The European Communities reserve-resource terminology has been modified to conform with the U.S.G.S.-U.S.B.M. classification. The horizontal scale depicts the total copper content of groups of possibly mineable deposits, the vertical scale the average grade of such groups. For the whole world, Brinck assumed a rock tonnage of 10^{18} tonnes (1.1 x 10^{18} tons) to an average depth of 2.5 km (1.6 miles) with an average grade of 70 ppm Cu. This is shown at the origin of the graph (left bottom). Reading the vertical scale, we find 70 ppm Cu and considering the total volume of 10^{18} tonnes (1.1 x 10^{18} tons) worldwide as a possible copper deposit, its content in Cu is 7 x 10^{13} tonnes (7.7 x 10^{13} tons) Cu on the horizontal scale.

Diagonally upwards from the origin, a line has been drawn to depict the expectable deposits of highest grade for any given copper content. According to the theory of de Wijs, used by the European Economic Communities, a resource tonnage R of average grade X and specific mineralizability* Q, can be split α times into 2^{α} equal parts of tonnage $\frac{R}{2^{\alpha}}$ each. As a result of this process, $\alpha + 1$ different grades of these parts can be expected, represented by one of the terms $X(1+Q)^{\alpha-K} \cdot (1-Q)^{K}$, with K an integer from 0 to α. Associated with each separate grade is a number of equal parts of that grade, represented by the tonnage $\frac{R\binom{\alpha}{K}}{2^{\alpha}}$ and, therefore, having a total metal content of $\frac{RX\binom{\alpha}{K} \cdot (1+Q)^{\alpha-K} \cdot (1-Q)^{K}}{2^{\alpha}}$ tons. Setting K=0, we find only one possible deposit of highest grade $X(1+Q)^{\alpha}$ and tonnage $\frac{R}{2^{\alpha}}$ having a metal content of $RX\left(\frac{1+Q}{2}\right)^{\alpha}$. To find the theoretical highest grade of a single copper deposit containing, for example, 1 million tonnes of Cu metal: Set: $RX\left(\frac{1+Q}{2}\right)^{\alpha} = 1{,}000{,}000$ or with $R = 10^{18}$ tonnes, X = 70 ppm, Q = 0.1981,

*This tongue-twisting word, not found in dictionaries and glossaries, has been coined by Brinck, to indicate the capacity of rock masses to contain metallic mineralization.

$$10^{18} \times 7 \times 10^{-5} \times \left(\frac{1.1981}{2}\right)^{\alpha} = 10^{6} \quad \text{or:} \quad \alpha = \frac{\log \frac{10^{-7}}{7}}{\log \frac{1.1981}{2}}$$

which yields $\alpha = 35.253$, meaning that the grade of this single deposit of $\frac{10^{18}}{2^{\alpha}}$ = 24.422 million tonnes (26.9 million tons) is expected to be

$$(1.1981)^{\alpha} \times 70 \text{ ppm} = 40{,}950 \text{ ppm or } 4.1\% \text{ Cu}$$

As can be seen on Figure 8, this is one of the points on the diagonal line through the origin.

By manipulating α the theoretical size and grade of deposits containing the same tonnage of metal (say 1 million tonnes), but of lower grades than those represented by the diagonal line, can be found. Clearly, the combined Cu-tonnage of these lower-grade deposits and the frequency of their occurrence are expectably higher than for the single highest-grade deposit. This trend towards more frequent occurrence with lowered grade is also expressed on Figure 8, and the entire system of lines has been labeled the "Iris-diagram" by Brinck.

In reading down along the 1 million tonne line of this diagram, at the intersection with the 10^7 tonne Cu line (up from the horizontal scale), for example, we find a grade (on the vertical scale) of 2.82 percent. The inference from this is that, worldwide, we may expect the presence of the equivalent of 10 ore bodies of 35.5 million tonnes (39.1 million tons) of an average of 2.82 percent Cu ore, each containing 1 million tonnes of Cu metal. This finding must be interpreted, not as saying that these 10 ore bodies will eventually be encountered as such, but only that the occurrence of such a body can be deemed to be 10 times as likely as that of the single 24.4 million tonne type ore body of 4.1 percent Cu of the diagonal line.

Close to the center of the Iris-line for one million tonnes, the graph shows a point marked 1963. This represents the total 1963 world inventory, established by the European Economic Communities, of Cu metal in past production and estimated reserves of 220 million tonnes (242 million tons) of Cu metal present in ore of an overall average grade of about 1.65 percent. From the data of this inventory, a specific mineralizability, Q = 0.1981, for copper was derived.

Another system of curved lines appears on Figure 8 and they attempt to delimit the economically viable combinations of tonnage and grade established for 1963 conditions, of interest to the industry and on which, therefore, exploratory activity should be focused.

The areas of economic interest delineated on the graph by limits of 40, 60, and 80¢/pound copper were arrived at by computing the expectable "unit cost" including operating, capital investment, and exploration costs per pound of copper, for each tonnage and grade combination, along any one of the lines of the Iris-diagram (1963 figures).

The area delineated by the 40¢/pound curve to the left and the vertical line through the point "1963" indicates that a total potential of about 0.98 x 10^{10} tonnes (1.08 x 10^{10} tons) of contained copper existed in 1963 for conditional and undiscovered resources. Even ignoring changes in price and amount mined, the total resources suggested by the Brinck method do not differ drastically from other estimates. They are less than an order of magnitude above the U.S.G.S. estimate and very close to the amount of recoverable copper suggested earlier from simple geochemical reasoning.

Unconventional Copper Resources: Deep Ocean Nodules

Besides conventional kinds of deposits, we must also be aware of the possibility that completely unconventional deposits, not heretofore exploited, may eventually be transferred from a resource category to a reserve. The only possibility we can identify is the copper contained in manganiferous nodules on the deep ocean floor. A great deal of attention has been focused on the nodules leading some to be optimistic, others skeptical about the possibility of future exploitation. There are four principal reasons for believing that a change from resource to reserve may soon be realistic:

1. Industry representatives predict that techniques allowing full scale mining of nodules from the floor of the Pacific Ocean at water depths between 3,650 and 4,580 m (12,000 to 15,000 feet) will be successfully developed and tested by 1976 (Dubs, 1974).

2. A viable regime of exploitations by the private mineral sector has been proposed to Congress (H.R. 12233 and S. 2878 introduced January 23, 1974 by Mr. Downing and Mr. Metcalf). The bill would provide the Secretary of the Interior with authority to regulate the hard mineral resources of the sea floor pending an international regime.

3. The First International Law of the Sea Conference, held in Caracas, Venezuela in June 1974, recognized copper-rich manganese nodules as a potentially valuable resource and one of the reasons for trying to settle ownership problems at future meetings.

4. At least two successful metallurgical procedures have been developed for processing the recovered nodules. Although the procedures now consume approximately five times as much energy to produce the same amount of refined copper as is consumed in production of copper by conventional methods, it can be anticipated that the energy consumption will decline with technological experience.

As a consequence of these measures, economic recovery of copper, nickel, cobalt, and possibly manganese and molybdenum has been predicted for 1980 at the level of 9,070 tonnes (10,000 tons) of nodules per day (dry weight) or about 2,730,000 tonnes (3,000,000 tons) of nodules per mining unit per year. If the various predictions have any validity, as indeed they seem to, the magnitude of the resource is such that nodules may be destined to play a significant role in future world copper production. It is possible to foresee a major role for the copper from nodules in the production of copper during this century.

Estimates of the magnitude of copper contained in nodules made in U.S.G.S. Professional Paper 820 under the classification of conditional resources are of the order of 45.4 million tonnes (50 million tons) of copper metal. The number is not explained in detail. Earlier estimates by Mero (1964) and by McKelvey and Wang (1969) have indicated that nodules are widespread and abundant on the floor of the world oceans, and suggest that the Professional Paper 820 figure is probably very conservative.

Size of Copper Resources in Nodules

Data for estimating the magnitude of the copper resource in nodules are, in the hands of the private sector, derived from privately-financed exploration and sampling of mining sites, and, in the public sector, derived from the National Science Foundation's IDOE program, which assembled data from various oceanographic institutions both U.S. and foreign. Unfortunately, there is little in the published material of the IDOE reports (1972, 1973) that can be directly applied to an estimate of the copper metal resource in manganese nodules and the private sector has so far been unwilling to make all of its information available.

Although nodules have been known since 1876, the quality of the public data bank remains low. Little statistical quality control can be applied to the data and the technology of sampling and exploration is still relatively primitive. Much progress also remains to be made in understanding the process of nodule genesis. Nonetheless, using such data as are available from IDOE reports, an empirical estimate of the copper nodule resource can be attempted either by a "geological control" or by an "area of influence" approach.

A. Geological Control Approach: By 1972 sufficient was known of the copper content of nodules from the world oceans for the data reported in Figure 9 to be compiled. Nodules containing the highest copper contents are found in a band on the Northwest Pacific Ocean floor (Figure 10) between 7°30 and 15° N; and 110° and 160° W covering an area of 375 square degrees or about 4,470,000 km^2 (1,720,000 square miles). This is a unique province of siliceous oozes with well-defined boundaries, characterized by low sedimentation rates. Although the word "average" is not defined in the reports, the copper contents of nodules shown on Figure 10 yield an "average" of 1.16 percent copper by weight.

Unfortunately, IDOE reports do not provide data on the abundance of nodules on the ocean floor, and we have been unable to obtain firm estimates from other sources. However, estimates by Schatz (1971) indicate a range of 5 to 20 kg/m^2 (1-4 lb/ft^2) with an "average" of 10 kg/m^2 (2 lb/ft^2) or 10,000 tonnes/km^2 (26,800 tons/mi^2) for ocean mining sites (Figure 11).

This suggests that the magnitude of the copper contained in this unique geological province could be as large as 4.5 x 10^{10} tonnes (5 x 10^{10} tons). If the North Pacific Province is unique, it represents roughly 4,470,000 km^2 (1,720,000 square miles) out of 450,000,000 km^2 (173,000,000 square miles) of total world ocean area, or, roughly, 1 percent. Because it is known that nodule densities vary greatly, but little is known quantitatively about the "patchiness" of nodule distribution, it would be safest to accept 4.5 x 10^{10} tonnes (5 x 10^{10} tons) as an unrecoverable maximum, and to assume that no more than about 1 percent of the predicted total is present and recoverable. This still suggests that at least 0.045 x 10^{10} tonnes (0.05 x 10^{10} tons) of Cu metal might be recovered (Tables 7 and 8). If the province is not unique, and if it can be hypothesized that even one comparable environment exists, copper in deep sea nodules might be assumed to increase proportionately to as much as 10^9 tonnes (1.1 x 10^9 tons).

B. Area of Influence Approach: If it is assumed that the oceans have been "thoroughly sampled" (i.e., no new sites remain to be found) and that each point with nodule abundances represents a location (on the map, Figure 9) arbitrarily comparable to that on Schatz's map (Figure 11), 51,500 km^2 (198,000 square miles) or four square degrees, the copper tonnage in nodules can be estimated as follows:

	No. of Stations	
	Cu 0.5 - 1.0%	Cu 1.0 - 20%
Pacific (Figure 9A)	66	25
Atlantic/Indian (Figure 9B)	11	0

By taking an "average" abundance of nodules of 10,000 tonnes per km^2 and median values for copper percentage, the tonnages are as follows:

	Copper Tonnage in Million Tonnes	
	@ 0.75% Cu	@ 1.5% Cu
Pacific (Figure 9A)	255	193
Atlantic/Indian (Figure 9B)	43	0
TOTAL	297	193

Because both estimates A and B are of the same order of magnitude, and because each is probably conservative, it appears that (i) the amount of copper in nodules may be of the same order of magnitude as copper in developed reserves of conventional ore deposits; (ii) because the nodules are an identified (discovered) resource, their role in future copper supplies will become increasingly significant; (iii) consideration should be given to research for enhancing knowledge of the deposits; and (iv) policy implications of nodules to future U.S. copper supply are highly significant.

Summary of World Copper Resources

However the world's copper resources are estimated, the recoverable tonnage, without having to extract the metal from silicate solid solution, seems to lie between 0.5×10^{10} and 1.4×10^{10} tonnes ($0.55 - 1.5 \times 10^{10}$ tons). The upper limit is an order of magnitude larger than the copper resource figure proposed by the U.S.G.S. in P.P. 820. World reserves of copper are large. World supply to the end of the century for this critical metal, so vital for the effective distribution of electricity, will apparently depend more on additional mining and processing capacity than on discovery of new deposits.

United States reserves of copper are 75 million tonnes (83 million tons), and the U.S.G.S. estimates that an additional 90 million tonnes (100 million tons) of undiscovered resources exist in known mining districts plus 109 million tonnes (120 million tons) in undiscovered districts (Table 6). Primary production of copper in the U.S. is 1.5 million tonnes (1.7 million tons) a year and growing. Viewed solely from a reserve in the ground point of view, domestic copper supplies thus seem to be a problem of the 21st, not the 20th century. However, economic factors intervene. These are discussed in the next section. Despite the seemingly sanguine reserve and resource pictures, it develops that additional discoveries at a significant rate will be necessary if U.S. production rates are to be maintained to the end of the century. The panel concludes, therefore, that there is a need to stimulate effective domestic mineral exploration and production.

On a more technical basis, the panel concludes that one specific aspect of copper resource appraisal which deserves special emphasis by the U.S.G.S. is a study of potential strata-bound copper provinces in the U.S., based on an analysis of strata-bound copper deposits of the world similar to the U.S.G.S. analysis of porphyry copper deposits.

UNITED STATES PRODUCTION CAPABILITY

U.S. and World Mine Production

Table 9A shows, according to the January 1973 U.S.B.M. Commodity Data Summary, primary mine production and reserves in the main copper producing countries of the world. The table suggests that U.S. reserves of copper are adequate to meet U.S. demands during the next 25 years. However, analyses of the supply of copper from primary U.S. sources presented by Wimpfen and Bennett of the U.S.B.M. before COMRATE, and published by Bennett et al. (1973), indicate that the future picture is not as clear as Table 9A suggests. Other factors intervene and make it most unlikely that self-sufficiency can be reached and maintained. COMRATE believes that these economic analyses, made using available information on reserves and conditional resources (identified subeconomic resources), have much validity. The analyses take into account the lead times required to expand an existing operation, or to start a new operation, and the declining production capacity of mines nearing exhaustion.

Potential Deficit in Domestic Primary Copper Supply

According to an analysis by Bennett et al., the apparent U.S. consumption of primary refined copper in 1972 was 1.7 million tonnes (1.9 million tons), which were supplied 86 percent from domestic mines, 10 percent from net imports, and 4 percent from a drawdown of stocks, such stocks being mostly of domestic origin. At an assumed, but widely accepted, average annual consumption growth rate of 4 percent, the demand for primary copper in 1985 would be nearly 2.6 million tonnes (2.9 million tons) (see Figure 12). However, projected mine production will be only 1.7 million tonnes (1.9 million tons) in 1985. This results in a predicted 900,000 tonne (1 million ton) deficit in domestic primary supply 12 years hence. For the immediate future, lack of domestic smelting capacity is more important than mine capacity. Due in part to environmental protection restraints, declining smelter capacity may also cause a deficit. Attempts to predict the possible longer term effects of environmental control regulations (see Figure 13) suggest a deficit of 410,000 tonnes (450,000 tons) by 1980.

Thus, the U.S.B.M. contemplates a reversal from self-sufficiency in copper to a growing dependency on imports (shown by the area between mine production and primary demand in Figure 13). The U.S.B.M. does point out, without endorsing its point, that mine production without any constraints, based on announced expansions in prior years and continuation of normal growth rate for the latter part of the period, could apparently satisfy U.S. primary demand. For this to be true, however, U.S. smelting and refining capacity would also have to be expanded in an environmentally acceptable manner to treat the mine production.

U.S. Production Capabilities Based on Identified Resources

An illustration of expectable comparative 1970 cost levels for a hypothetical, large, new U.S. operation, comprising a mine, mill and dump leach facilities is given in Figure 14, derived from data by the U.S.B.M. The chart shows a "Net Smelter Return" royalty of 5 percent on the resulting production: copper sulfide concentrates; copper cement precipitates and electrowon copper. The chart also shows the relative magnitude of taxes, profit required for a 12 percent DCF-ROI after tax, and postconcentrator costs of transportation, smelting and refining to suggest a "minimum selling price of copper."

Each "cost" element in the graph includes depreciation and amortization, allowable depletion, local and federal taxes, and net profit required for a 12 percent discounted cash flow return on investment (DCF-ROI) after taxes. However, the tax and profit contributions to the minimum selling price for the mining and milling operations only are separately broken out in the two left-hand bars of the chart; similarly for the leaching methods these contributions are separately shown in the two right-hand bar charts. The transportation, smelting, and refining costs are total charges (including tax and profits) for those separate operations.

It is to be noted that, as dump leaching can be considered a scavenging operation, the costs of mining and transportation to the dump are fully charged to the (normal) flotation operation (two left-hand bars in chart) and only the costs of bedding, circulation and recovery are debited to leaching (two right-hand bars in chart).

The distribution of U.S. primary copper production, according to metallurgical extraction method, is shown in Table 9B. It shows that there has been little change over the last 15 years in the proportion of hydrometallurgical to pyrometallurgical copper recovery. Announcements have been made by 10 companies for pilot plants to test new "chemical" smelting, i.e., hydrometallurgical process for treatment of copper sulfides. However, only one new, sizable metallurgical plant has been announced recently; a smelter for 90,000 tonnes (100,000 tons) metal per year in New Mexico. It would appear that new metallurgical approaches will have an impact on production only 15 years hence.

In 1950 and 1960, 25 percent of ore was extracted in underground operations; the increasing amount of open-pit extraction is reflected by a decrease of underground extraction in 1970 to 16 percent, 1973--27 percent.

Breakthroughs in rock shattering (fracturing) technology and the development of lixiviants that would allow in situ solution mining of low-grade sulfide ores at low costs would considerably increase the reserves of copper in the U.S. However, even though considerable work is being done along these lines, no low-cost methods have been developed to date for deep, low-grade, sulfide ores. Again, such approaches, if they are successfully developed, will affect production of copper in 20 years or more.

Using the costs shown in Figure 14, the U.S.B.M. has determined the amount of domestic recoverable copper resources, i.e., measured and indicated economic resources (reserves) available at different prices of copper. Such resources were computed also for different rates of return (DCF-ROI after all taxes). The results, graphed in Figure 15, show how price-sensitive copper resources are. At a price of $4.40 per kilogram ($2 per pound) of copper, about 162 million tonnes (180 million tons) of copper could be economically recovered. This is to be compared with 1972-1973 statements of U.S. reserves in the range 73 to 81 million tonnes (81 to 90 million tons) of copper, at 50 to 60¢ price. These estimates allow a credit for the by-products of copper production: gold, molybdenum, and silver.

The U.S.B.M. determined that the potential annual production rate, at 12 percent DCF-ROI and 50¢ price, is about 2.3 million tonnes (2.5 million tons) of copper per year (see Figure 16). However, this assumed that no restrictions are placed on the scheduling of production from the deposits; no time is allotted for expansion, development, and startup. But, in reality, some copper operations justify a long projected life, whereas others can be reasonably projected for only 10 to 15 years. Also, in reality, time allowances must be made for preproduction development, startup, and expansion (Bailly, 1966). For a new operation, feasibility and environmental studies will take from one to three years and the subsequent preproduction development from two to seven years. Startup period needed to reach decision capacity will be six to eighteen months.

Using the 1970 capacity as a base and applying realistic time factors, it then turns out that the increased theoretical capacities shown in Figure 16 could not be reached at all, as shown in Figure 17. If domestic operators could double their production rate (from two percent to four percent of reserves per year), the theoretical capacities would be reached; such an approach is not economically wise in the eyes of the mine operators. In any case, such an approach would deplete domestic reserves by the year 2000. In addition, the maximum capacity that could be reached at different prices of copper could not be maintained for long, unless a continuous replacement of reserves takes place through discoveries and/or improvement in extraction technology. The projections shown in Figure 17 will, most likely, prove incorrect because the industry will react to circumstances and search for, discover, and develop new deposits and develop new extractive technology, in response to anticipated demand.

As for most resource industries, one of the limiting shortages may turn out to be the availability of capital. It has been estimated that, by the year 2000, the U.S. copper industry will have to replace 50 percent of its 1.8 million tonne (2 million ton) metal annual capacity now installed. To this must be added new capacity of 3.6 million tonnes (4 million tons) by 2000. The investment required "based on $3,630 per annual tonne ($3,300 per annual ton) would approximate $16.5 billion or $600 million each year" (Lawrence, 1974). This capital requirement is very sizeable when compared to the sales value of $2 billion of the primary copper produced currently per year, especially since recently started operations cost much more than $3,600 per annual tonne ($3,300 per annual ton) capacity--actually $4,400 to $5,500 per annual tonne ($4,000-$5,000 per annual ton).

One conclusion resulting from the U.S.B.M. study is that additional discoveries must be made which will transfer undiscovered resources into the identified economic resource (reserve) category. COMRATE's copper resource estimates, and those of the U.S.G.S., indicate that the resource base most likely does exist in the U.S. To obtain this transfer, copper exploration must be continued successfully.

Need for Copper Discoveries

Additional production needed to satisfy the U.S.B.M. demand forecast is equivalent to putting on stream, each year, an operation processing 36,000 tonnes (40,000 tons) per day of ore, averaging 0.6 percent copper. Assuming a minimum economic life of 20 years (at a high production rate of five percent of reserves per year), this is about equivalent to a 230 million tonne (250 million ton) deposit, i.e., a Class #3 deposit. In addition, just to stay even on its current ore reserves, the U.S. must discover each year a 230 million tonne (250 million ton) deposit containing 0.8 percent copper, or a 300 million tonne (330 million ton) deposit containing 0.6 percent copper, i.e., another Class #3 deposit. In other words, to maintain its position and avoid future depletion by maintaining a steady state of reserve readiness, the U.S. alone needs to find two Class #3 deposits each year, or one Class #2 (giant) deposit, the equivalent of 530 million tonnes (580 million tons) of 0.6 percent copper ore containing 3.2 million tonnes (3.5 million tons) metal. To achieve the same reserve readiness, the world must discover each year a billion ton ore deposit containing 0.8 percent copper, i.e., a large Class #2 (giant) deposit. The record indicates this has not been achieved in the U.S. in recent years through known new discoveries. Reserve development has largely been in areas of old discoveries. As discussed, deposits have limits to their sizes, and it is becoming more prevalent to test a body completely before mining starts. Future maintenance of production must depend to an increasing extent on new discoveries. It is clear from the following discussion that the cost of discovery is high.

COST OF MINERAL DISCOVERY

Copper is one of several mineral commodities which are used in sufficient volume, and the reserves of which are so limited, that discovery-oriented exploration programs for it by industry and/or government are economically justified.

Exploration Phases

In any given area containing, or judged to be favorable to the occurrence of copper deposits, the search effort over time follows an evolution pattern including three generations (or phases) of exploration.

The first generation exploration effort consists of surface prospecting which results in discovery of outcropping deposits. This has been, to date, the most productive approach for copper as well as for most other metal commodities. For instance, the major part of the pre-1960 porphyry coppers was found through prospecting.

The second generation exploration effort consists of geological mapping and instrumental (geochemical, geophysical, airborne, etc.) surveys around known deposits, or known metal showings, resulting in discovery of new deposits in known mineralized districts. During that phase the extrapolation is pragmatic, rather timid, and with limited conceptual content. This approach, however, results in substantial increases in reserves. For instance, as shown in Figure 18, the reserves created by 21 years of second-generation exploration of porphyry coppers in North America during the period 1950-1971 amount to 40 million tonnes (44 million tons) of metal contained in 39 deposits, as of 1971. This can be compared with reserves, as of the same year, of 41 million tonnes (45 million tons) in 17 deposits found before 1950, in the same area.

The third generation exploration effort consists of using geological occurrence models (sometimes with genetic content) of deposits and of their settings. Such models allow the formulation of predictions about mineral provinces and mineral occurrences in the absence of mineralized outcrops and in the absence of anomalies resulting from instrumental surveys. This conceptual approach has not yet produced many reserves, but it is expected that it will in the future, thanks to the major improvements in recent times in the conceptual, genetic content of the science of ore deposition. This approach is essentially geological remote sensing. It is followed, of course, by geological and instrumental field surveys.

For each generation, the first discovery in any province is usually followed by a surge of new discoveries. Then this generation (or phase) ends slowly, with a few more discoveries which show a continued increase in time and money for each new discovery.

Discovery Costs

The discovery cost per deposit, or per unit of commodity is negligible during the first generation effort. It increases considerably during the second generation effort. As shown in Table 10, this cost (including all expenditures previous to development as defined for tax purposes) during the 20-year period, 1951-1970, increased in Canada, from 0.6 percent to two percent of the gross value of metals contained in the discoveries. During the 15-year period, 1955-1969, it increased in the western U.S., from one percent to 2.2 percent of the gross value of metals discovered. The year 1950 marks approximately the transition throughout the world from first to second generation exploration efforts.

The efficiency of the exploration dollar varies considerably with different generations of exploration, different regions, and different targets; this is clearly shown by Table 11.

Over the last 20 years, the efficiency of the exploration dollar has decreased by about two-thirds. It appears that this decrease will continue. Table 10 also shows that the cost for each discovery has increased by a factor of seven in Canada; even though no reliable data are available for the U.S., the increase is probably of the same order. No data are available for costs of third generation discoveries; COMRATE estimates that it will be at least 10 times the cost of the second phase.

Adequacy of Domestic Exploration

For large to medium-sized copper deposits in Arizona, the average investment to equip a deposit for production, including smelter and refinery capacity, is presently about 100 times the actual discovery costs (excluding cost of failures at other sites) and the average value of annual production derived therefrom, about 30 times the discovery cost. The U.S.B.M.'s economic study of copper supply from domestic resources indicated that the copper industry in the U.S. should each year equip one new mine extracting daily 40,000 tons of ore averaging 0.6 percent copper, just to satisfy the expected increase in demand. With the recovery of five kilograms of copper per tonne (10 pounds per ton) of ore, this represents a gross value of metal production of about $66,000,000 annually, and a total investment of $200,000,000. In this light, maintaining the 1970 level of exploration expenditures, in Arizona and New Mexico, is apparently adequate to satisfy the additional demand. However, knowing that exploration costs have increased very rapidly in the last 20 years and will, most probably (barring cost-effective breakthroughs in exploration science and technology), continue to increase drastically, it appears that the current domestic exploration level is not adequate to satisfy the expected continued increase in demand. It is definitely not adequate to provide for a replacement of reserves at the producing mines.

The need for increasing (a) exploration effectiveness--"doing the right things" which lead to discovery; (b) exploration efficiency--"doing such things well and at low cost"; and (c) the intensity of the exploration effort, is becoming critical. Because of this realization, COMRATE is planning to convene several future workshops on mineral exploration science and technology to investigate the following topics:

1. R & D on geological models for discovery planning and resource appraisal;

2. R & D for improving detection technology: (a) indirect detection (geological mapping, geophysical and geochemical techniques, remote sensing) and (b) direct detection (sampling technology, with emphasis on drilling technology).

LIST OF TABLES

TABLE 1 Past Production + Ore Reserves of World Copper Deposits, Excluding U.S.S.R., as of 1970 (Data from AMAX)

Units: Short Tons

Tonnage Limits (1,000 tons metal)	Class of Deposit	Number of Deposits in each class	Total Copper contained in each class (1,000 tons metal)	Average Copper Content per Deposit (1,000 tons metal)	Cumulative Percentage of Total Number of Deposits	Cumulative Percentage of Total Copper Contained
31,623						
10,000	#1	6	103,849	17,308	1.9	24.2
3,162	#2	40	173,474	4,448	14.3	64.7
1,000	#3	45	86,313	1,876	28.9	84.8
316	#4	81	48,157	595	54.6	96.0
100	#5	68	13,623	200	76.2	99.2
32	#6	52	3,021	58	92.7	99.8
10	#7	22	519	24	99.7	99.9
3	#8	1	9	9	100.0	100.0
	Total	315	428,965			

TABLE 2 Main Geologic Types of Copper Deposits (After Pelissonier, 1968, 1972) and Their Geologic Ages (After C. Meyer, unpublished)

Geologic Type	Cu Distribution by Type % 1968	1972	Tonnage Class (from Table 1)	Geologic Age Range (million years)
Porphyry Coppers	51.8	52.4	Class 1 - Class 6	Mostly 30-130. Some in Paleozoic: 300-500. Some as recent as 1.
Strata-Bound Coppers (In Sedimentary Rocks)	26.5	26.9	Class 1 - Class 5	Mostly 500-1,800. A few 200-500. One recent.
Massive Sulfides (In Volcanic Assemblages)	10.7	9.9	Class 3 - Class 7	Mostly 1,800-3,200. None between 500 and 1,800. Some younger than 500. A few very recent.
Subtotal	89.0%	89.2%		
In Basic and Ultrabasic Rocks	2.9	4.7		
In Intermediate-Basic Lavas	2.7	2.3		
Other Types	5.4	3.8		
Total	100.0%	100.0%		

TABLE 3 Median Tonnage, Grade and Copper Content of Porphyry Copper Deposits (Including Past Production + "Reserves") (After U.S.G.S.)

Units: million tonnes (million short tons)

	Number of Deposits	Size in million tonnes ore*	Grade % Cu	Copper Content in million tonnes copper contained*
All Porphyry Copper Deposits	78	222 (246)	.73	1.6 (1.8)
U.S. and Mexico	39	244 (268)	.72	1.7 (1.9)
Canada	15	188 (206)	.53	1.0 (1.1)
Andes and Pacific	24	216 (238)	.93	2.0 (2.2)
Pre-1960 Porphyry Copper Deposits	19	450 (495)	.84	3.8 (4.2)

* Million short tons shown in parentheses.

TABLE 4-1 Production and Reserves (1) (2)
Bingham Canyon Porphyry Copper Mine, Utah
Owned and Operated by Kennecott Copper Corporation

(Tonnage in short tons - See References in Table 4-2)

	Sept. 1899	Jan. 1, 1915	Jan. 1, 1930	Dec. 31, 1970
Annual Production (prior year)				
Tonnage of Waste Rock	---	1,252,961 (9)	19,821,357	97,103,000
Tonnage of Ore Mined	---	6,470,166	17,724,100	40,147,500
Average Cu grade %	---	1.4%	0.99%	0.68%
Tonnage of Copper Metal Recovered	---	91,876	175,469	273,003
Cumulated Production from start in July 1904				
Tonnage of Waste Rock	---	12,529,611	213,195,619	1,286,629,883 (10)
Tonnage Ore Mined	---	34,053,723	192,732,074	1,170,042,091
Average Cu grade %	---	1.464%	1.191%	0.927%
Tonnage Gross Copper Contained	---	498,696	2,296,078	10,851,559
Tonnage Copper Metal Recovered	---	309,636	1,658,464	9,745,560
Reserves - description	Measured Ore (3)	Ore (4)	Ore (5)	Proven Ore (6)
Tonnage of Ore	12,385,000 (7)	342,500,000	640,000,000	1,773,300,000 (8)
Average Cu grade %	2.0%	1.45%	1.066%	0.71%
Tonnage Gross Copper Contained	247,700	4,966,000	6,822,000	12,588,000
Tonnage Estimated Copper Recoverable	---	---	---	11,011,000

TABLE 4-1 Continued.

Size of Deposit = Past Production + Reserves				
Tonnage of Ore	12,385,000 (7)	376,533,723	832,732,074	2,943,342,091 (8)
Average Cu grade %	2.0%	1.452%	1.094%	0.796%
Tonnage Gross Copper Contained	247,700	5,468,696	9,118,078	23,439,559
Tonnage of Recovered + Estimated Recoverable Copper Metal	---	---	---	20,756,560

TABLE 4-2

1. Arrington, Leonard J. and G.B. Hansen, 1963, "The Richest Hole on Earth": a History of the Bingham Copper Mine: Utah State University, Monograph Series, Vol. XI, No. 1, October 1963, 103 pp.

2. Kennecott Copper Corporation, Annual Reports for years 1963 through 1970 and March 10, 1971 Registration Statement with Securities and Exchange Commission. Production estimated for 1963 and 1967.

3. Parsons, A.B., 1933, Utah--The Prospect: in The Porphyry Coppers, 1933. On pages 54-55, the author summarizes the Jackling-Gemmell evaluation report written at Bingham in September, 1899.

4. Kennecott Copper Corporation, 1916, Report for the period May 27, 1915 to December 31, 1915: New York, April 7, 1916, p. 11.

5. Boutwell, J.M., 1935, Copper deposits at Bingham, Utah: in Copper Resources of the World, 16'th International Geological Congress, Vol. 1, p. 347-539; see p. 358.

6. Kennecott Copper Corporation, 1971, Stock Prospectus filed with Securities and Exchange Commission, April 22, 1971, p. 11.

7. In the report mentioned in (3) above, the authors indicate the "great extent" of the deposit, thus implying additional reserves would be found.

8. To these reserves should be added Anaconda's Carr Fork reserves which, geologically, are part of the same ore body and have been described in Anaconda's 1972 Annual Report as a substantial tonnage of 1-3% copper ore, for which feasibility studies for a large scale, underground mining operation are under way.

9. Average for the 10 prior years.

10. This tonnage of waste from start through 1961; actually a considerable amount of waste rock has been dump-leached for several decades and has contributed up to 25% of the recovered copper metal in recent years.

TABLE 5 Main Copper Provinces of the World
Percent Distribution of Copper Metal (after Pelissonnier, 1972)*

Copper Provinces (Pelissonnier, 1972)	Percentage of Copper in All Provinces	Main Types of Deposits
Andean	28.8	Porphyry coppers
North American Cordilleran	22.7	Porphyry coppers
South Central Africa	19.3	Strata-bound
North American Great Lakes	8.2	Massive sulfides, strata-bound, and basic rocks
Northern Europe	8.5	Strata-bound and massive sulfides
Ural-Kazakstan	4.1	Porphyry coppers, massive sulfides, and strata-bound
East Mediterranean	2.2	Massive sulfides and porphyry coppers
Huelva, Spain	1.8	Massive sulfides
Philippines	1.3	Porphyry coppers
Japan	1.9	Massive sulfides
Eastern Australia	1.2	Strata-bound and massive sulfides
	100.0%	

* This represents 98% of copper inventoried in the world by Pelissonnier

TABLE 6 Copper Resources

Units: million tonnes (short tons) metal contained (after USGS and USBM)

	Identified Resources — Includes Measured+ and Indicated Resources		Undiscovered Resources — Assuming economic extractability			
	Economic ("Reserves")	Subeconomic ("Conditional" Resource)	Adjacent to Identified Resources*	In known mining districts	In undiscovered mining districts	Total Resources
U.S.A.	75 (83)	100 (110)	?	90 (100)	110 (120)	375 (413)
World total, U.S.A. included	340 (370)	300 (330)	?	360 (400)	290 (320	1,300 (1,420)

* Sometimes referred to as "inferred" and included in "Identified Resources." Such occurences, however, are speculative projections which have not been sampled; they are not truly identified and their economics cannot be evaluated.

\+ Estimates of Measured Reserves are computed with a margin of error of 20% or less. The reader should keep in mind that numerical expressions of other resource categories are guesses with margins of error greater than 20%. The probable errors increase from left to right in the table.

TABLE 7 Probable Copper Resources in Manganese Nodules, All Oceans

	Atlantic and Indian Oceans	N.E. Pacific Nodules Belt	Rest of Pacific	Total
Nodules: tonnes/km^2	10^4	10^4	2.5×10^3	-----
(tons/mi2)	(28,500)	(28,500)	(7,100)	-----
% Cu	0.75	1.50	0.75	
Copper: tonnes/km^2	75	150	19	-----
(tons/mi2)	(215)	(430)	(54)	-----
km^2/station	51.4×10^4	51.5×10^4	51.5×10^4	
(mi2/station)	(19.8×10^4)	(19.8×10^4)	(19.8×10^4)	-----
Copper: tonnes/station	3.85×10^6	7.71×10^6	0.98×10^6	-----
(tons/station)	(4.24×10^6)	(8.54×10^6)	(1.08×10^6)	-----
No. of stations	11	25	66	102
Copper: tonnes	42×10^6	193×10^6	65×10^6	300×10^6
(tons)	(46×10^6)	(212×10^6)	(72×10^6)	(330×10^6)

TABLE 8 Probable Copper Resources, Pacific Ocean Nodule Belt

Latitude range	7.5 - 15°N	(7.5°)
Longitude range	110 - 160°W	(50°)
Area, square degrees	375	
km^2 (mi^2)	4.5×10^6	(1.73×10^6)
Nodules, tonnes/km^2(tons/mi^2) median	10^4	(28,500)
Nodules, tonnes (tons)	4.5×10^{10}	(4.95×10^{10})
Copper Resources @ 0.75%	338×10^6	(372×10^6)
tonnes (tons) @ 1.5%	675×10^6	(773×10^6)
@ mean value	432×10^6	(475×10^6)

TABLE 9A World Copper Mine Production and Reserves (1000's short tons)
(From U.S.B.M. - Commodity Data Summaries - Jan. 1973 and Jan. 1974)

Country	1971	1972	1973 (est.)	Estimate of Reserves* Jan. 1973	Jan. 1974
United States	1,522	1,665	1,717	81,000	83,000
Canada	720	801	900	30,000	33,000
Chile	791	799	800	56,000	58,000
Peru	235	248	240	22,000	23,000
Zaire	449	473	510	20,000	20,000
Zambia	718	791	800	27,000	29,000
Other Free World	1,231	1,437	1,533	64,000	82,000
Communist countries (except Yugoslavia)	999	1,100	1,200	40,000	42,000
World total	6,665	7,314	7,700	340,000	370,000

* Reserves include "Measured" reserves computed with a margin of error of 20% or less, and "Indicated" and "Inferred" reserves estimated within a much larger margin of error.

TABLE 9B U.S.A. Primary Copper Mine Production
(Distribution according to metallurgical recovery method)

		1950	1960	1970	1971	1972	1973 (Est.)	1975 (Projected)
Pyrometallurgical Recovery (Smelting, Refining)								
Concentrates from sulfide flotation	%	92.0	87.0	86.5	86.4	86.3	86.0	84.0
Cement copper obtained by precipitation of iron								
(Dump leaching	%	6.0	8.5	9.0	9.0	9.0	9.0	10.0
(Heap leaching	%	1.0	1.0	1.2	1.2	1.3	1.5	1.5
(In-place leaching	%	----	1.5	1.0	1.0	1.0	1.0	1.5
Hydrometallurgical Recovery (Solvent Extraction, Electrowinning)								
(Dump leaching	%	----	----	.2	.3	.3	.3	.5
(Heap leaching	%	----	----	.1	.1	.1	.2	.3
(Vat leaching	%	1.0	2.0	2.0	2.0	2.0	2.0	2.2
Total U.S. Primary Copper Mine Production 1000's short tons		910	1,082	1,545	1,522	1,665	1,717	1,850

TABLE 10 Mineral Exploration Costs and Results in North America (After Statistics Canada 1973, and AMAX 1971)

	Value of Metals Discovered		Average per Discovery in $ Millions	
	Annual Average in $ Billions	In $'s per each $* of Expl. Expend. - Total Industry	Exploration Expenditures	Value of Metals Discovered
A. Canada				
Nonferrous metals (and asbestos) 1971 $ - 1971 metal prices.				
Period				
1951 - 55	4.4	160	2	320
1956 - 60	5.4	110	6	670
1961 - 65	5.2	95	6	570
1966 - 70	5.4	55	14	770
B. Western U.S.A.				
Metals excluding Uranium 1970 $ - 1970 metal prices.				
Period				
1955 - 59	2.8	80	NA	NA
1960 - 64	3.2	59		
1965 - 69	4.1	45		

* Equivalent to efficiency of Exploration Dollar. The expenditures include expenditures for all exploration programs whether successful or not.

TABLE 11 Cost of Mineral Discovery

- Percentage of Total Exploration Cost Resulting in Discovery
- Efficiency of Exploration Expenditures
  (After AMAX, 1970 and P. A. Bailly, 1973)

Discovery of one deposit in Province named below	Total Cost to Industry in $ Millions	Actual Cost to Discoverer in $ Millions	Percentage of Successful Exploration	Average Gross Value of Deposit in Province $ Millions	Efficiency of Exploration Dollar ($ value of Metals Discovered per each $ of Exploration Expenditure) For Discoverer	For Industry
Arizona Porphyry Copper	20	2	10%	1,600	800	80
British Columbia Porphyry Copper	33	2	6%	750	375	23
Mexico Porphyry Copper*	10	2	20%	2,700	1,350	270
Eastern Canada Massive Sulfides	100	1	1%	1,000	1,000	10

* Based on limited data

LIST OF ILLUSTRATIONS

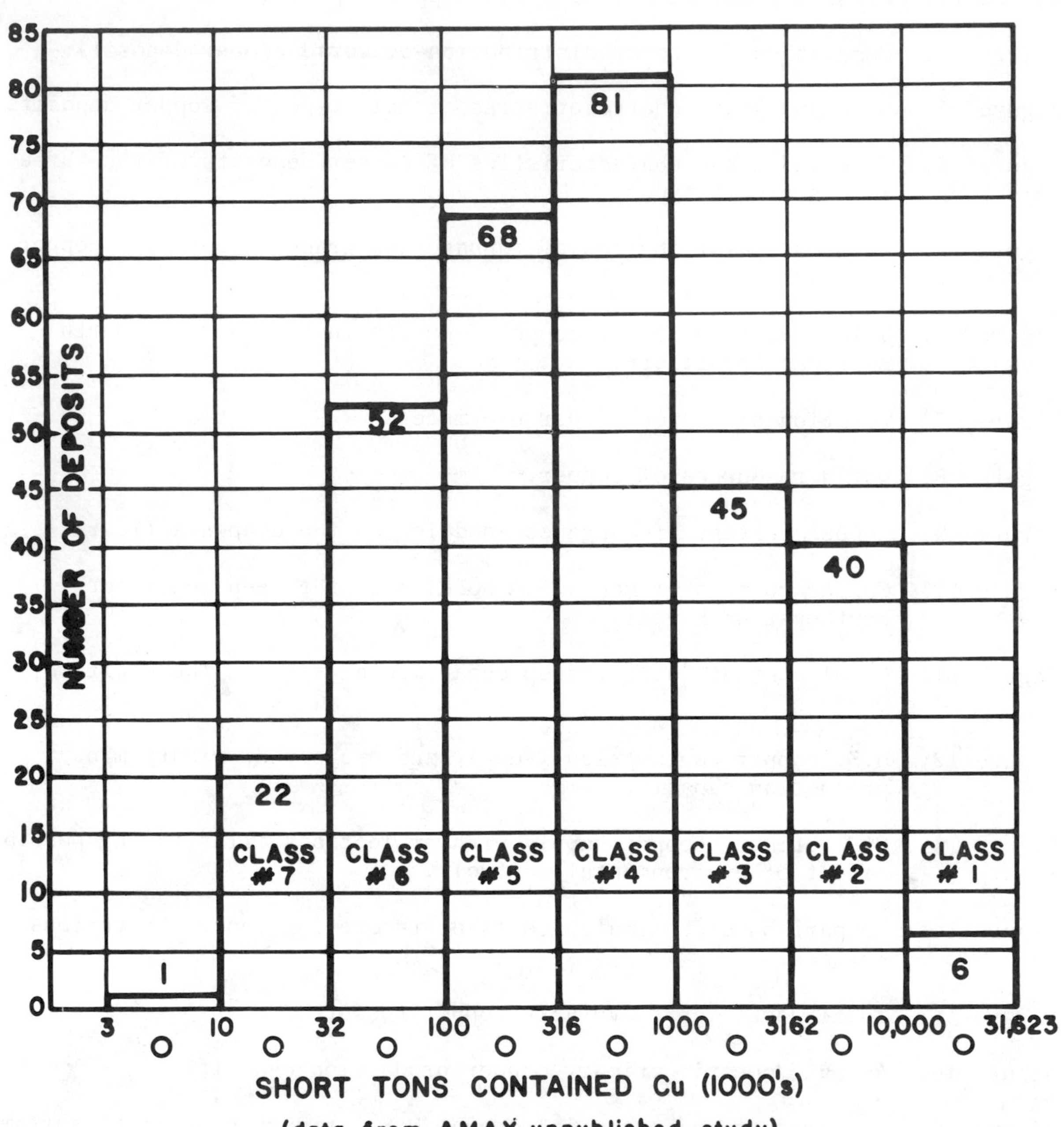

SOURCE: Courtesy of S. K. Hamilton.

FIGURE 1: Frequency Distribution, World Cu Deposits (Excluding U.S.S.R.)

315 Deposits as of 1970

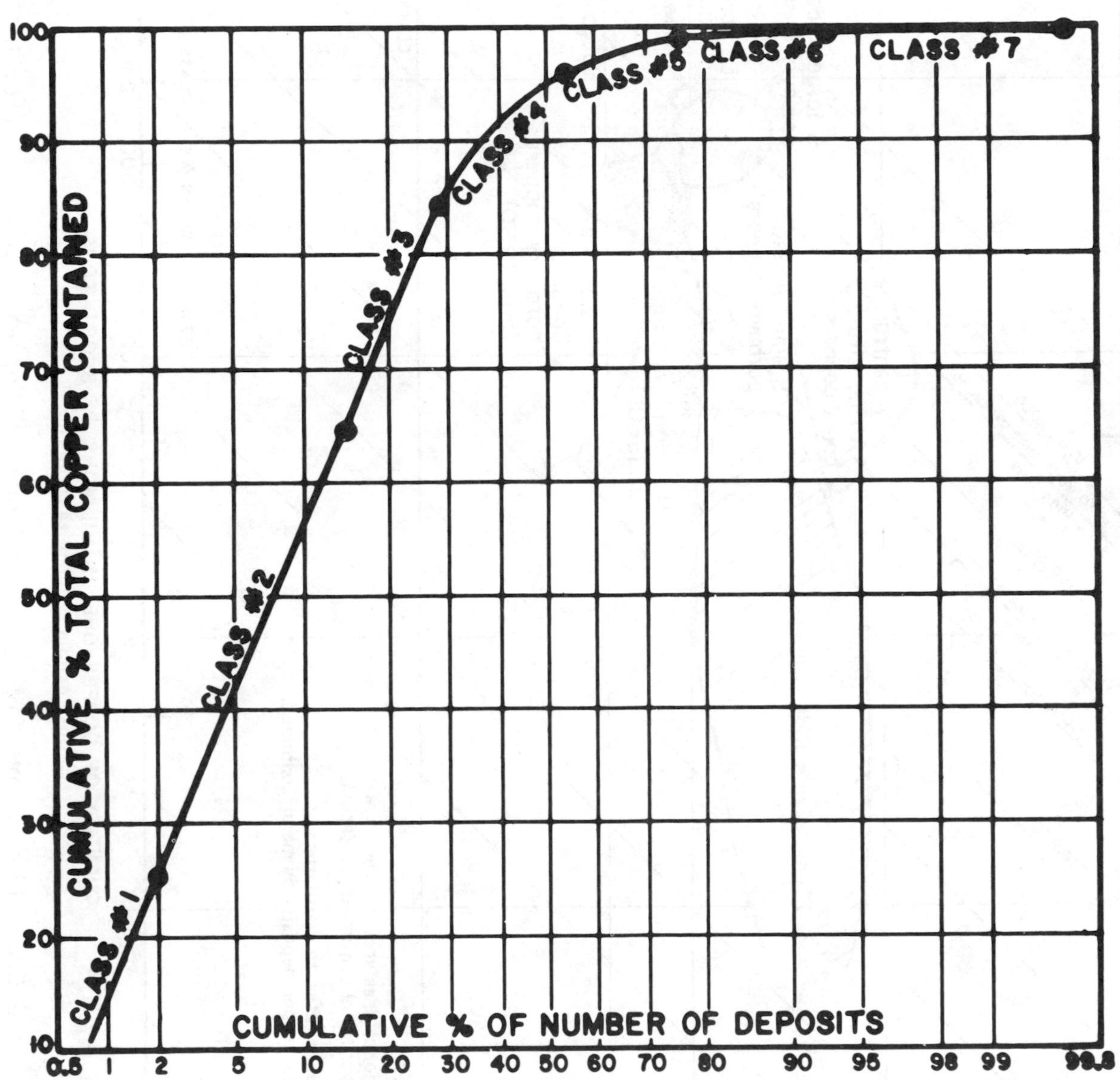

SOURCE: Courtesy of S. K. Hamilton.

FIGURE 2: Cumulative Frequency Distribution of World Copper Deposits (Excluding U.S.S.R.)

315 Deposits as of 1970

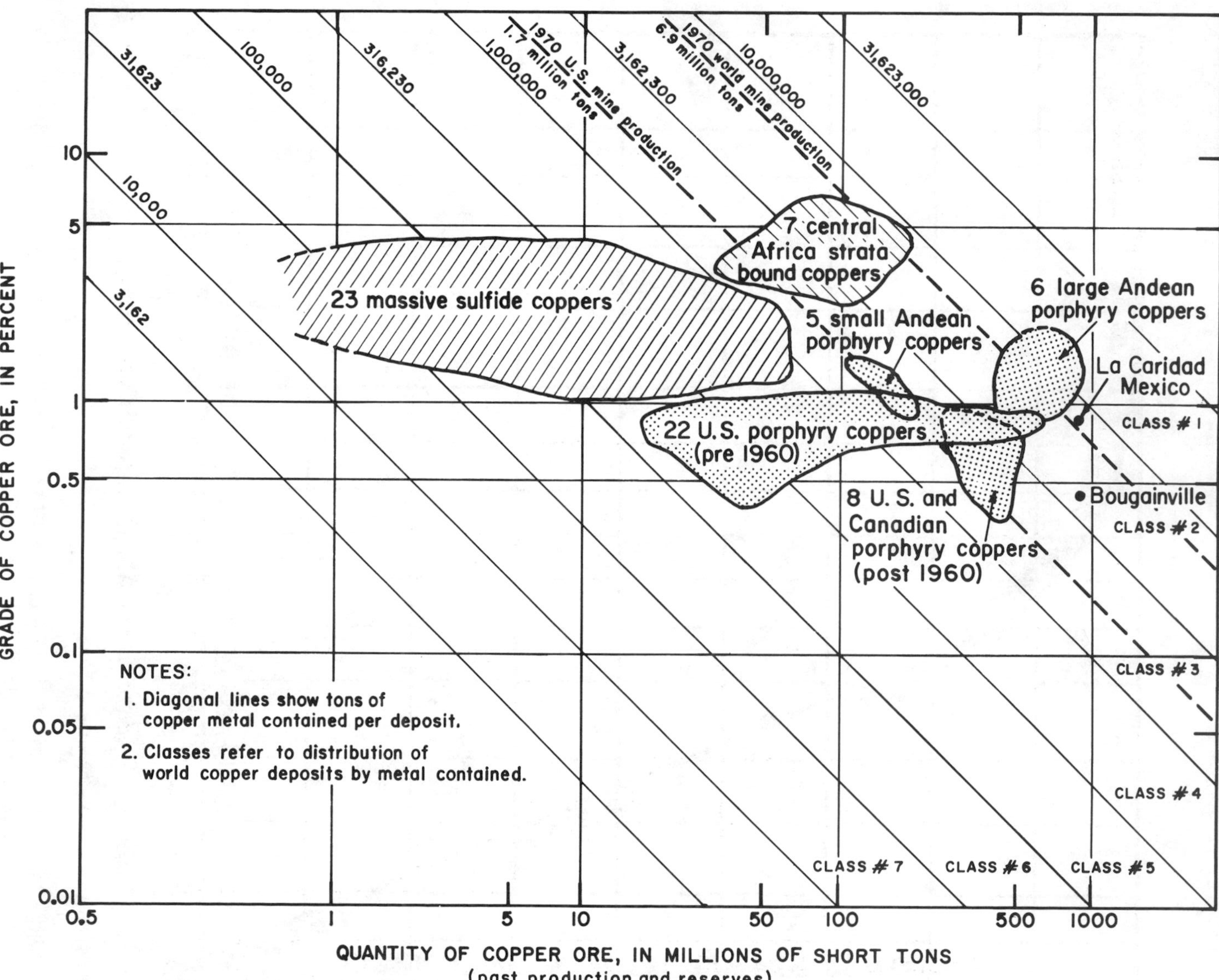

FIGURE 3: Size and Grade Characteristics of Main Types of Copper Deposits (after U.S.G.S.)

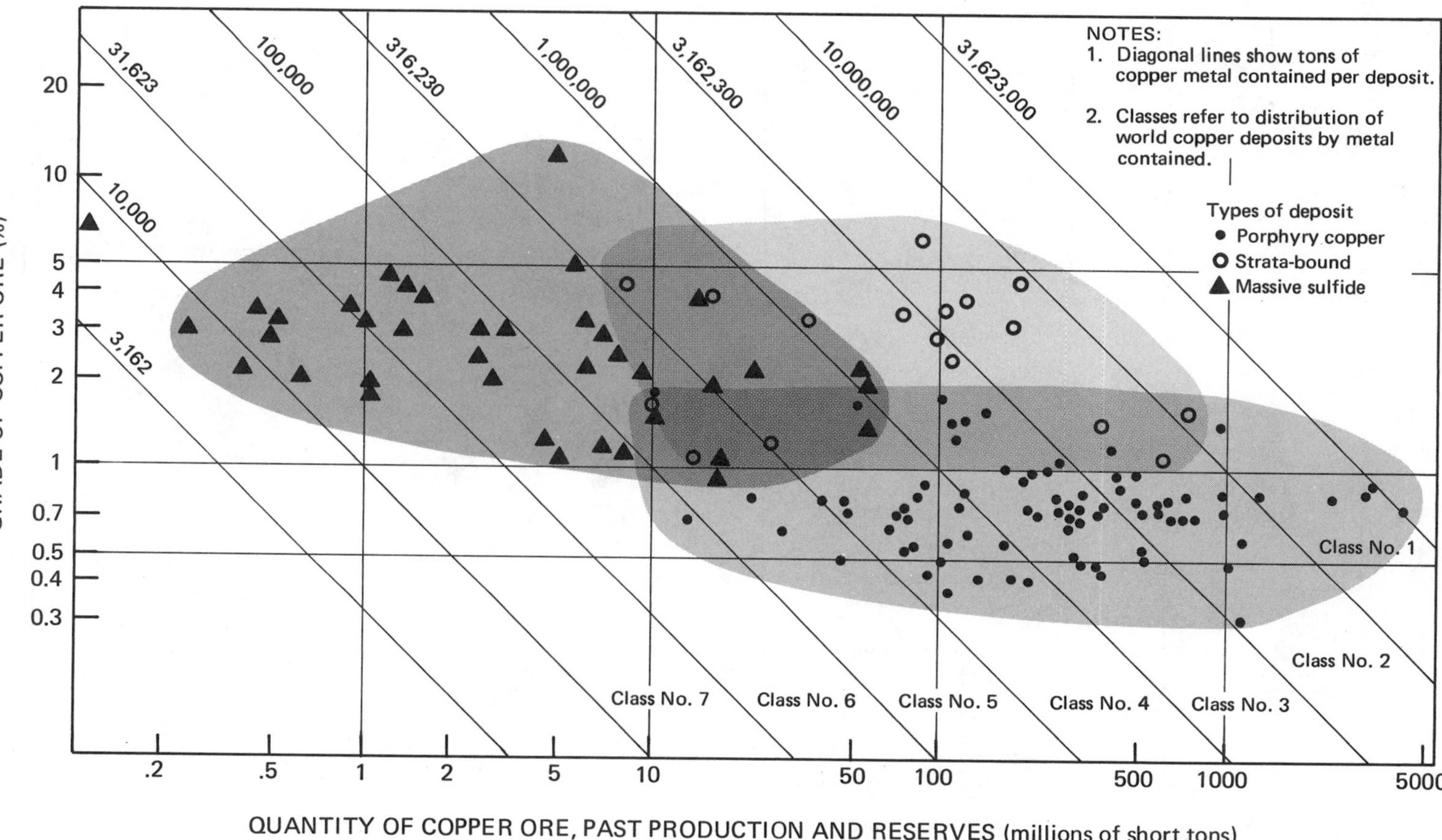

FIGURE 4: Size and Grade Characteristics of Copper Deposits of the Three Main Geological Types (after U.S.G.S.)

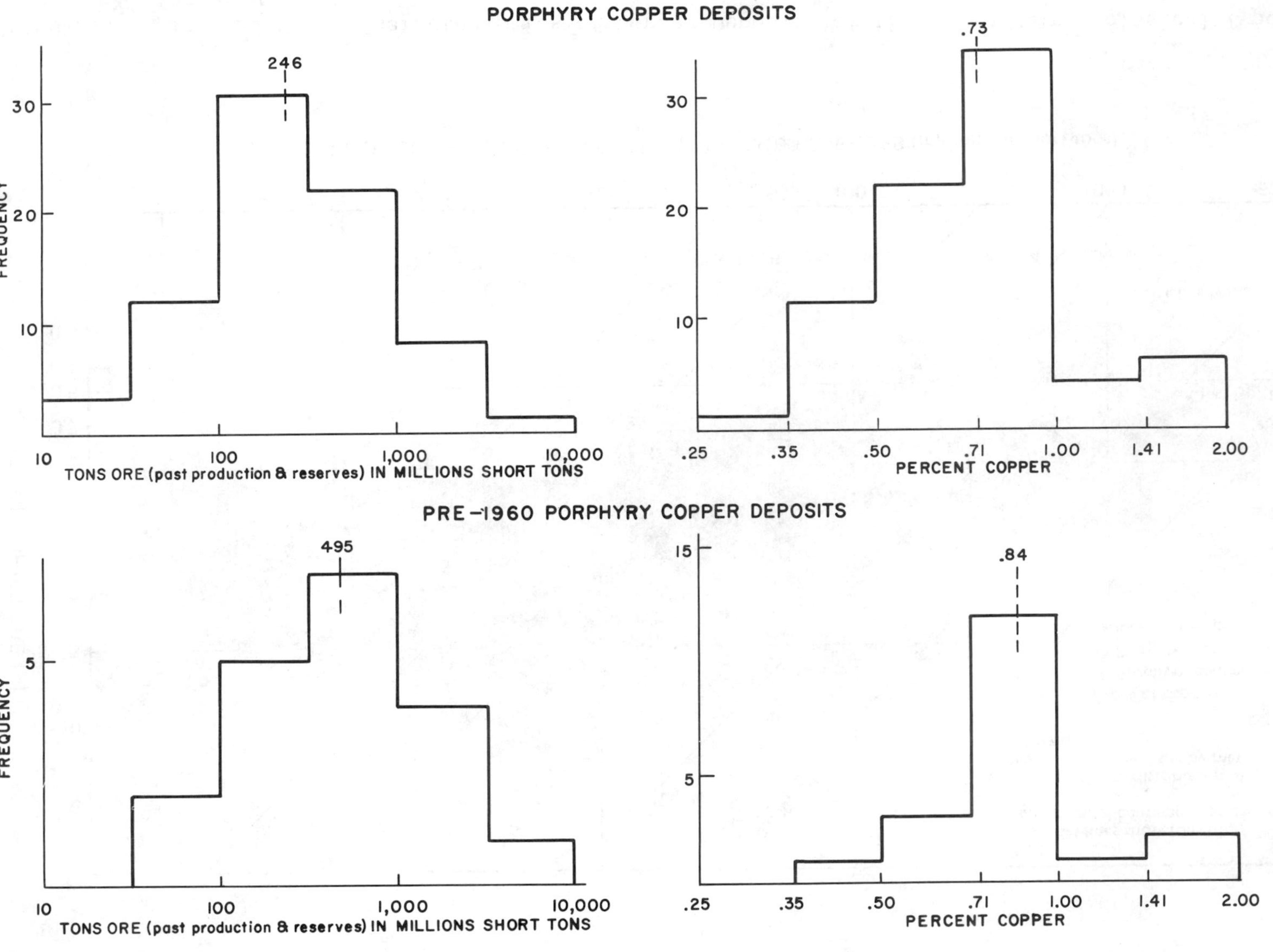

FIGURE 5: Frequency Distribution of Tonnage and Grade of Porphyry Copper Deposits (after U.S.G.S.), Median Values are Indicated

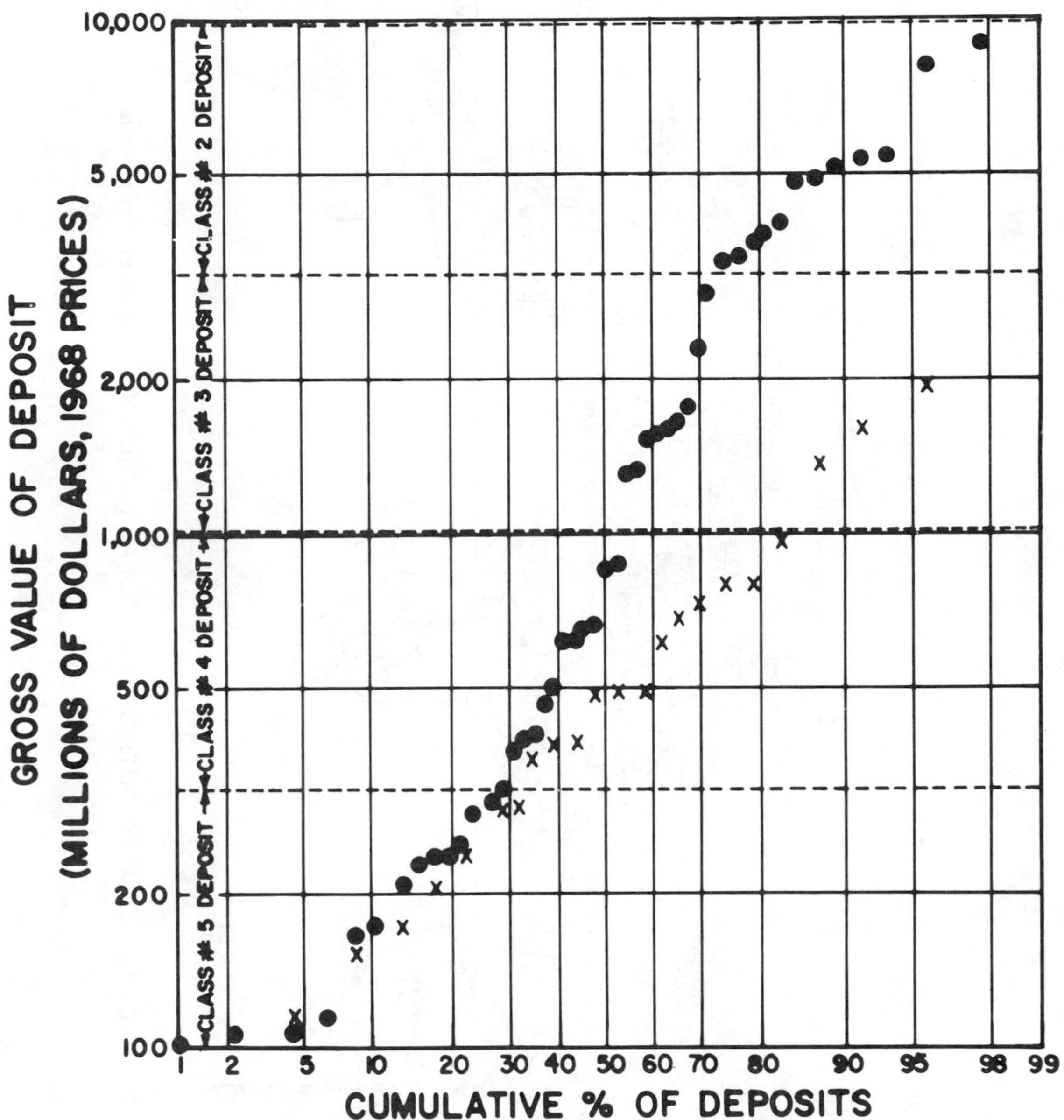

SOURCE: Courtesy of S. K. Hamilton.

FIGURE 6: Value Distribution of Copper Deposits in 2 Parts of the North American Cordilleras Copper Province (1968)

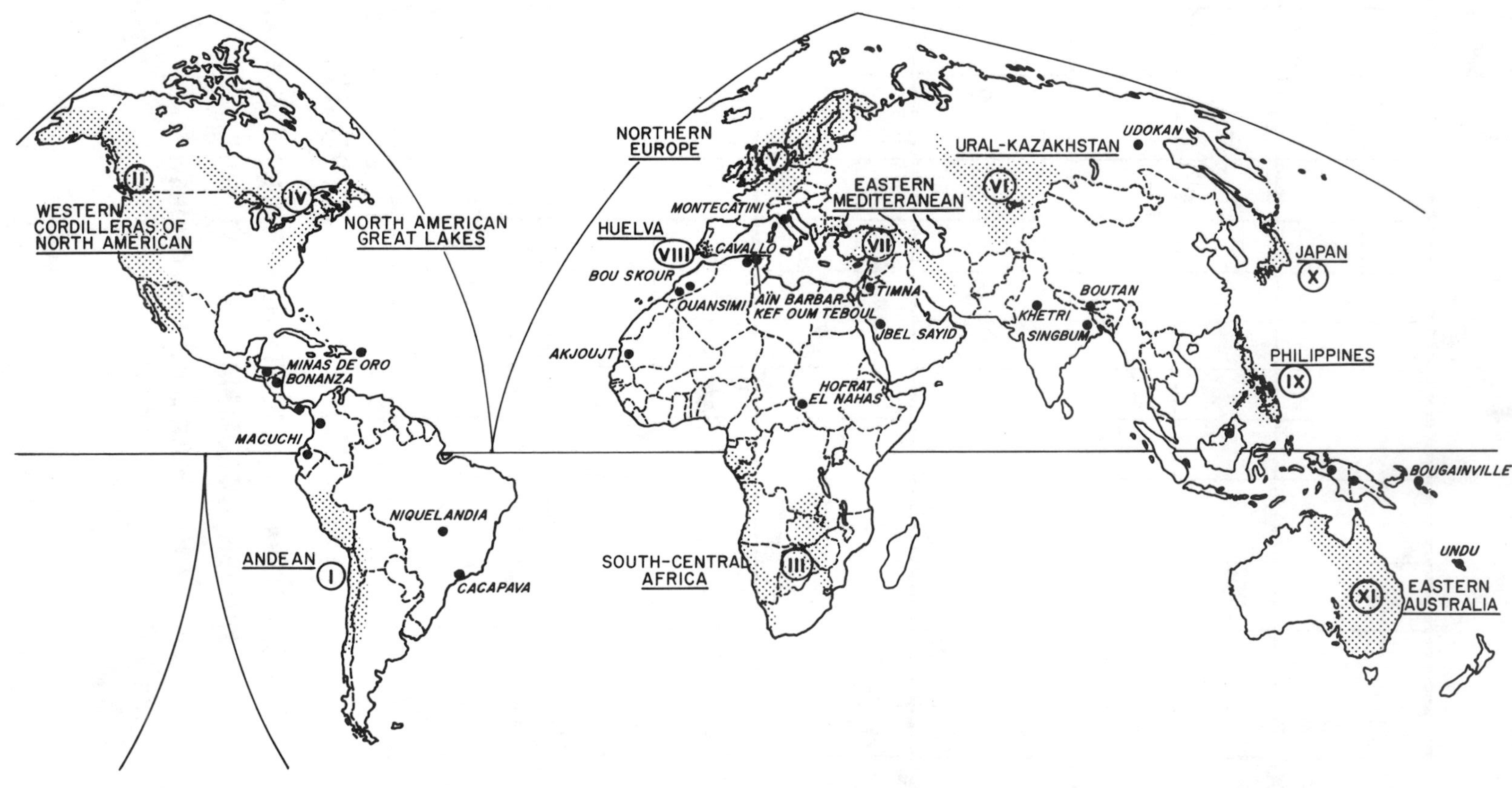

FIGURE 7: Location of Eleven Copper Provinces (I through XI)

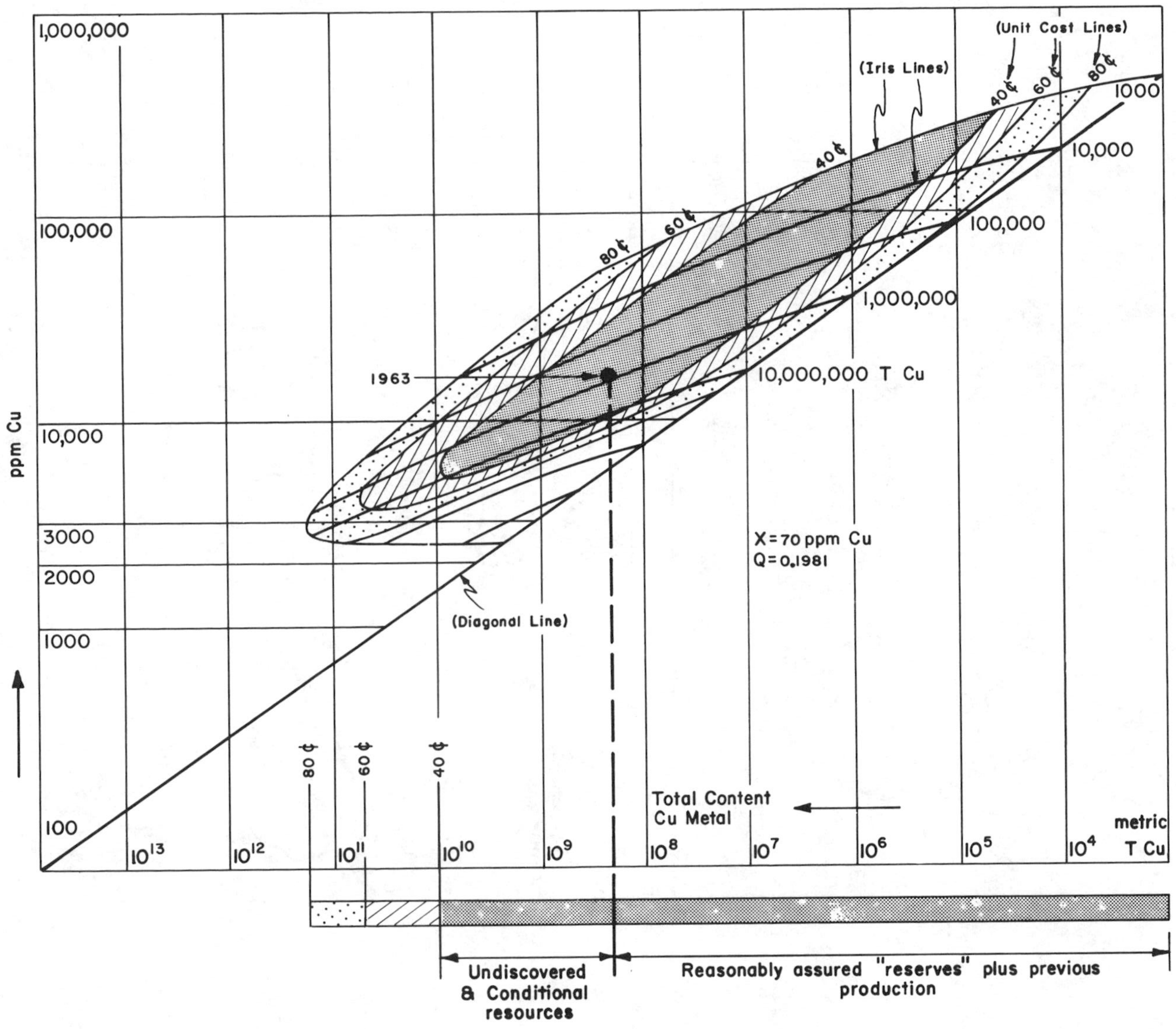

FIGURE 8: World Resources of Copper (From article "MIMIC" by J.W. Brinck, Eurospectra, Vol. X, No. 2, June 1971)

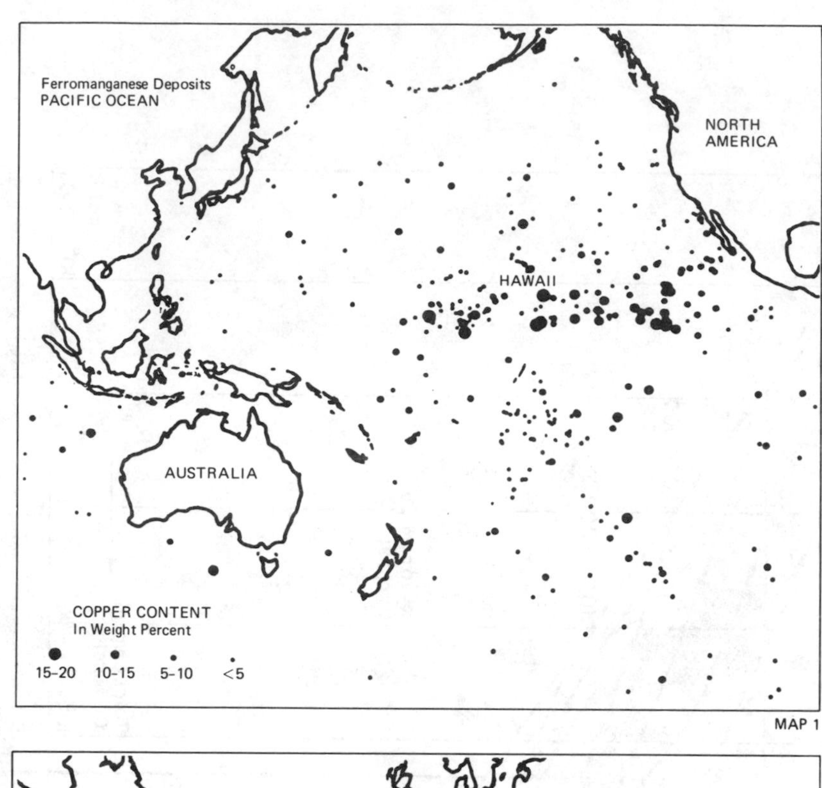

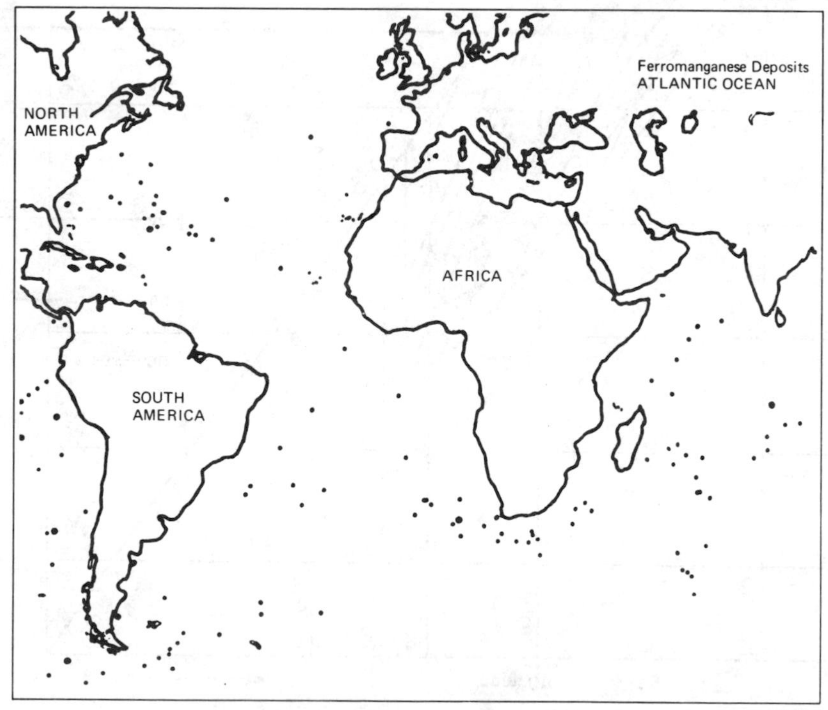

SOURCE: IDOE Technical Report No. 4, 1973, NSF/GX 33616, p. 32.

FIGURE 9: Copper Content of Manganese Nodules of the Deep Sea Floor

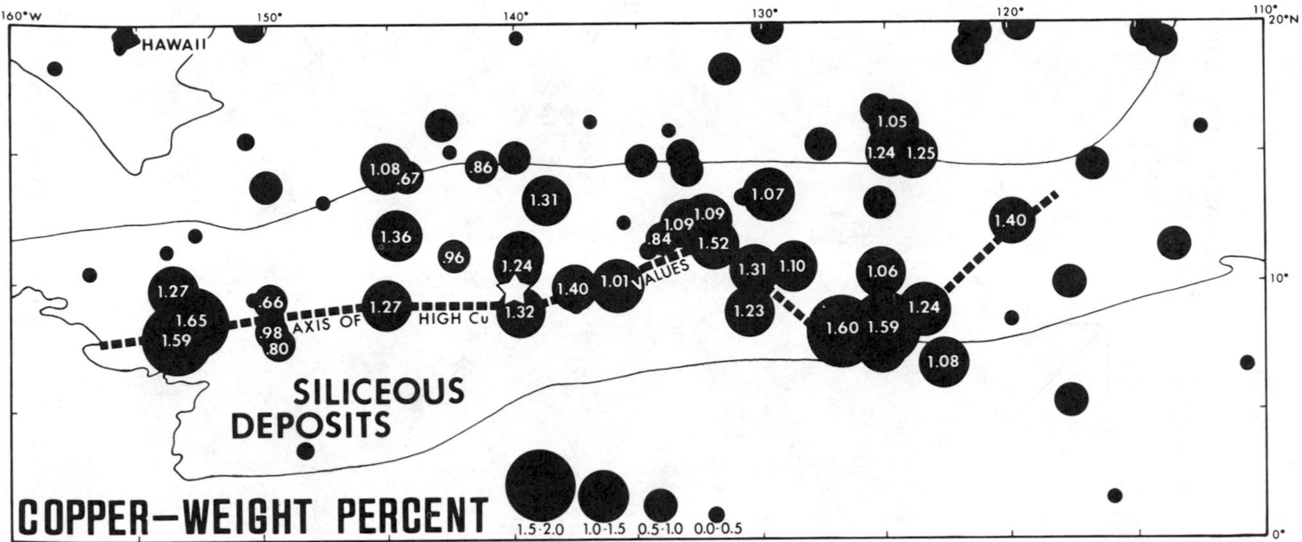

SOURCE: The Origin and Distribution of Manganese Nodules in the Pacific and Prospects for Exploration, Edited by Maury Morgenstein, Honolulu, Hawaii, July 23–25, 1973, p. 82. (Hom *et al.,* 1973).

FIGURE 10: Copper Content of Manganese Nodules on the Deep Sea Floor Southeast of Hawaii

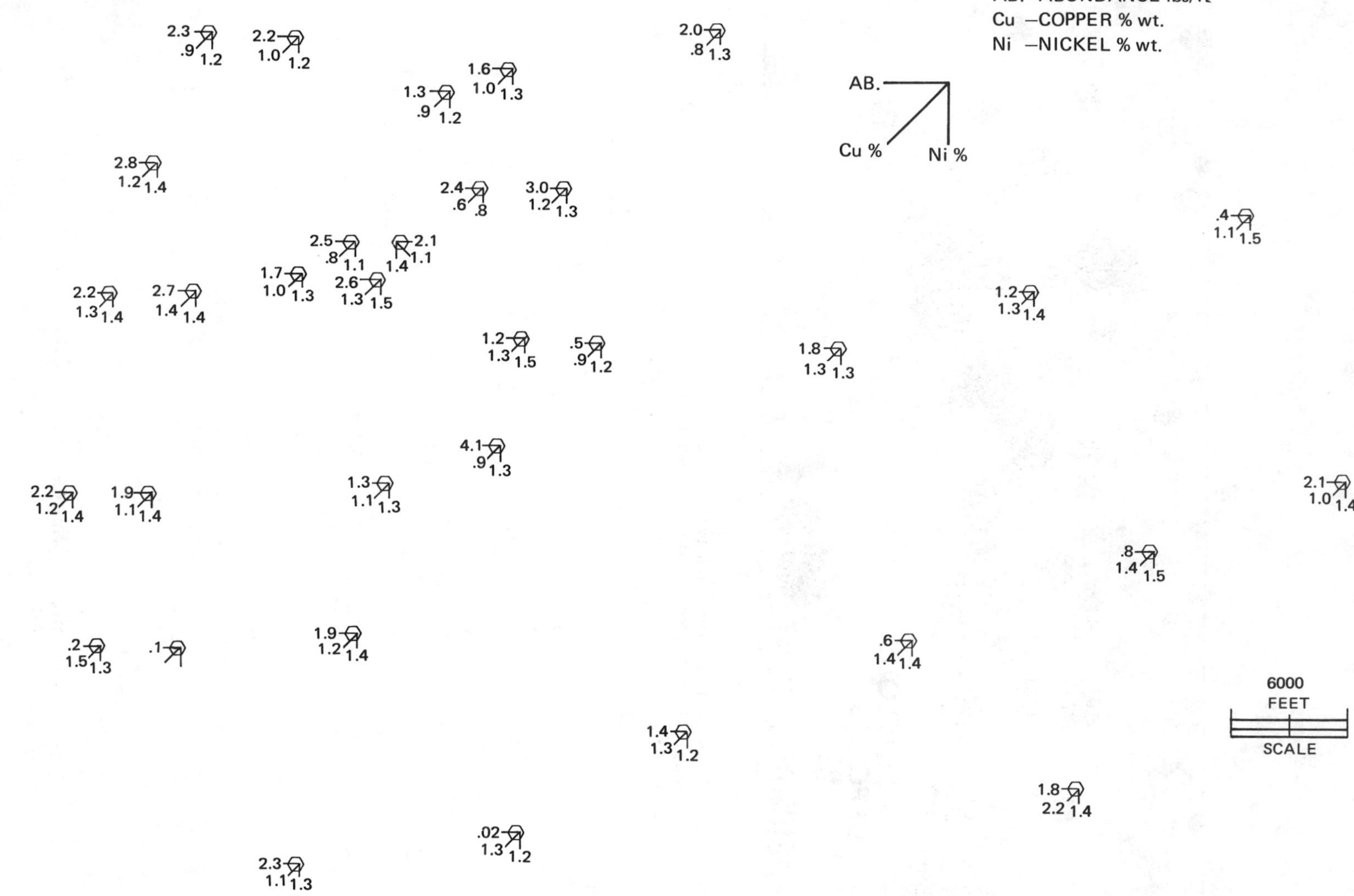

SOURCE: Metal Contents at the Deep Ocean Mining Site in the Pacific Ocean (Schatz, C. E., 1971).

FIGURE 11: Metal Contents at the Deep Ocean Mining Site in the Pacific Ocean

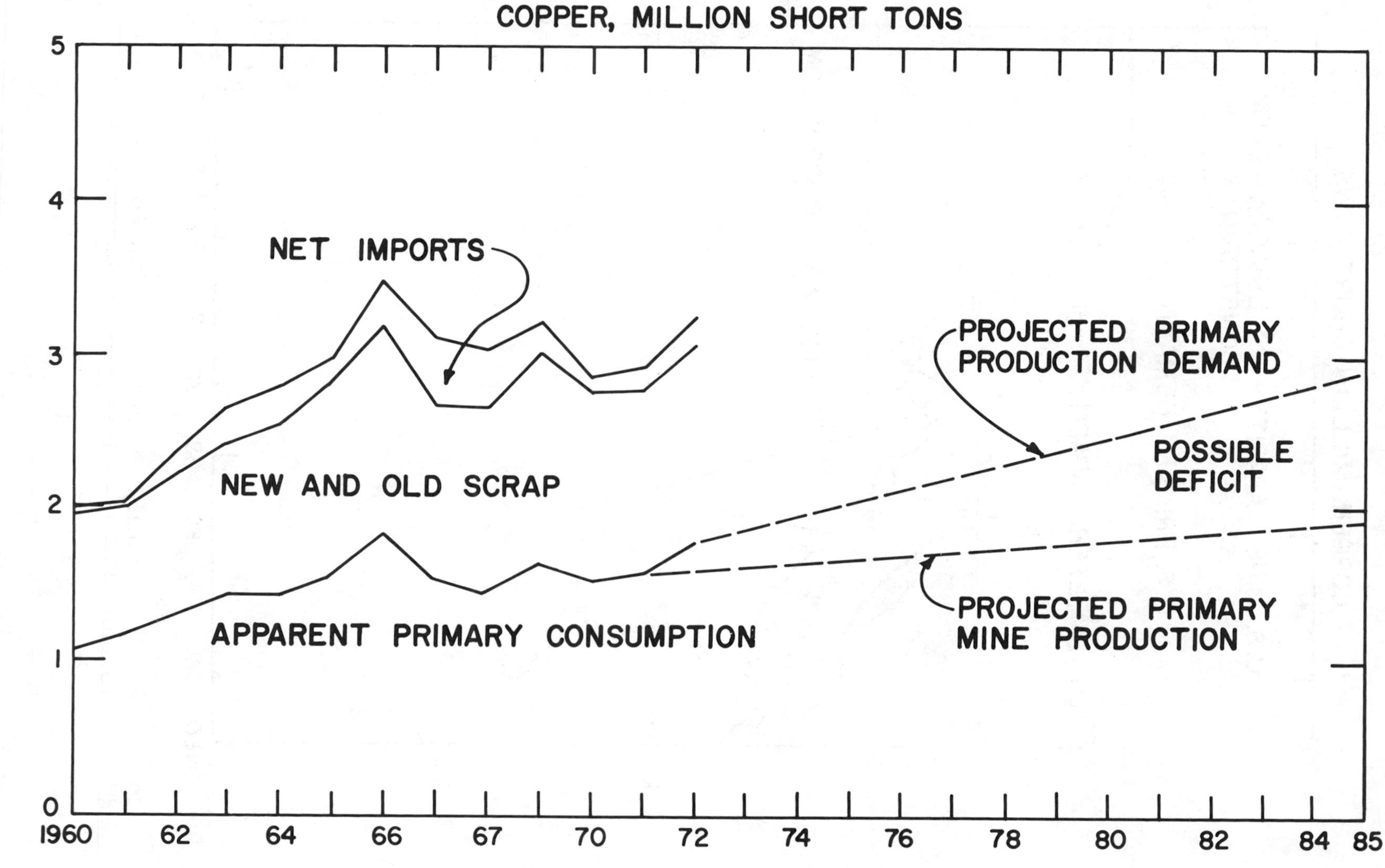

FIGURE 12: U.S. Copper Consumption; Supply and Projected Primary Mine Production Capacity (after U.S.B.M.)

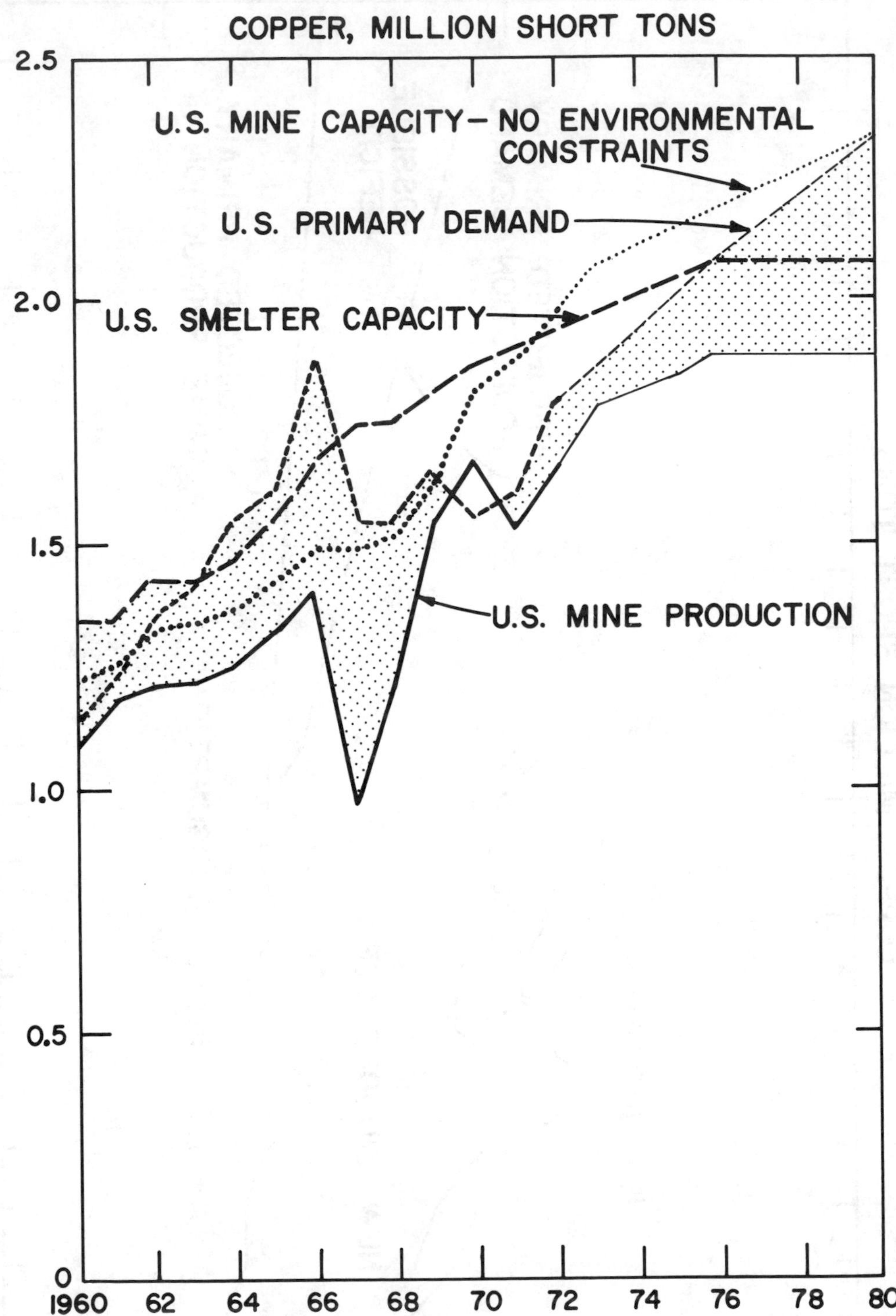

FIGURE 13: The U.S. Primary Copper Supply with Projections Illustrating Possible Effect of Environmental Controls (after U.S.B.M.)

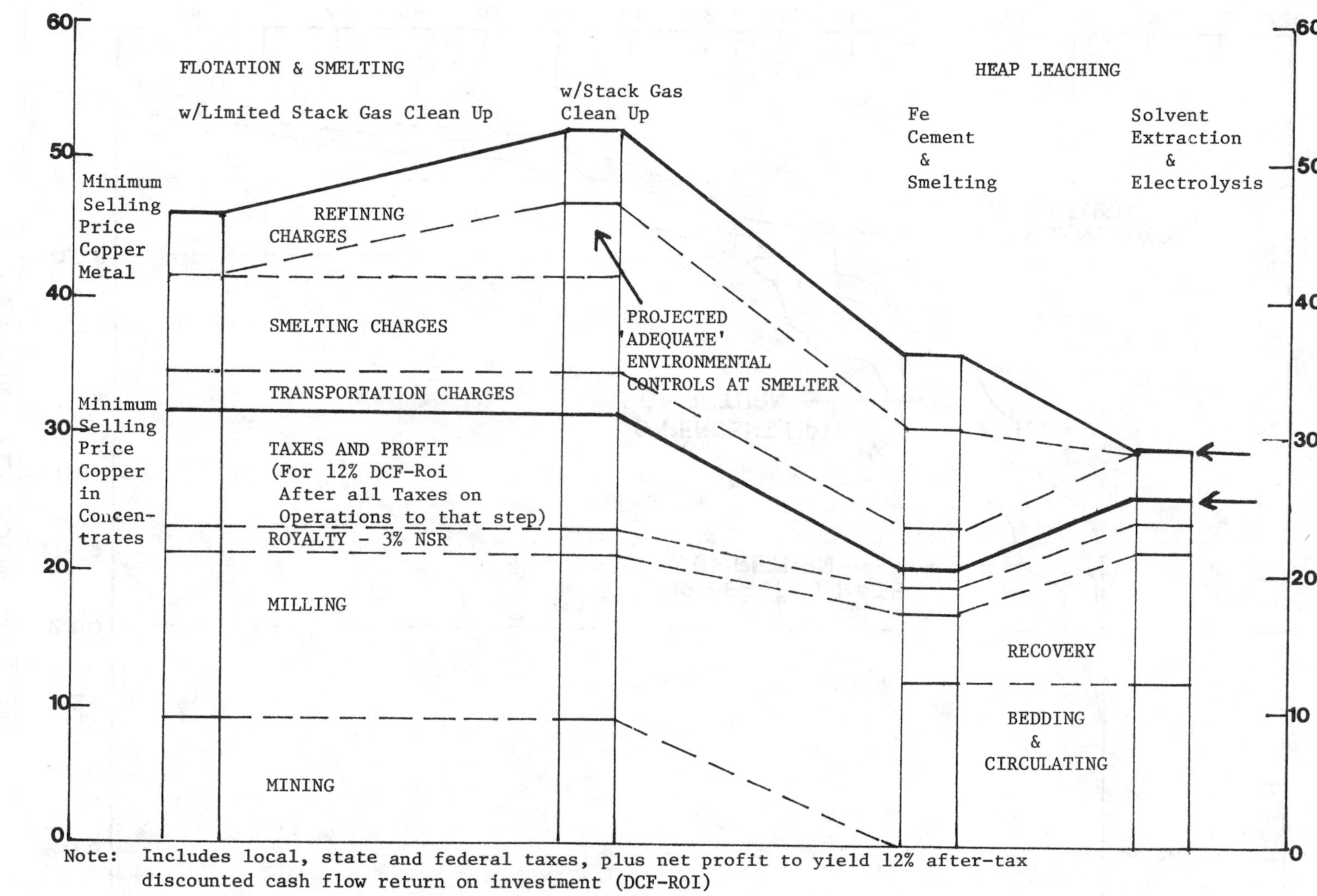

Note: Includes local, state and federal taxes, plus net profit to yield 12% after-tax discounted cash flow return on investment (DCF-ROI)

FIGURE 14: Elements of the Price of Copper By Method of Production: 1970 Estimates for Large New Mine-Mill Operations in U.S.A. (Source: Minerals Yearbook - U.S.B.M., 1973)

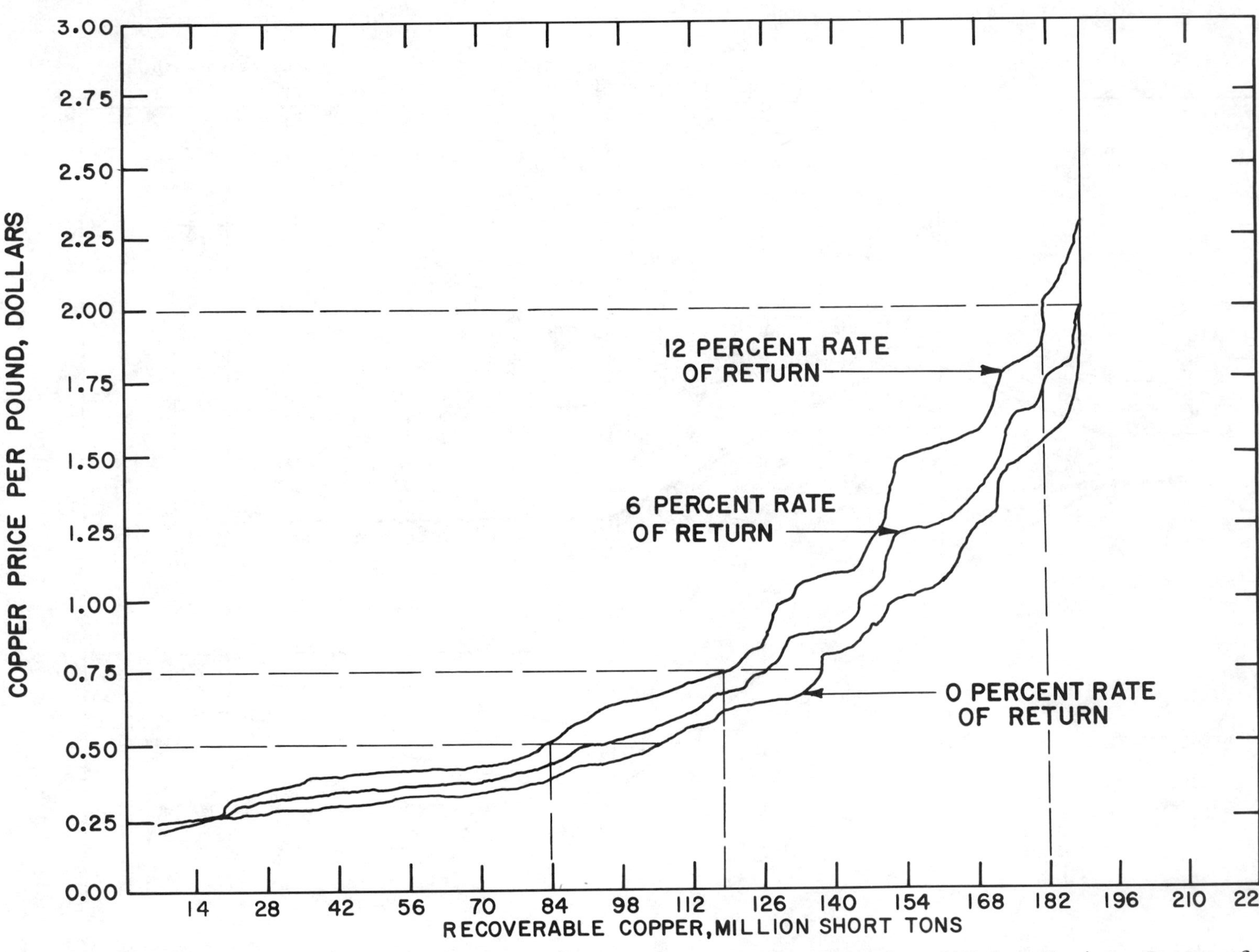

FIGURE 15: Total Domestic Recoverable Copper Resources--Price Calculated Using Various Rates of Return (DCF-ROI after tax). (After U.S.B.M.)

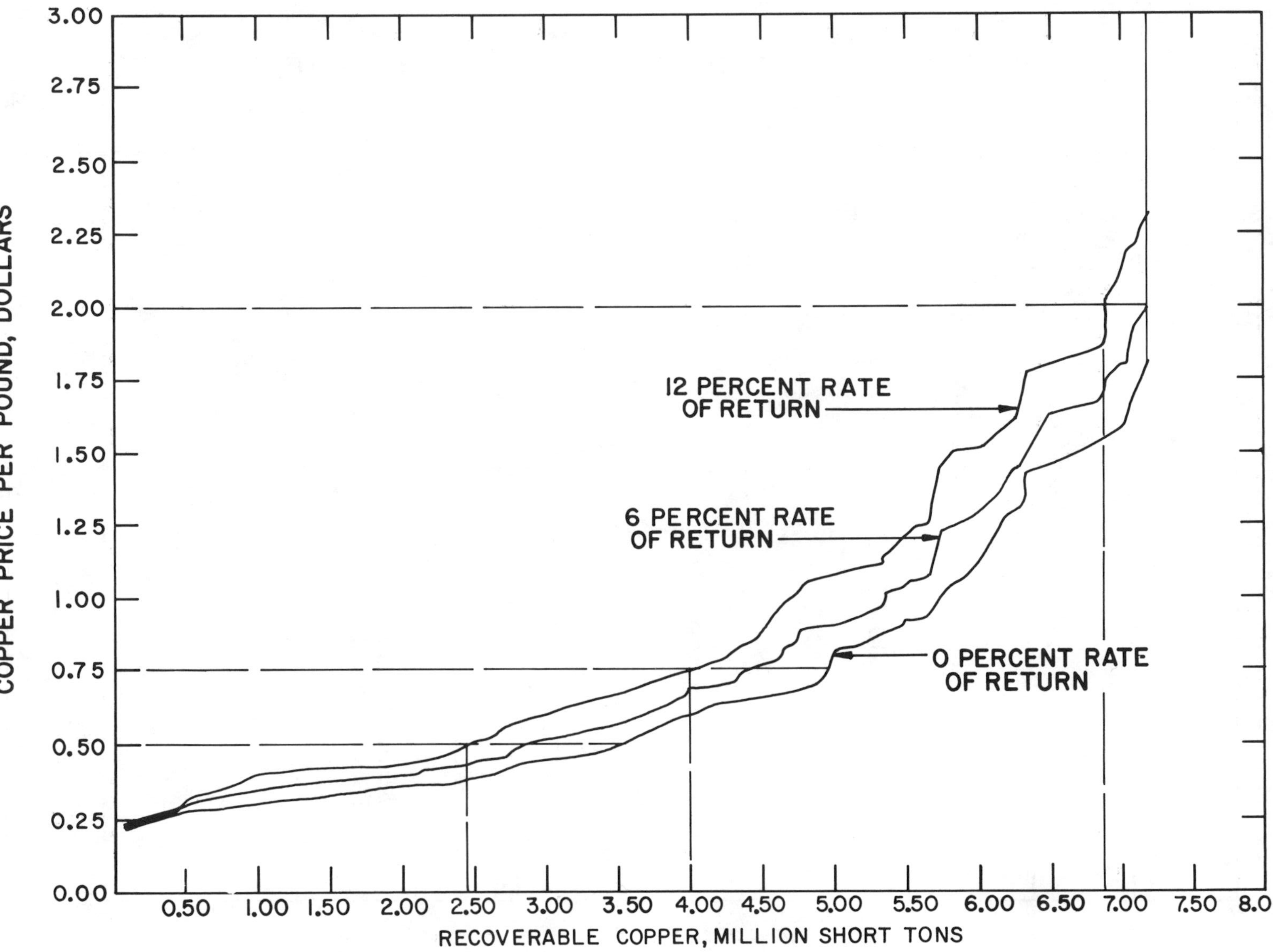

FIGURE 16: Annual Domestic Primary Copper Production Capacity--Price Calculated Using Various Rates of Return (DCF-ROI after tax). (After U.S.B.M.)

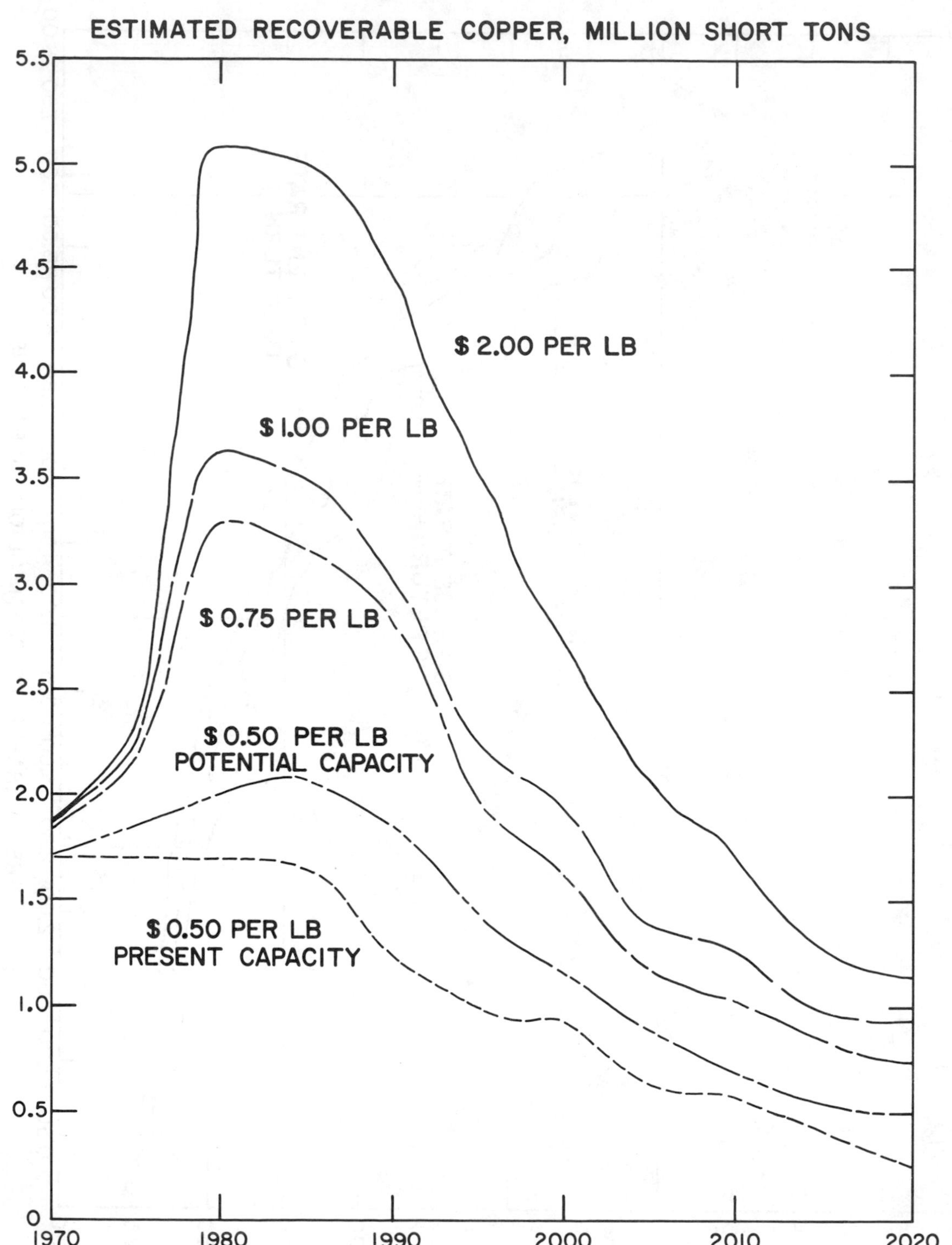

FIGURE 17: The Present and Potential Annual U.S. Primary Capability From Domestic Resources Based on Price of Copper (after U.S.B.M.)

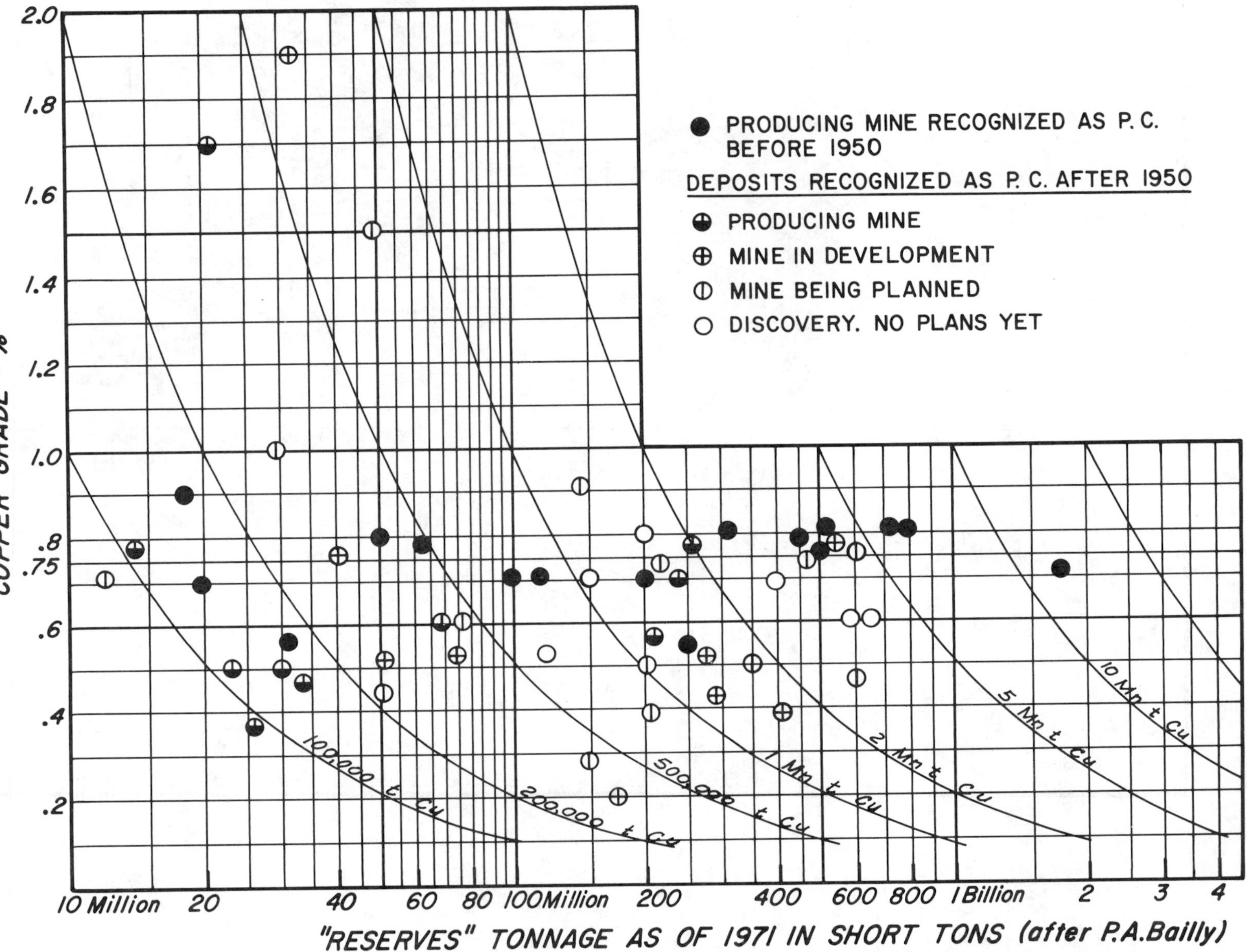

FIGURE 18: Porphyry Copper Deposits in North America (Reserves as of 1971)

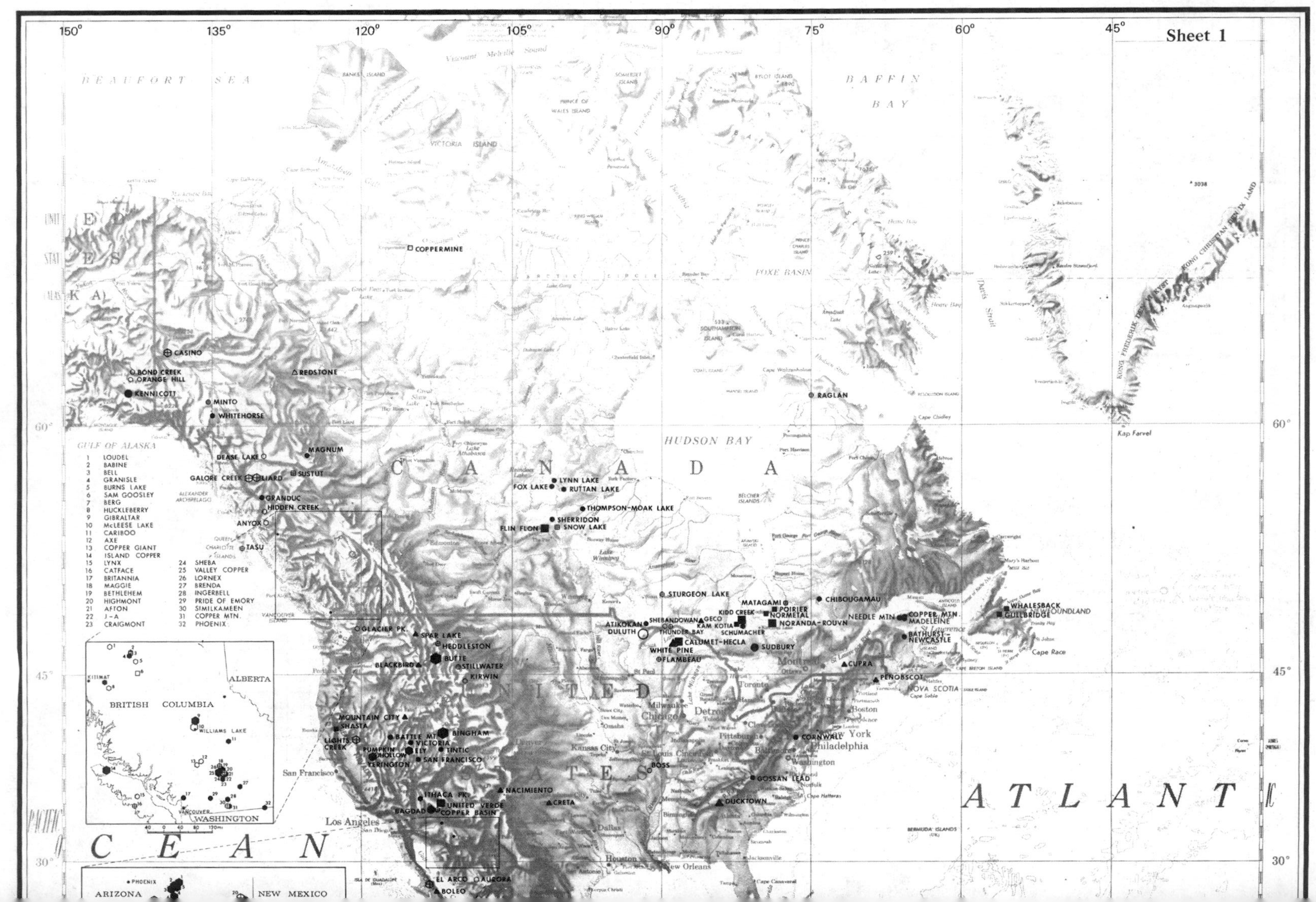
Sheet 1
BEAUFORT SEA
BAFFIN BAY
FOXE BASIN
Davis Strait
HUDSON BAY
VICTORIA ISLAND
CANADA
UNITED STATES
ATLANTIC
PACIFIC OCEAN
GULF OF ALASKA
Kap Farvel
Cape Race
NOVA SCOTIA
COPPERMINE
CASINO
BOND CREEK
ORANGE HILL
KENNICOTT
MINTO
WHITEHORSE
REDSTONE
MAGNUM
DEASE LAKE
GALORE CREEK
LIARD
SUSTUT
GRANDUC
HIDDEN CREEK
ANYOX
TASU
LYNN LAKE
FOX LAKE
RUTTAN LAKE
THOMPSON-MOAK LAKE
SHERRIDON
FLIN FLON
SNOW LAKE
RAGLAN
STURGEON LAKE
CHIBOUGAMAU
MATAGAMI
POIRIER
KIDD CREEK
NORMETAL
KAM KOTIA
NORANDA-ROUYN
SCHUMACHER
GECO
SHEBANDOWAN
ATIKOKAN
DULUTH
THUNDER BAY
WHITE PINE
CALUMET-HECLA
FLAMBEAU
SUDBURY
NEEDLE MTN
COPPER MTN.
MADELEINE
BATHURST-NEWCASTLE
WHALESBACK
GULLBRIDGE
NEWFOUNDLAND
CUPRA
PENOBSCOT
CORNWALL
GOSSAN LEAD
DUCKTOWN
BOSS
GLACIER PK.
SPAR LAKE
HEDDLESTON
BLACKBIRD
BUTTE
STILLWATER
KIRWIN
MOUNTAIN CITY
SHASTA
LIGHTS CREEK
BATTLE MT
VICTORIA
BINGHAM
TINTIC
PUMPKIN HOLLOW
ELY
YERINGTON
SAN FRANCISCO
ITHACA PK.
UNITED VERDE
BAGDAD
COPPER BASIN
NACIMIENTO
CRETA
EL ARCO
AURORA
BOLEO
1 LOUDEL
2 BABINE
3 BELL
4 GRANISLE
5 BURNS LAKE
6 SAM GOOSLEY
7 BERG
8 HUCKLEBERRY
9 GIBRALTAR
10 McLEESE LAKE
11 CARIBOO
12 AXE
13 COPPER GIANT
14 ISLAND COPPER
15 LYNX
16 CATFACE
17 BRITANNIA
18 MAGGIE
19 BETHLEHEM
20 HIGHMONT
21 AFTON
22 J-A
23 CRAIGMONT
24 SHEBA
25 VALLEY COPPER
26 LORNEX
27 BRENDA
28 INGERBELL
29 PRIDE OF EMORY
30 SIMILKAMEEN
31 COPPER MTN.
32 PHOENIX
BRITISH COLUMBIA
ALBERTA
WASHINGTON
WILLIAMS LAKE
KITIMAT
VANCOUVER
ARIZONA
NEW MEXICO
PHOENIX
San Francisco
Los Angeles
Houston
New Orleans
Dallas
Chicago
Detroit
Milwaukee
Boston
New York
Philadelphia
Washington
Toronto
Montreal

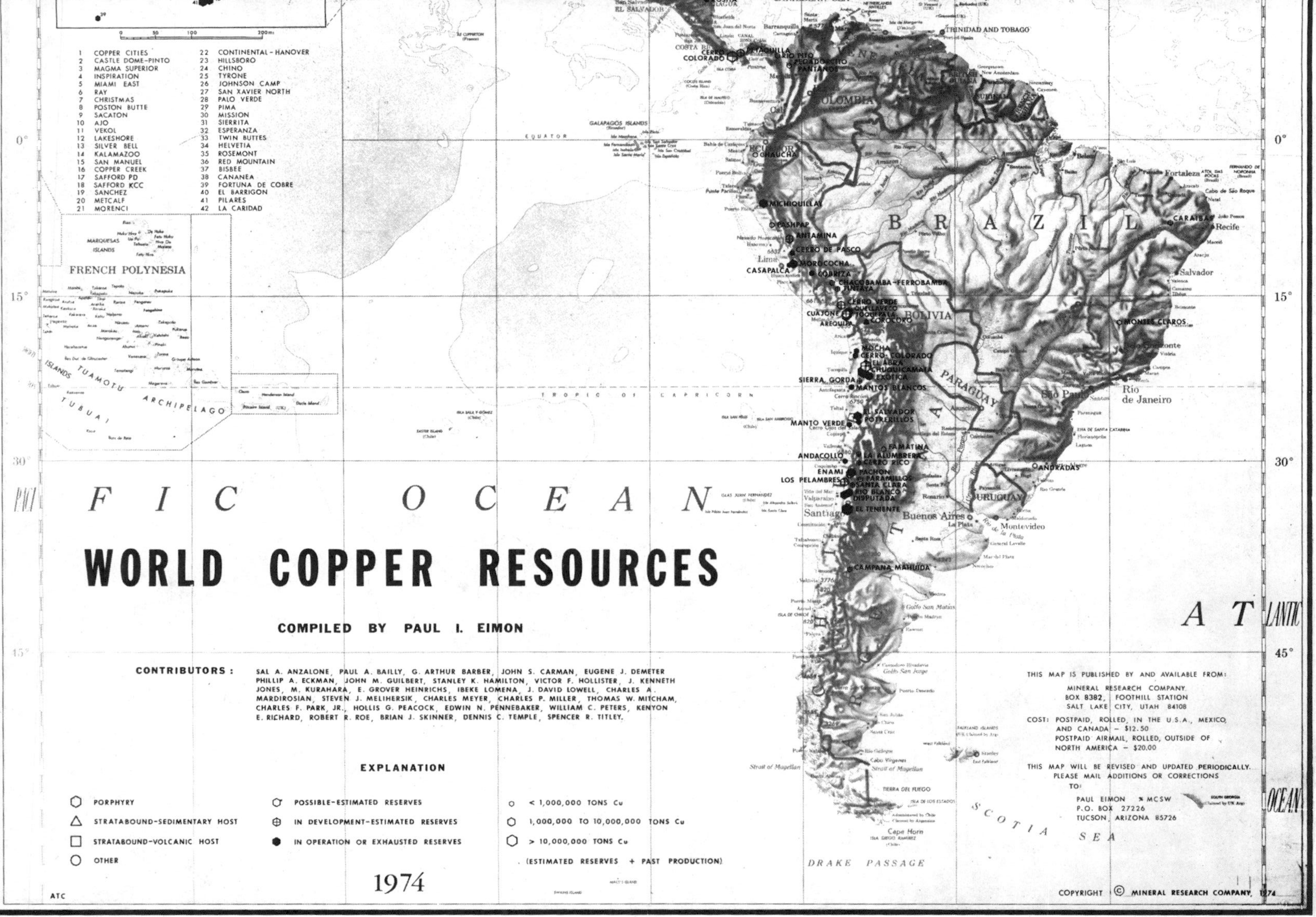

FIGURE 19: Map of the World--Western Hemisphere

Sheet 2
NORWEGIAN SEA
UNION OF SOVIET SOCIALIST REPUBLICS
RUSSIAN SFSR
KAZAKH SSR
UNITED KINGDOM
FRANCE
POLAND
ALGERIA
LIBYA
CHINA
BLACK SEA
CASPIAN SEA
ICELAND
FAEROE ISLANDS
SHETLAND ISLANDS
YAMAL PENINSULA
GYDANSKIY PENINSULA
KOLA PENINSULA
White Sea
Moscow
Leningrad
Stockholm
Copenhagen
Berlin
Hamburg
Rome
Naples
Barcelona
Istanbul
Alexandria
Tashkent
Delhi
NORIL'SKOYE
TALNAKH
PECHENGSKAYA
MONCHEGORSKAYA
OUTOKUMPU
PITKARANTA
AITIK
SULITJELMA
BOLIDEN
FALUN
GARPENBERG
JOMA
TURINSKAYA
KIROVGRADSKAYA
DEGTYARSKOYE
KARABASHKAYA
SIBAYEVSKOYE
GAYSKOYE
BLYAVA
DZHEZKAZGAN
KOUNRAD
ALMALYK
SAR CHESMEH
TIMNA
MAJDANPEK
BOR
RUDNANY
MANSFELD
RAMMELSBERG
AVOCA
PARYS MTN.
THARSIS
CERRO COLORADO
SANTIAGO
CAVALLO
15°
30°
45°
60°

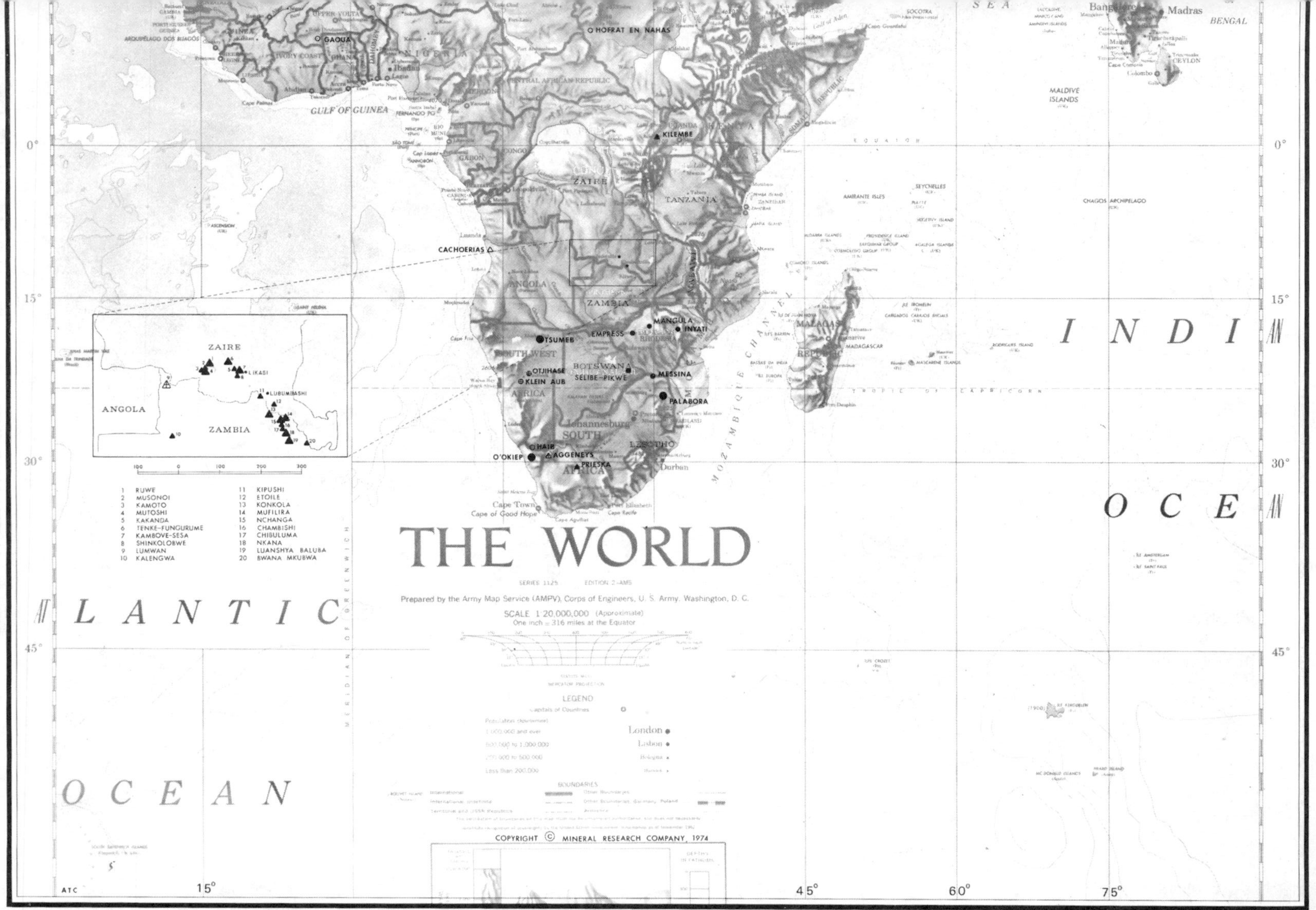

FIGURE 19 (continued): Map of the World--Europe and Africa

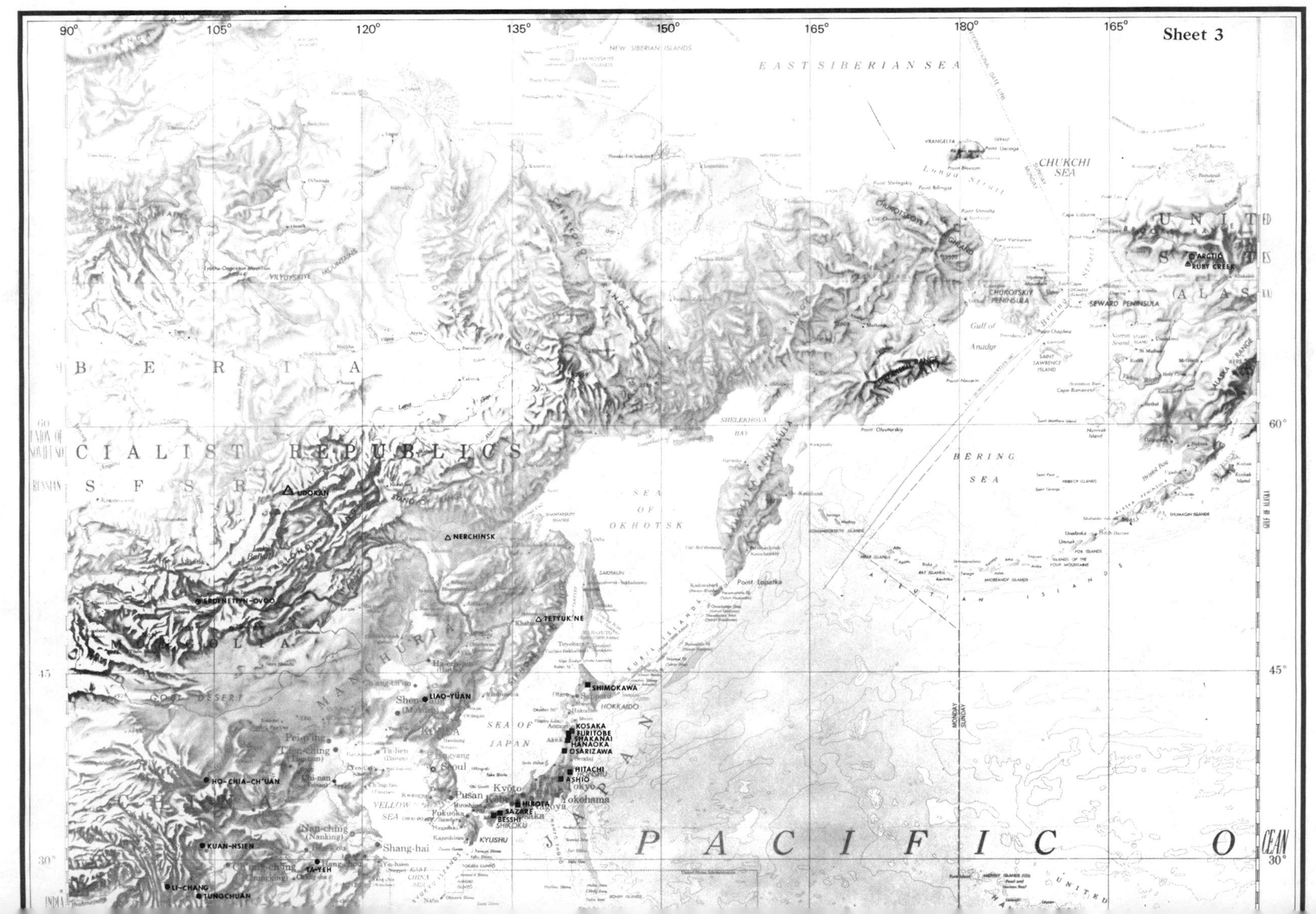
Sheet 3
90°
105°
120°
135°
150°
165°
180°
165°
60°
45°
30°
EAST SIBERIAN SEA
NEW SIBERIAN ISLANDS
CHUKCHI SEA
Longa Strait
CHUKOTSKIY PENINSULA
SEWARD PENINSULA
Gulf of Anadyr
Bering Strait
ALASKA RANGE
ARCTIC
RUBY CREEK
SAINT LAWRENCE ISLAND
BERING SEA
ALEUTIAN ISLANDS
SHELEKHOVA BAY
SEA OF OKHOTSK
Point Lopatka
KURIL ISLANDS
SIBERIA
SOCIALIST REPUBLICS
S F S R
UDOKAN
NERCHINSK
ERDENETIYN-OVOO
TETYUK'NE
MANCHURIA
GOBI DESERT
LIAO-YUAN
SHIMOKAWA
HOKKAIDO
KOSAKA
FURITOBE
HANAOKA
OSARIZAWA
HITACHI
ASHIO
HIBOTA
SAZARE
BESSHI
SHIKOKU
KYUSHU
SEA OF JAPAN
JAPAN
Seoul
Pusan
Kyoto
Yokohama
Tokyo
YELLOW SEA
EAST CHINA SEA
Shang-hai
Nan-ching (Nanking)
HO-CHIA-CH'UAN
KUAN-HSIEN
TA-YEH
LI-CHANG
TUNGCHUAN
MONDAY
SUNDAY
P A C I F I C O CEAN
UNITED STATES (ALASKA)

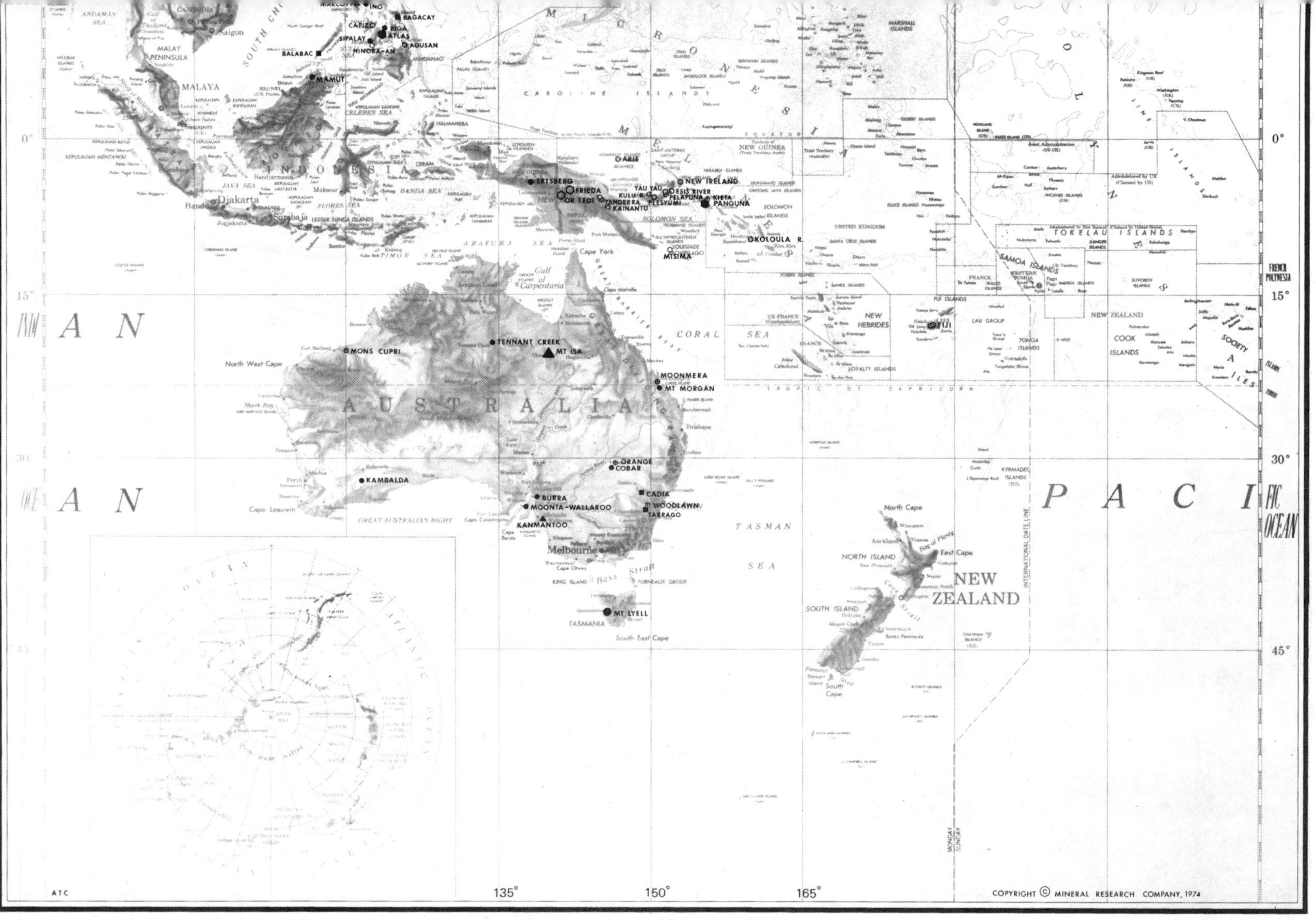

FIGURE 19 (concluded): Map of the World--Asia and the Pacific

SECTION III

THE IMPLICATIONS OF MINERAL PRODUCTION FOR HEALTH AND THE ENVIRONMENT

The Case of Coal

The following report is dedicated to the memory of Hubert E. Risser, member of this Panel and co-author, who died in September, 1974.

CHAPTER VII
INTRODUCTION TO SECTION III

Resources, estimated to range from 1,400 to 2,900 billion tonnes (1,600 to 3,200 billion short tons), make coal our nation's most abundant domestic source of fuel. Most projections of our needs for energy call for an increased reliance on coal. According to recent estimates, our annual consumption by the year 2000 may range from 1.2 to 2.7 billion tonnes (1.3 to 3 billion tons)*, or about two to three times the present rate of use. Increases of this magnitude imply large impacts on many features of the physical environment, on human health, and on other biological species. Early in its deliberations, this panel selected coal to illustrate some of the numerous problems related to minerals, e.g., extraction, processing, use, and waste disposal. The fuel crisis of late 1973-early 1974, and the strong consideration now being given to relaxing the Clean Air Act, ostensibly to encourage greater dependence on coal, underscore the timeliness of this selection.

The consequences for the physical environment of mining and burning coal, and the costs imputed to these impacts, have been studied intensively and are matters of public record. (CEQ, 1973; Washington Post, 1974.) We have not attempted to review or judge this material comprehensively. Instead, we have focused mainly on the implications for health, both of the miner and of the general public. The hazard to the miner is direct and awesome. Among all major occupations, his is probably most susceptible to accidents and disease.

The hazards to the general public arising from the burning of coal are less obvious, and more difficult to isolate and weigh. But they become even more compelling because of the vast numbers of citizens who are at risk, and because of ramifications that air pollution has for rainfall, vegetation, and segments of our food chain. In the annex to this section we have added a brief description of the global cycling of sulfur to show the contribution made by man.

---

*References: CEQ, 1973; Dupree and West, 1972; NAS/NAE, 1974b; Occupational Health, 1974a; Ray, Dixy Lee, 1973; Washington Post, 1974. (See Bibliography of References.)

EXTRACTION

The projected increases in coal production will require a greater number of miners. Unless the current efforts by government and industry to improve safety and prevention are successful, the problem of mining accidents and respiratory illness from the inhalation of coal dust, so-called "black-lung" or coal workers' pneumoconiosis (CWP), are likely to be magnified. Owing to improvements in technology, the number of fatal and non-fatal accidents per ton of coal mined has decreased, more notably in surface than in deep mines. Nonetheless, mining remains one of the most hazardous of occupations. This fact is implicit in the continuing and substantial efforts by the Federal Government, mining industry, and unions to improve the safety records. The most important factor in reducing accidents in the short-term appears to be task experience: the shorter the experience, the greater the risk of accident. The essential elements of prevention are: training for new miners before they begin work; ongoing training especially for those moving to new tasks; and strict safety regulations that are enforced. In the long run technological improvements should reduce the risks.

Compensation for CWP now exceeds one billion dollars annually (Occupational Health, 1974a). The amount has risen steadily since the enactment of the Federal Coal Mine Health and Safety Act of 1969. Any illness so costly, in which the disability (shortness of breath) is not unique but may be mimicked by other illnesses that are not necessarily related to mining, is bound to cause controversy. We take this opportunity to recommend that a study be undertaken to determine the full social and economic costs of CWP. Such a study should consider the problems of etiology, diagnosis, prevalence, preventions, compensation, and adjudication.

USE

SO_2 is a by-product of the combustion of fossil fuels. In the atmosphere, the gas undergoes a variety of chemical transformations. The types of transformations and the rates at which they occur are subject to many influences: sunlight, relative humidity, other pollutants such as ozone and oxides of nitrogen, trace metals that act as catalysts, and dusts. Sulfates and sulfuric acid are formed in these reactions, and become incorporated in submicronic* aerosols. The distances over which the reactions take place are probably of the order of 1,000 kilometers (620 miles). Consequently, the air pollution arising from sulfur emissions is regional and not simply urban or confined to the immediate proximity of the coal-fired power plants.

The importance of SO_2 as a hazard to public health is probably due to these chemical end-products, not to the parent gas itself. Among the segments

---

*1 micron or micrometer = one-thousandth of a millimeter
= 39 millionths of an inch

of the population apparently most vulnerable to this form of pollution are the aged, individuals already ill with cardiac or pulmonary disease, and infants and children.

Air polluted with sulfur oxides has two significant adverse ecological effects. One is damage to certain plants by gaseous SO_2. The second results from the production of acid rain. The impact of acid rain on forests and bodies of fresh water is coming under increasing study.

More information about aero-chemistry is required if effective control of air pollution is to be achieved. Almost certainly, technology will have to be directed toward minimizing the emissions of SO_2. The dispersion of emissions from tall stacks (over 152 meters, or 500 feet), while reducing ground-level concentrations over relatively short distances, is likely to ensure a more complete oxidation of gas in the atmosphere and a wider distribution of the end-products.

CHAPTER VIII
PROJECTED DEMANDS, RESOURCES

In the years from 1953 to 1973, the estimated national consumption of total gross energy surged from 9.5×10^{18} cal. to 1.9×10^{19} cal. (37.6×10^{15} British Thermal Units [BTU] to 75.6×10^{15} BTU) (Dupree and West, 1972; U.S.B.M., 1974). Accompanying this growth was a major shift in the relative contributions of the different sources of energy (Figure 1). The use of coal, which constitutes about 95 percent of the fossil fuel resources within the United States, rose only 3.7 percent. At the same time, the use of petroleum increased almost 104 percent and the use of natural gas more than 185 percent.

Over recent decades, the use of coal has declined for nearly all activities other than the generation of electrical power. For electrical power alone, consumption rose from 267 million tonnes (295 million tons) in 1968 to about 352 million tonnes (388 million tons) in 1973, presumably because coal was the least expensive fuel available. At present, the generation of electrical power and manufacture of coke* together account for about 85 percent of the total tonnage (Table 1). By far the largest fraction is utilized east of the Mississippi River; in 1972, this fraction amounted to about 94 percent. Forecasting our demands for energy has become commonplace. A representative forecast of continued exponential growth of energy consumption would give a value for 1980 in the neighborhood of 2.4×10^{19} cal. (95×10^{15} BTU), increasing to 3.15 to 3.18×10^{19} cal. (125 to 130×10^{15} BTU) in 1985, and to over 5×10^{19} cal. (200×10^{15} BTU) in 2000. It is worthy of note that in the last 20 years, such predictions have tended to underestimate demands beyond the initial few years. (Committee on Interior and Insular Affairs, 1974.)

Not unexpectedly, estimates of the quantities of coal that may be needed to satisfy these projections are themselves subject to wide variation. They are clouded by such uncertainties as (1) how rapidly nuclear power will grow; (2) how much coal will be used to supplement dwindling supplies of natural gas and petroleum; (3) to what extent the United States will provide coal

---

*Consumption has remained relatively stable for the manufacture of coke for several decades.

abroad; (4) what are to be the domestic and worldwide rates of economic growth; (5) the possibilities for conserving power through increased efficiencies of use; and (6) the ability of the coal industry to increase coal production. Allowing for the uncertainty of any projections beyond a few years, our discussion will proceed on the following assumptions for the use of coal:

1985: 900 million tonnes (1 billion tons) (68 percent above 1972 level); cumulative total from 1973 to 1985: 8.6 billion tonnes (9.5 billion tons)

2000: 1.8 billion tonnes (2 billion tons) (233 percent above 1972 level); cumulative total from 1985 to 2000: 20.4 billion tonnes (22.5 billion tons)

Therefore, cumulative total from 1973 to 2000: 29 billion tonnes (32 billion tons).

Coal is found in many regions of the country. The resources as determined by exploration and mapping are estimated to exceed 1400 billion tonnes (1550 billion tons). The varieties of coal and their distribution are shown in Table 2. While approximately half of this amount is estimated to be recoverable, only one third, or 236 billion tonnes (260 billion tons), is within the reach of current technology at current economic costs. Estimates of the coal reserves in beds lying close to the surface exceed 127 billion tonnes (140 billion tons). About 60 percent of these surface deposits lie west of the Mississippi River, mostly in Montana, North Dakota, and Wyoming.*

PROPERTIES OF COAL

The thermal value of coal varies considerably. Bituminous coals provide the most heat per unit of weight (about 6,700-8,300 cal/g, or 12,000-15,000 BTU/pound) followed by sub-bituminous coals (about 4,400-6,200 cal/g or 8,000-11,000 BTU/pound) and lignite (3,900-4,400 cal/g, or 7,000-8,000 BTU/pound). Eastern coals are generally richer thermally than those of the Rocky Mountains and Northern Great Plains. Nonetheless, the western coals offer several distinct advantages which ultimately render them less costly to mine: for example, they possess thicker seams (up to 33 meters, or 100 feet) and thinner overburdens (0-50 meters, or 0-150 feet) than the eastern varieties. In 1972, the estimated average cost per ton of coal at the mines was $3.74 in Wyoming, $2.03 in Montana for bituminous and lignite coal, and $2.02 in North Dakota for lignite, compared with a national average of $7.66. To some

---

*The figures relating to coal reserves are taken from U.S. Geological Survey Bulletin 1275, Coal Resources of the United States, January 1, 1967 and U.S. Geological Survey Bulletin 1322, Stripping Coal Resources of the United States, January 1, 1970, and recent updating of western coal reserves by state governments and U.S.B.M.

degree, however, the lower price of the western coal also reflected their lower thermal content. (U.S.B.M., 1972.)

Western coal has another advantage over eastern coal that is of immediate relevance to this report: generally, its sulfur content is lower.* Indeed, an administrative rule (EPA, 40 CFR60.43) reflecting the increasing concern over air pollution disqualifies a major portion of the coal reserves of the U.S. from use until the sulfur can be removed (adequately and economically), before or during combustion. Current EPA standards for new installations which are scheduled to apply to all fuels by 1975, limit the maximum emission of SO_2 to 2.16 Kg per million Kcal (1.2 pounds per million BTU) of fuel energy. Accordingly, unless the sulfur is removed or the blending of coal is widely adopted, a bituminous coal of 6,700 cal/g (12,000 BTU/pound) will be limited to no more than 0.72 percent sulfur, while a lignite of 3,900 cal/g (7,000 BTU/pound) will be permitted 0.42 percent sulfur. The sulfur content of most coal being used at present exceeds these limits. The tabulations of the producers of 67 percent of the total national output of coal in 1972 show that only 8 of the 23 coal producing districts in the country yielded coal with an average sulfur content of 0.7 percent or less.** The 8 districts accounted for about 11.5 percent of the entire output. Even so, some of these low sulfur coals, because of poor thermal content, could not have met the 2.16 Kg per million Kcal (1.2 pounds per million BTU) standard without the further removal of sulfur.

The implications are obvious. If projected increases in the use of coal are realized, i.e., 68 percent by 1985 and 233 percent by 2000, we will need commensurate reductions in the emission of SO_2 just to prevent further deterioration of the atmosphere (see Chapter X: Use: Sulfur Emissions, Atmospheric Reactions). Two basic strategies would be available: either we move to increased reliance on limited reserves of low sulfur coal and/or we institute stricter control of emissions. Neither alternative appears to receive administrative favor at present. The problem might be exacerbated if there were delays in setting into operation new nuclear power plants# or if a trend developed toward the conversion of oil and gas-fired power plants to coal. Under these circumstances, the demand for coal could exceed present projections.

---

*Smaller but significant amounts of low sulfur, bituminous, surface coal are present in the central and eastern parts of the U.S. These reserves are estimated to be between 2.3 and 2.7 billion tonnes (2.5 and 3 billion tons). However, the availability of the land for mining is in question.

**Located in southern Appalachia, northern and southern Colorado, New Mexico, Wyoming, Utah, and some mines in Montana and North Dakota.

#Not unlikely in view of the AEC's recent shutdown of 21 of a total of 50 nuclear plants to check the cooling systems for leaks (<u>Congressional Record</u>, 1974).

Since midwestern and eastern coals often cannot meet present emission standards without additional fuel or flue gas desulfurization (FGD), large quantities of coal from Wyoming and Montana are being shipped into the midwestern markets. The total combined production for the two states, which was about 14.5 million tonnes (16 million tons) in 1972, is expected to reach about 32 million tonnes (35 million tons) by 1974. Other western states having low sulfur coal are also expected to show accelerating production. The trend could be stalled or reversed either if the power plants were to add FGD equipment or if the technology for the pre-combustion removal of sulfur were to become available.

AREA OF LAND NEEDED

One Km^2 of coal contains about 1.3 million tonnes per meter (one acre contains about 1,800 tons per foot) (thickness) of seam. In underground mining about 50 percent of the coal that is in situ can be extracted, while in surface mining, this fraction generally ranges from 80 to 85 percent. The area needed to produce any fixed quantity of coal is therefore influenced by both the type of mine and the thickness of the coal bed (Figure 2). Averitt (U.S.G.S., 1970) estimated that the average thickness of coal recovered by surface mining in 1970 was 1.5 meters (5 feet), while the seams ranged from under 0.61 meters to 18.3 meters (2 feet to 60 feet) or more. If 80 percent of this coal is to be extracted, a million tons can be mined from less than 12 acres of a 60-foot seam, whereas the same tonnage in a 2-foot seam would require almost 350 acres. Because the beds of coal are thicker in the Rocky Mountains and Northern Great Plains than elsewhere, less acreage is needed in these regions to yield a specified amount of energy, even after allowances are made for differences in thermal content. It should be emphasized, however, that uncertainty exists over how successfully these areas can be rehabilitated following surface mining (CEQ, 1973; NAS/NAE, 1974a; See Chapter X: Use: Reclamation). Major problems in rehabilitation are likely wherever the overburden is high in acidity or salinity, the soil cover is thin, and its elements, including micro-organisms, are delicately balanced, and--this being of fundamental importance--the rainfall is sparse.

It is not possible to estimate with reasonable assurance the area of land that will be used in coal mining during the remainder of the century. Currently, approximately 50 percent of the tonnage is being mined by surface methods. It will be recalled that the projected cumulative production of coal is about 8.6 billion tonnes (9.5 billion tons) by 1985, and 29 billion tonnes (32 billion tons) by 2000. If half of these tonnages 4.3 and 14.5 billion tonnes, or (4.75 and 16 billion tons) were to be obtained from surface mines with seams averaging 1.6 meters (5 feet) in thickness and with an extraction rate of 80 percent, the areas directly involved would represent about 2,700 Km^2 (2,250,000 acres) by 2000. Naturally, if the average thickness of the seam were greater, these requirements would be reduced. Thus, for an average seam 12.3 meters (40 feet) thick, the estimated acreages would be only 82,500 (334 Km^2) by 1985 and 300,000 (1,210 Km^2) by 2000. To the extent that reliance on western (or eastern) surface-mined coal is increased in the future, it is vital that in selecting the mining sites we be cognizant of a number of factors: the potential for rehabilitating the land, the water that is available, and the social, aesthetic, and economic impacts that will be felt in the regions that are directly involved.

MANPOWER NEEDS

The same projections for use, (i.e., 0.91 billion tonnes, or 1 billion tons in 1985, 1.8 billion tonnes, or 2 billion tons in 2000) represent increases of 68 percent and 233 percent above the amount mined in 1972. In 1972, about 112,000 workers were engaged in underground mines, and an additional 34,000 in surface mines. Should underground mines produce the same percentage of coal in the future as they have done recently (about 50 percent of the total), 75,000 additional underground workers would be required by 1985 and 224,000 by 2000 (Figure 3). These numbers will be subject to lowering as improvements in the efficiency of production take place. Nonetheless, it must be recognized that such projections, if they are to be met, will require considerable expansion of one of our most hazardous occupations (see Chapter IX: Extraction: Accidents, CWP).

CAPITAL INVESTMENT

To satisfy the projected increases in production, a large number of new mines would be needed. The capacity for increasing production in the mines now in operation is too small to sustain much growth. To increase production 363 million tonnes (400 million tons) annually by 1985 would require about 80 new mines, each capable of producing 4.5 million tonnes (5 million tons) of coal per year (Figure 4). At an estimated average capital investment for both surface and underground mines of $15 per annual ton of production, $8 billion of capital investmant would be required. An additional $6 billion in capital for 200 new mines would be necessary to achieve the projected goal for the year 2000. Capital markets probably could provide these funds. The delays that are likely to occur in obtaining adequate equipment, and in providing all the supplementary needs of new mining sites, make it most unlikely that such a schedule of growth could be met unless concerns for the environment were circumvented.

TABLE 1 Coal Consumption by Sector, 1973

Consumer Sector	Million Tonnes	(Million Tons)	Percent of Consumption
Electrical Power Generation	349.6	(388.4)	64.8
Industrial	145.4	(161.6)	33.2
Household and Commercial	12.6	(14.0)	2.0
Totals	507.6	(564.0)	100.0

Source: U.S.B.M. 1974

TABLE 2 Total Estimated Remaining Coal Resources of the United States, January 1, 1967

[In millions of short tons. Figures are for resources in the ground, about half of which may be considered recoverable. Includes beds of bituminous coal and anthracite 14 in. or more thick and beds of subbituminous coal and lignite 2.5 ft. or more thick]

	Overburden 0-3,000 ft. thick							Overburden 3,000-6,000 ft. thick	Estimated total remaining resources in the ground 0-6,000 ft. overburden
	Resources determined by mapping and exploration					Estimated additional resources in unmapped and unexplored areas[1]	Estimated total remaining resources in the ground	Estimated resources in deeper structural basins[1]	
State	Bituminous coal	Subbituminous coal	Lignite	Anthracite and semianthracite	Total				
Alabama------	13,518	0	20	0	13,538	20,000	33,538	6,000	39,538
Alaska-------	19,415	110,674	(2)	(3)	130,089	130,000	260,089	5,000	265,089
Arkansas-----	1,640	0	350	430	2,420	4,000	6,420	0	6,420
Colorado-----	62,389	18,248	0	78	80,715	146,000	226,715	145,000	371,715
Georgia------	18	0	0	0	18	60	78	0	78
Illinois-----	139,756	0	0	0	139,756	100,000	239,756	0	239,756
Indiana------	34,779	0	0	0	34,779	22,000	56,779	0	56,779
Iowa---------	6,519	0	0	0	6,519	14,000	20,519	0	20,519
Kansas-------	18,686	0	(4)	0	18,686	4,000	22,686	0	22,686
Kentucky-----	65,952	0	0	0	65,952	52,000	117,952	0	117,952
Maryland-----	1,172	0	0	0	1,172	400	1,572	0	1,572
Michigan-----	205	0	0	0	205	500	705	0	705
Missouri-----	23,359	0	0	0	23,359	0	23,359	0	23,359
Montana------	2,299	131,877	87,525	0	221,701	157,000	378,701	0	378,701
New Mexico---	10,760	50,715	0	4	61,479	27,000	88,479	21,000	109,479
North Carolina----	110	0	0	0	110	20	130	5	135
North Dakota-	0	0	350,680	0	350,680	180,000	530,680	0	530,680
Ohio---------	41,864	0	0	0	41,864	2,000	43,864	0	43,864

TABLE 2 Continued.

Oklahoma------	3,299	0	([4])	0	3,299	20,000	23,299	10,000	33,299
Oregon--------	48	284	0	0	332	100	432	0	432
Pennsylvania--	57,533	0	0	12,117	69,650	([5]) 10,000	79,650	0	79,650
South Dakota--	0	0	2,031	0	2,031	1,000	3,031	0	3,031
Tennessee-----	2,652	0	0	0	2,652	2,000	4,652	0	4,652
Texas---------	6,048	0	6,878	0	12,926	14,000	26,926	0	26,926
Utah----------	32,100	150	0	0	32,250	48,000	80,250	35,000	115,250
Virginia------	9,710	0	0	335	10,045	3,000	13,045	100	13,145
Washington----	1,867	4,194	117	5	6,183	30,000	36,183	15,000	51,183
West Virginia-	102,034	0	0	0	102,034	0	102,034	0	102,034
Wyoming-------	12,699	108,011	([2])	0	120,710	325,000	445,710	100,000	545,710
Other States--	[6]618	[7]4,057	[8]46	0	4,721	1,000	5,721	0	5,721
Total----	671,049	428,210	447,647	12,969	1,559,875	1,313,080	2,872,955	337,105	3,210,000

[1]Estimates by H.M. Belkman (Washington), H.L. Berryhill, Jr. (Virginia and Wyoming), R.A. Brant (Ohio and North Dakota), W.C. Culbertson (Alabama), K.J. England (Kentucky), B.R. Haley (Arkansas), E.R. Landis (Colorado and Iowa), E.T. Luther (Tennessee), R.S. Mason (Oregon), F.C. Peterson (Kaiparowits Plateau, Utah), J.A. Simon (Illinois), J.V.A. Trumbull, (Oklahoma), C.E. Wier (Montana), and the author for the remaining States.

[2]Small resources and production of lignite included under subbituminous coal.

[3]Small resources of anthracite in the Bering River field believed to be too badly crushed and faulted to be economically recoverable. (See Barnes, 1951.)

[4]Small resources of lignite in beds generally less than 30 in. thick.

[5]After Ashley (1911).

[6]Arizona, California, Idaho, Nebraska, and Nevada.

[7]Arizona, California, and Idaho.

[8]California, Idaho, Louisiana, Mississippi, and Nevada.

Source: U.S.B.M., 1974.

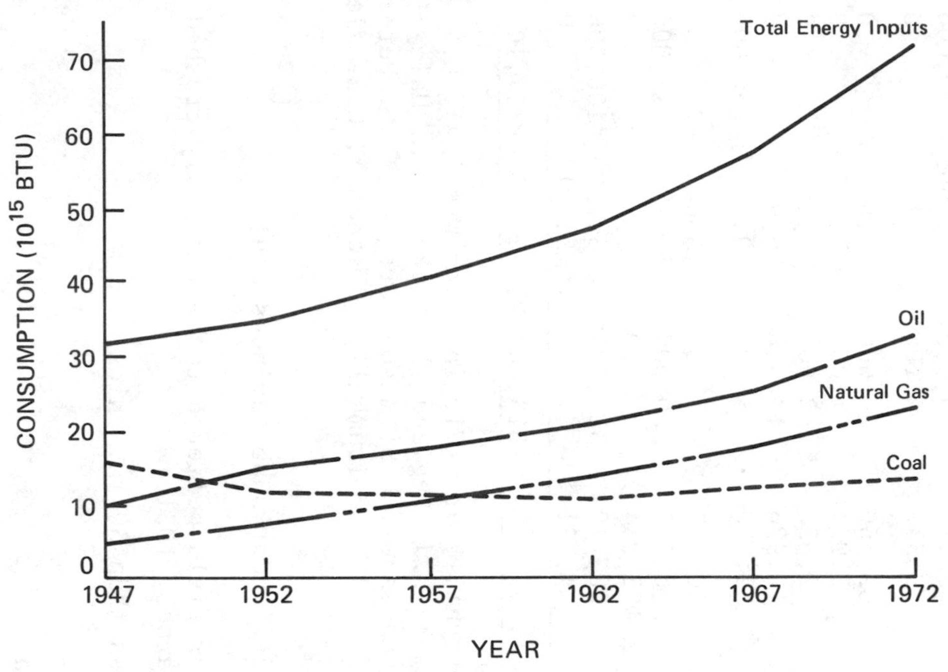

	YEAR					
	1947	1952	1957	1962	1967	1972
Total Energy Inputs (10^{12} BTU)	33,035	36,458	41,706	47,422	58,265	72,091
Oil Consumption as a Percentage of Total Energy (%)	34.4	42.1	44.5	44.8	43.5	45.5
Natural Gas as a Percentage of Total Energy (%)	13.7	21.3	25.0	29.8	31.3	32.3
Coal as a Percentage of Total Energy (%)	47.9	32.6	26.8	21.5	21.0	17.2

SOURCE: Dupree and West (1973).

FIGURE 1: United States Oil, Natural Gas, and Coal Consumption, 1947-1972

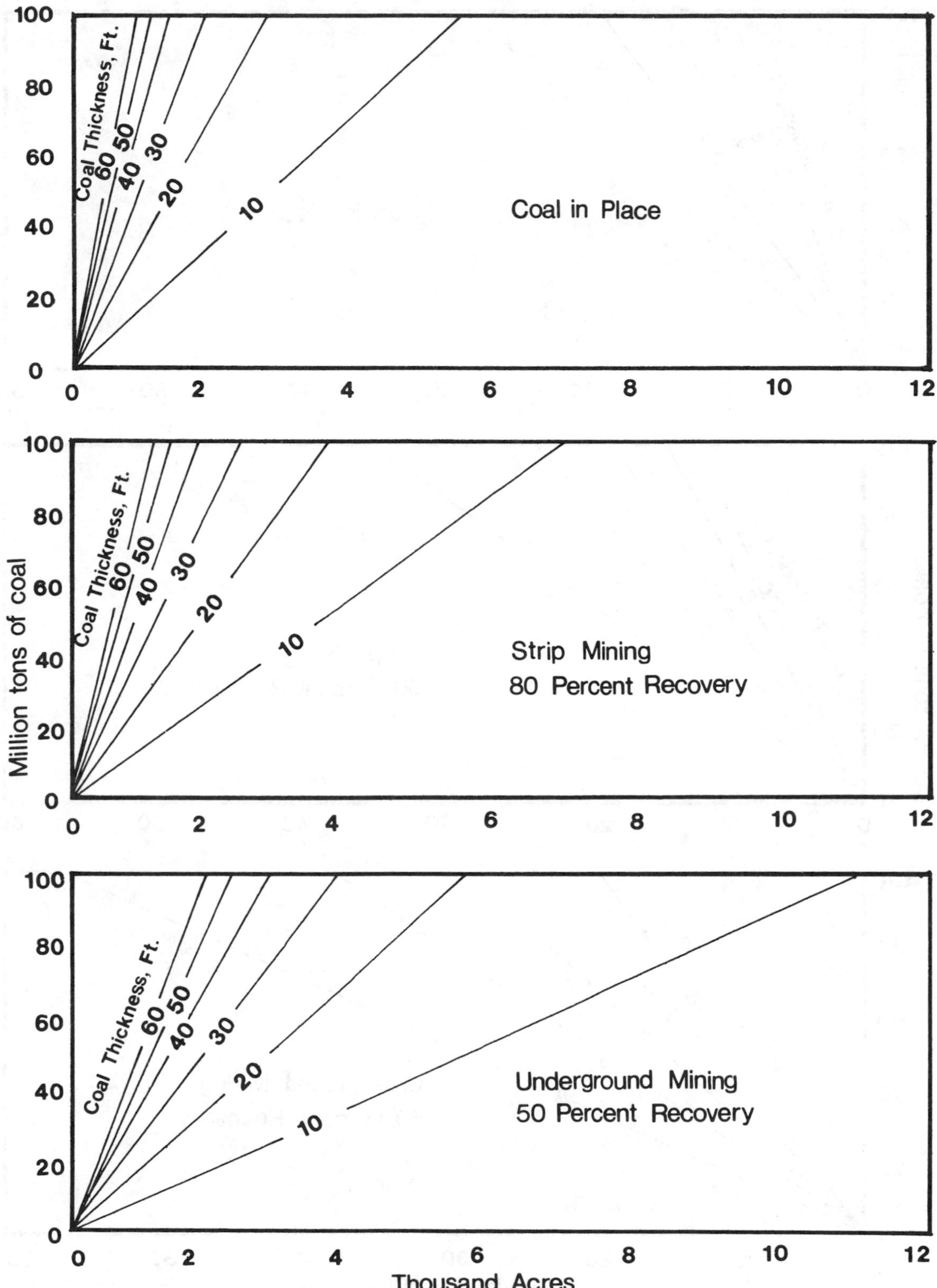

FIGURE 2: Acres Per Given Quantity of Coal--By Thickness

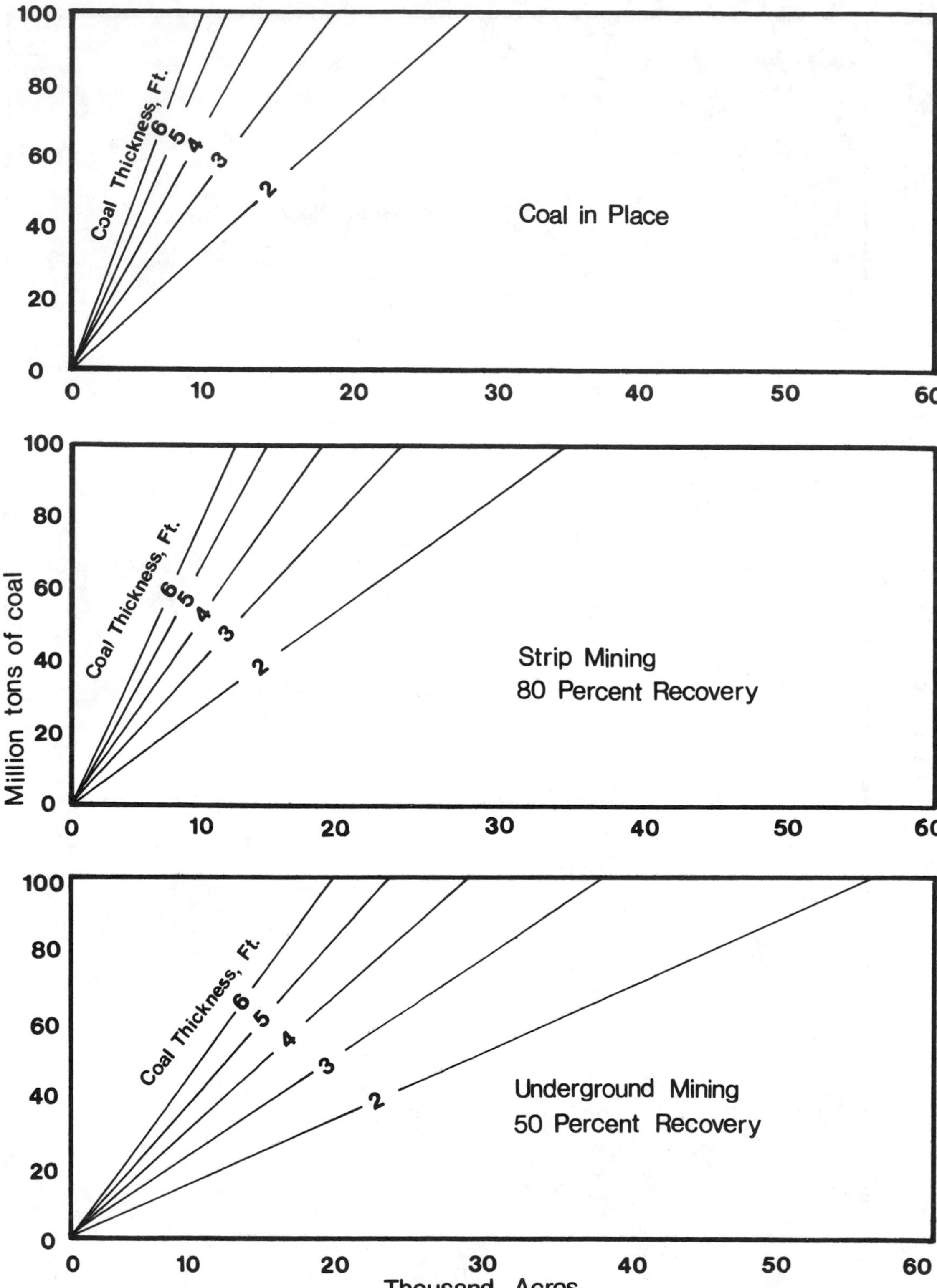

FIGURE 2: Acres Per Given Quantity of Coal--By Thickness

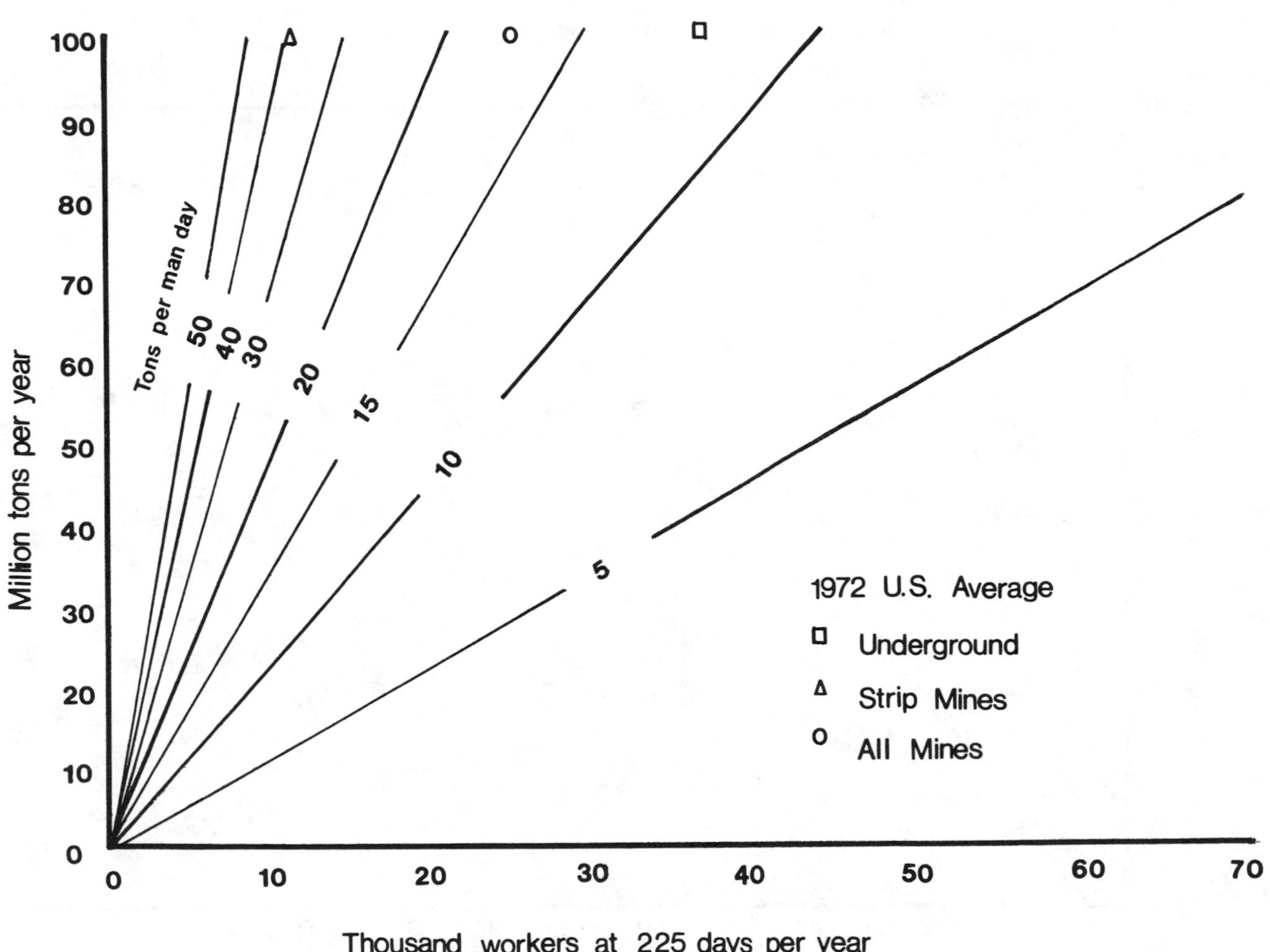

FIGURE 3: Coal Mine Workers Required for Given Outputs

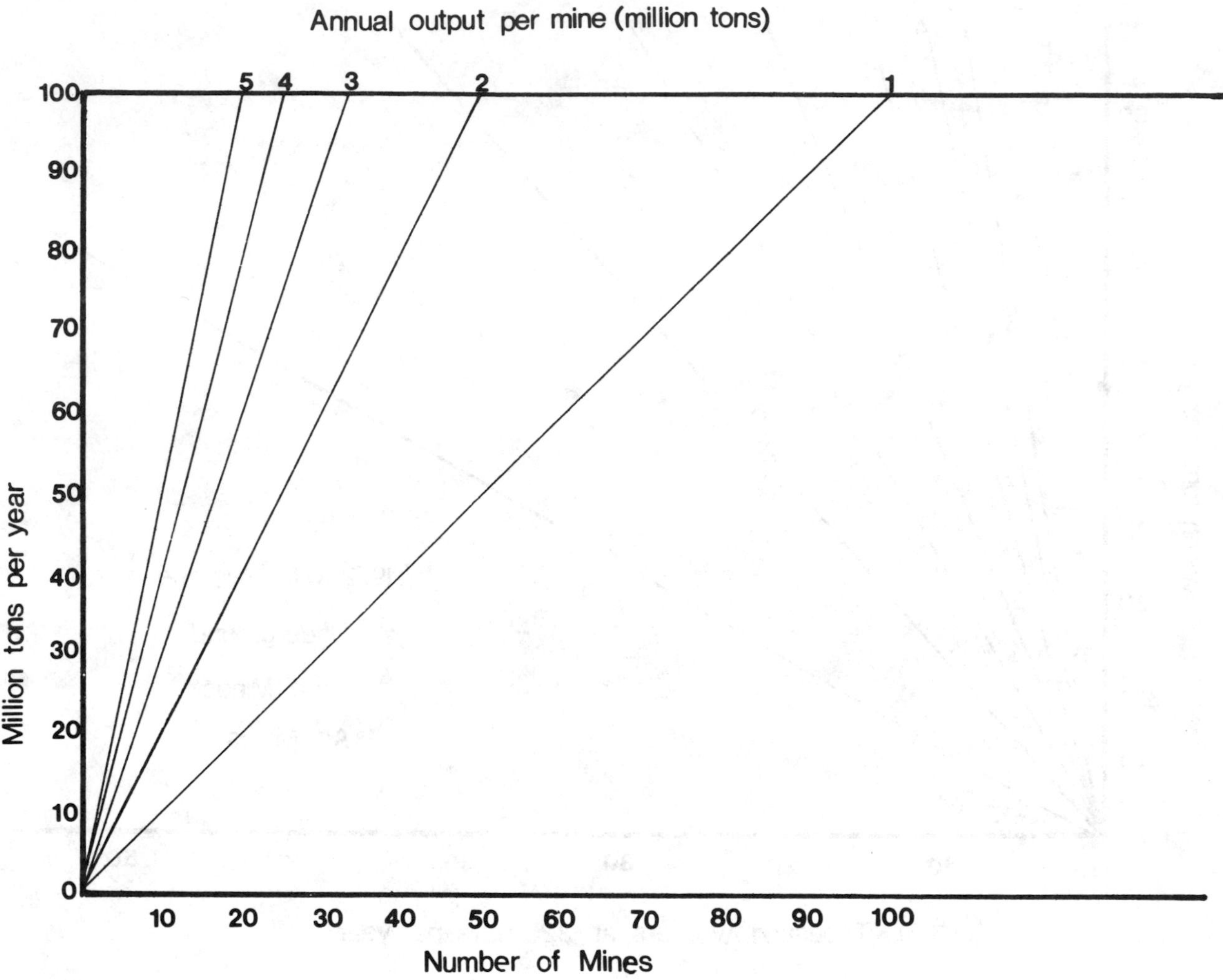

FIGURE 4: Number of Coal Mines Required to Provide Given Annual Tonnages

CHAPTER IX
EXTRACTION

ACCIDENTS

Mining Accidents

Coal mining in the United States has a grim record of safety which has improved in the past 25 years principally by reason of fewer miners being at risk. Because more miners will be needed to meet the projected increases in coal production, it is timely to review the two major hazards of mining: accidents and coal workers' pneumoconiosis (CWP).

In 1920, almost all of our coal came from underground mines. Starting in the 1940s, surface mining began to increase while production from underground mines fell off, so that by the early 1970s each method contributed about half of the total (Figure 5).

The total number of coal miners employed annually in the United States began to fall rapidly in the 1920s (Figure 6). The decline has continued with periodic fluctuations until about the present decade. Since then the number has remained relatively stable (Figure 6). Virtually the entire earlier decline in employment occurred in underground mines. It ran parallel with the fall in coal production in the 1920s and 1930s, but thereafter continued independently of production. Improvements in the methods of both underground and surface mining which appeared in the 1940s allowed production to be maintained with fewer miners.

Fatal Accidents

In the past four decades, over 90 percent of the fatal accidents in mining have occurred underground (Figure 7). The decline in fatal underground accidents has essentially mirrored the decline in employment; that is, there have been fewer miners at risk. The number of annual deaths among surface miners has remained remarkably stable, averaging 28 from 1930 to 1973.

There were one to two underground fatalities per million manhours from 1930-1971 (Figure 4). Therefore, the rate declined progressively to about 0.5 fatalities per million manhours in 1973. While the improvement probably reflects implementation of the Federal Coal Mine Health and Safety Act of 1969, judgment should be suspended since the recent statistics are preliminary and incomplete. From the 1930s on, the number of fatal accidents per million tons of coal (Figure 5) has fallen progressively owing to improvements in mining techniques. These improvements have permitted greater rates of productivity at no risk to safety. Nonetheless, both methods of evaluation clearly demonstrate the superior safety of surface mining.

Non-fatal Accidents

The annual occurrences of non-fatal mining accidents involving a loss of time from work are shown in Figure 10. The shape of the changes in occurrence among underground miners and the relative constancy of such accidents among surface miners are reminiscent of the findings for fatalities. The trends are reasonably similar when the numbers of accidents are related either to millions of hours worked (Figure 11) or to millions of tons of coal produced (Figure 12). Again, surface mining is distinctly safer. In the 1970s, the frequency of both fatal and non-fatal accidents, when corrected for the number of hours worked or coal produced, has repeatedly been greater in the underground mines by a range of two to eight times.

Prevention

Efforts to improve coal mine safety are underway. As a result of the Federal Coal Mine Health and Safety Act of 1969, and the Federal Metal and Nonmetallic Mine Safety Act of 1966, the U.S. Bureau of Mines began implementation in 1970, the Secretary of the Interior created the Mining and Safety Administration (MESA) in May 1973. MESA's stated goals call for the development and enforcement of strong mining safety and health standards, the development of safety training for all miners and supervisors, the provision of assistance for the solution of technical problems, and the direction of research towards the reduction of disease and injuries.

A study by Theodore Barry and Associates (1972) for the U.S. Bureau of Mines in June 1972, clearly showed that experience with a particular mining task is a critical factor influencing accidents, i.e., the less experience with the task, the greater the risk of injury. Limited efforts are underway to apply this information. West Virginia, in cooperation with MESA, now provides forty hours of pre-entry training for new coal miners. These programs, even in the absence of any final judgment of their efficacy, ought to be encouraged, to become more widespread.

Recommendation

To implement rapidly both pre-entry and ongoing training programs in specific mining tasks.

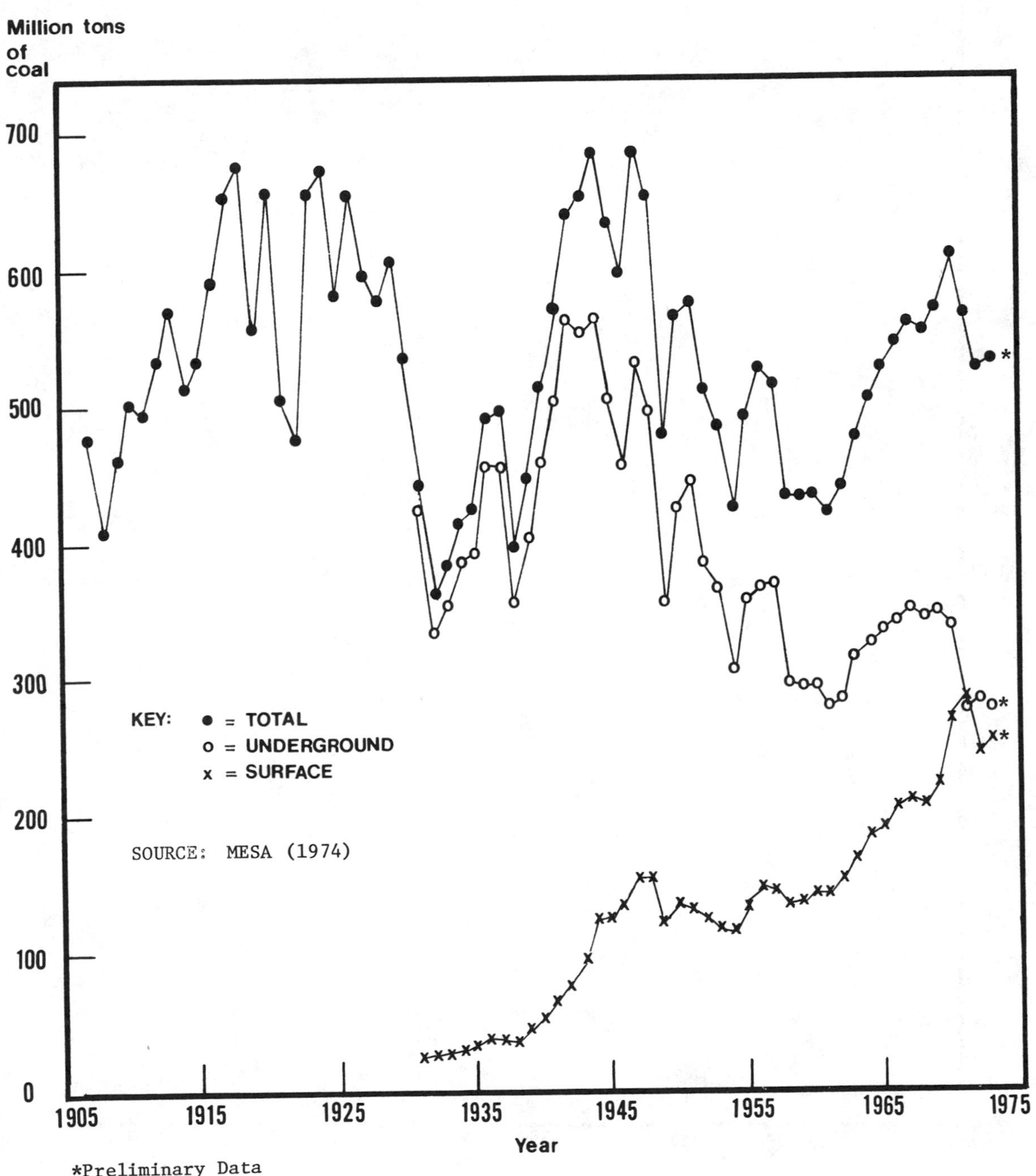

*Preliminary Data

FIGURE 5: Coal Production U.S.A. 1907-1973

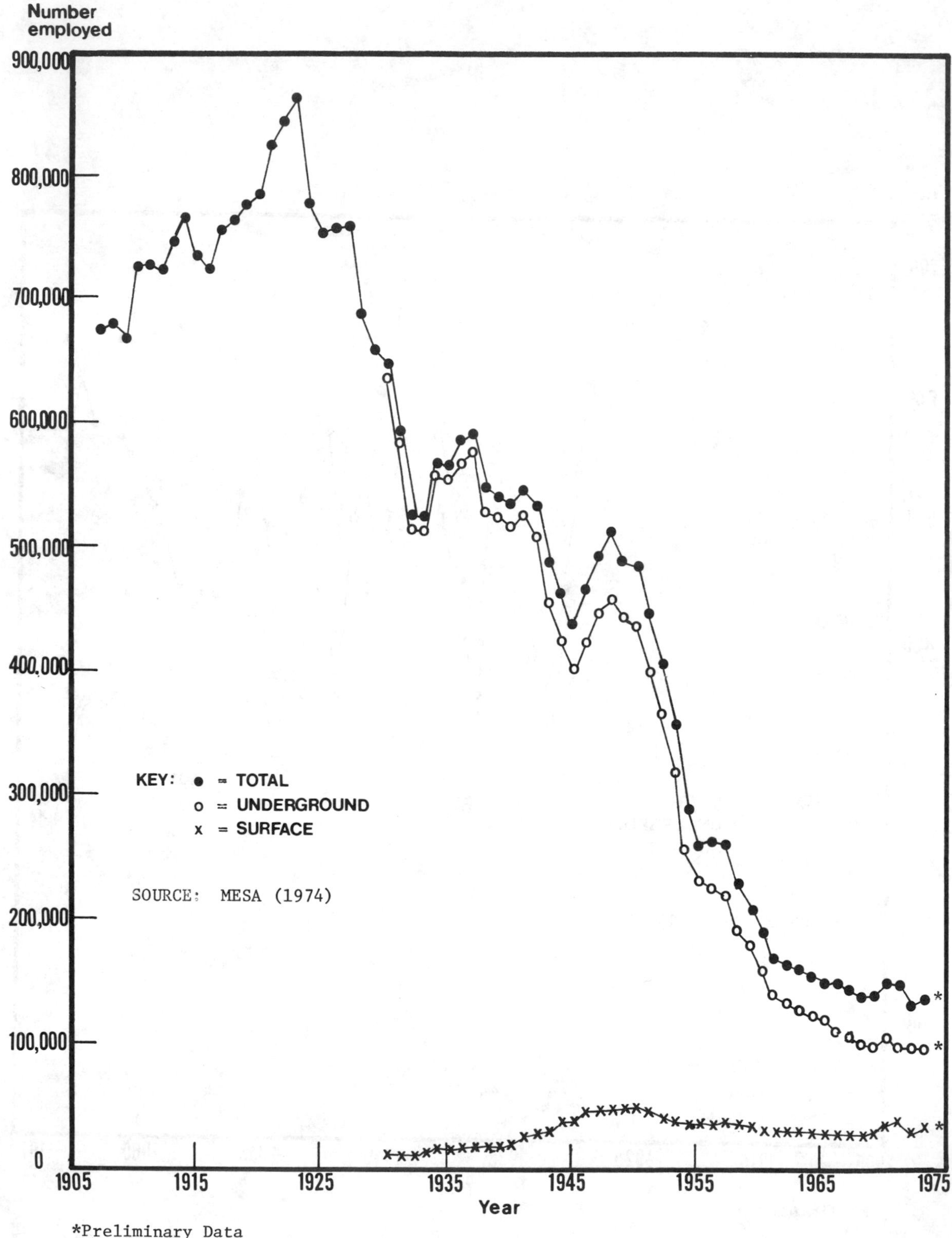

*Preliminary Data

FIGURE 6: U.S. Coal Miners Employed 1907-1973

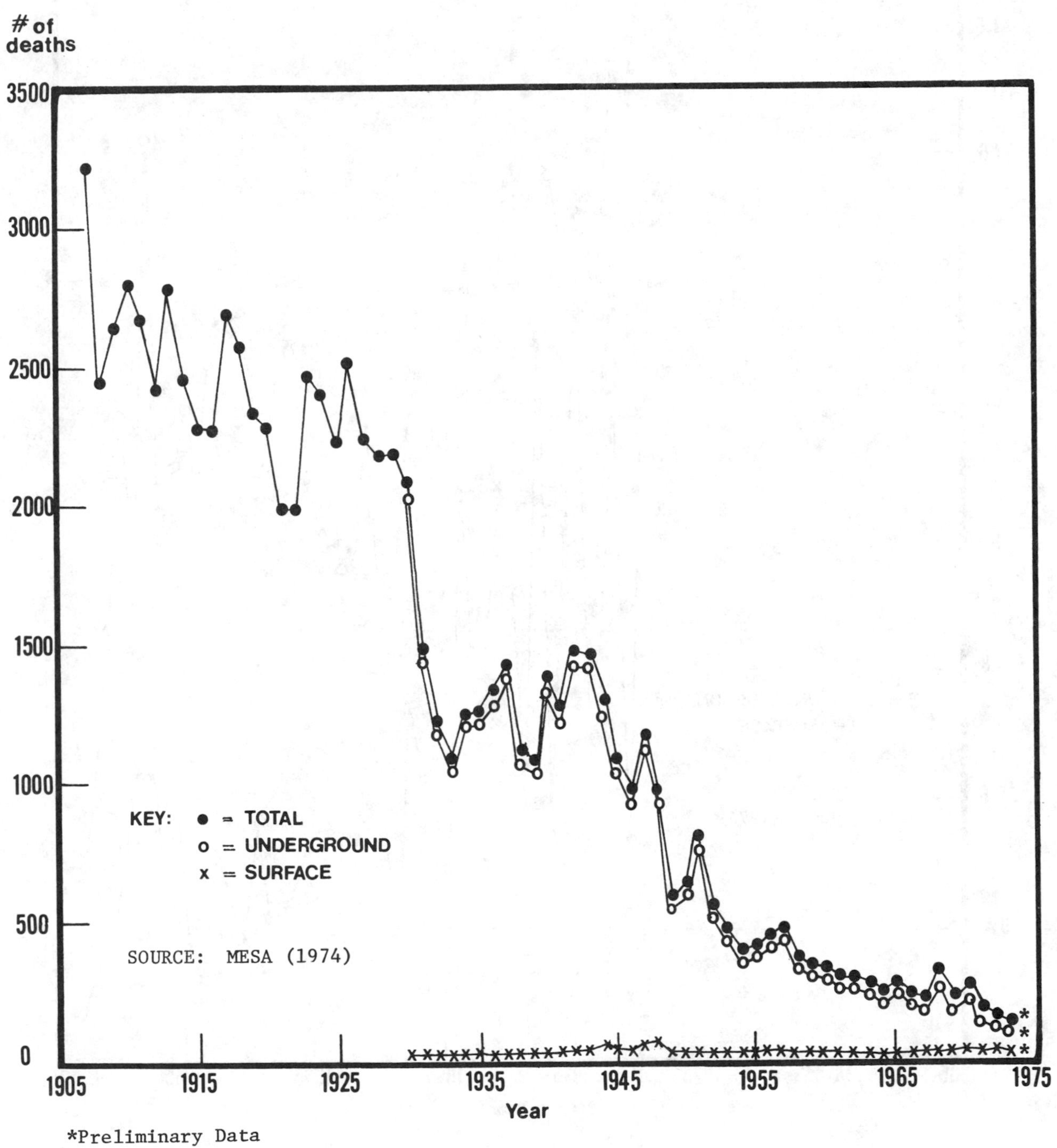

FIGURE 7: Total U.S. Coal Mine Deaths from Accidents 1907-1973

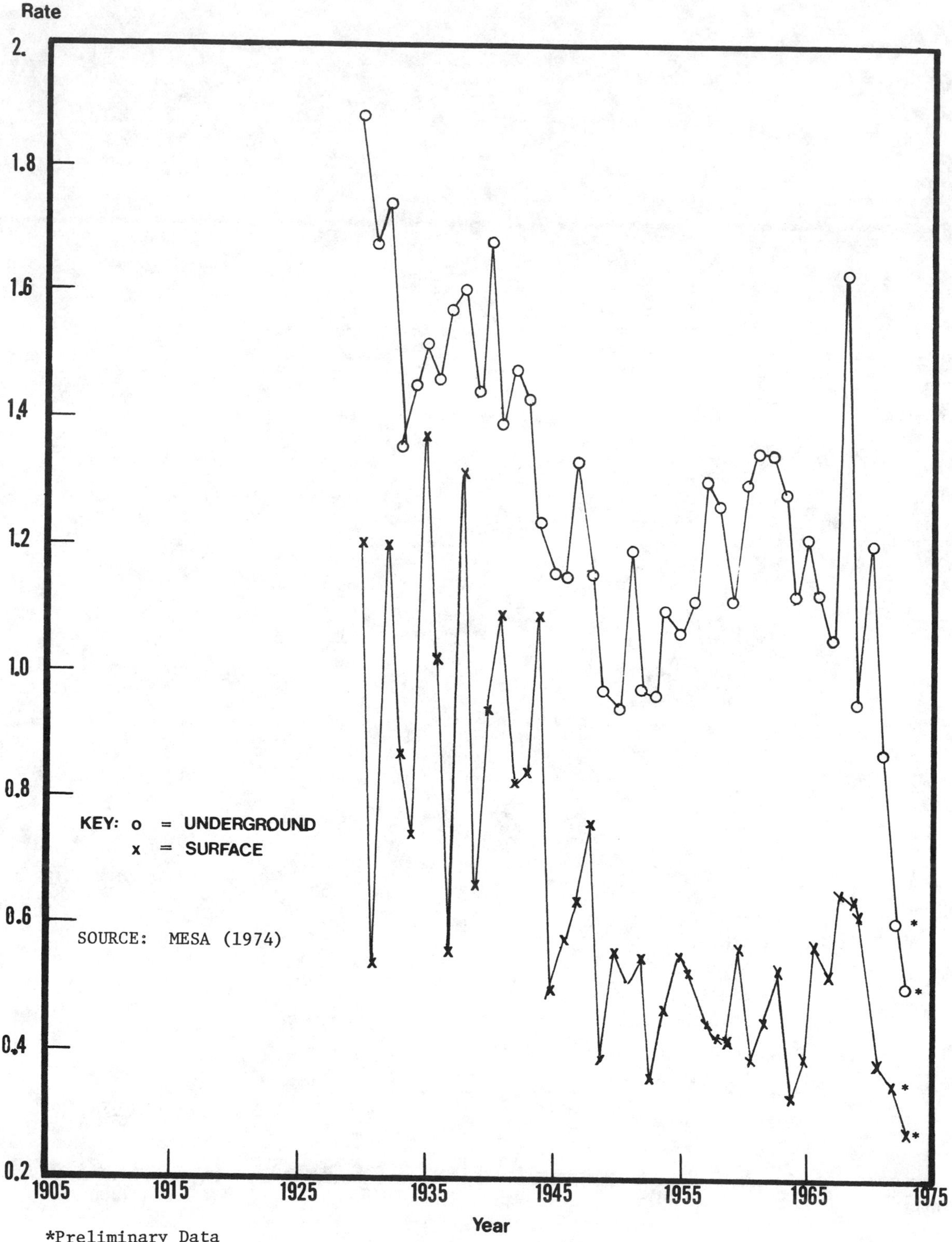

FIGURE 8: Frequency Rate of Fatal Accidents in U.S. Coal Mines 1907-1973 Per Million Man-Hours Worked

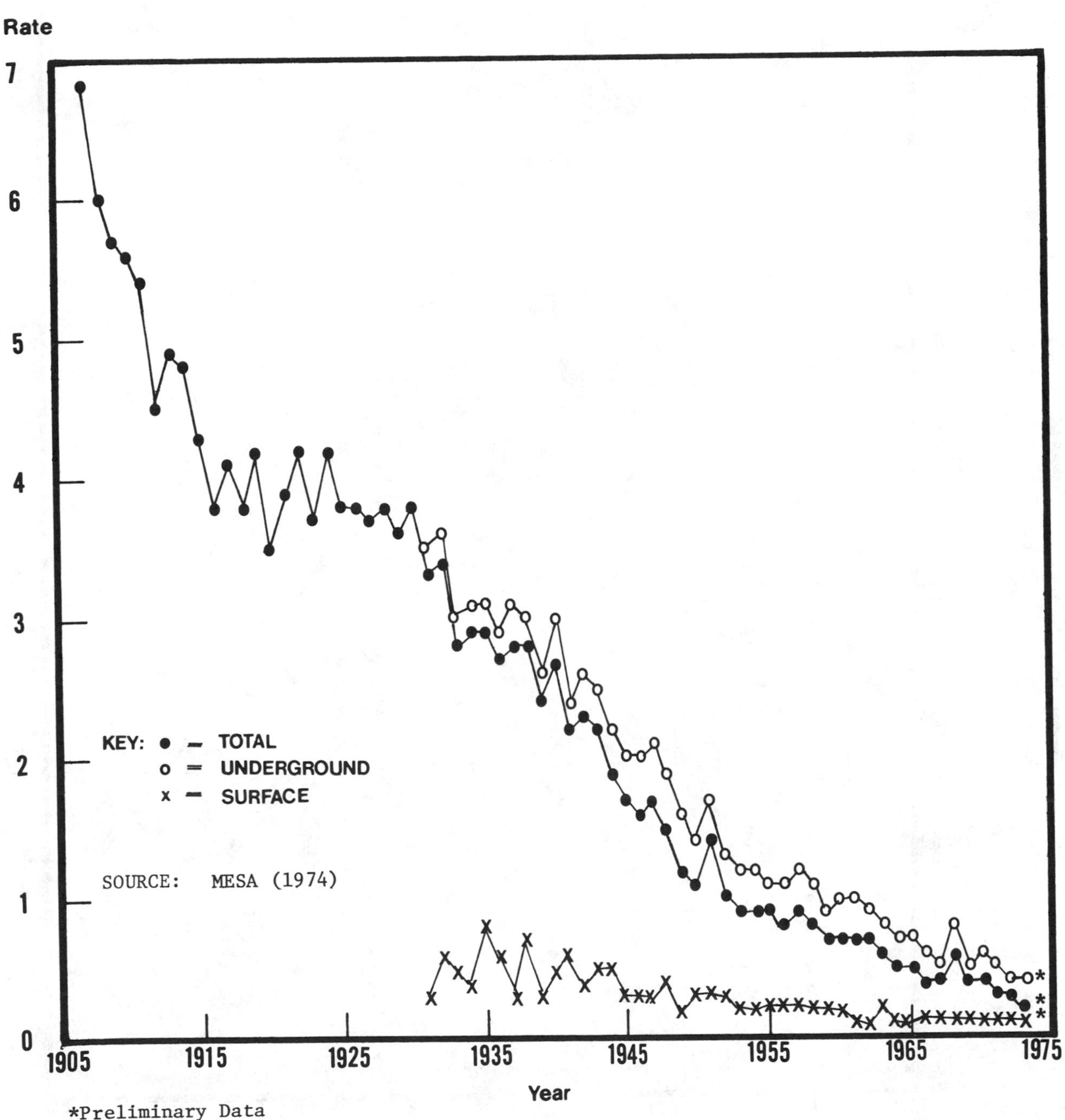

*Preliminary Data

FIGURE 9: Accidental Fatalities Per Million Tons of Coal Mined in U.S. 1907-1973.

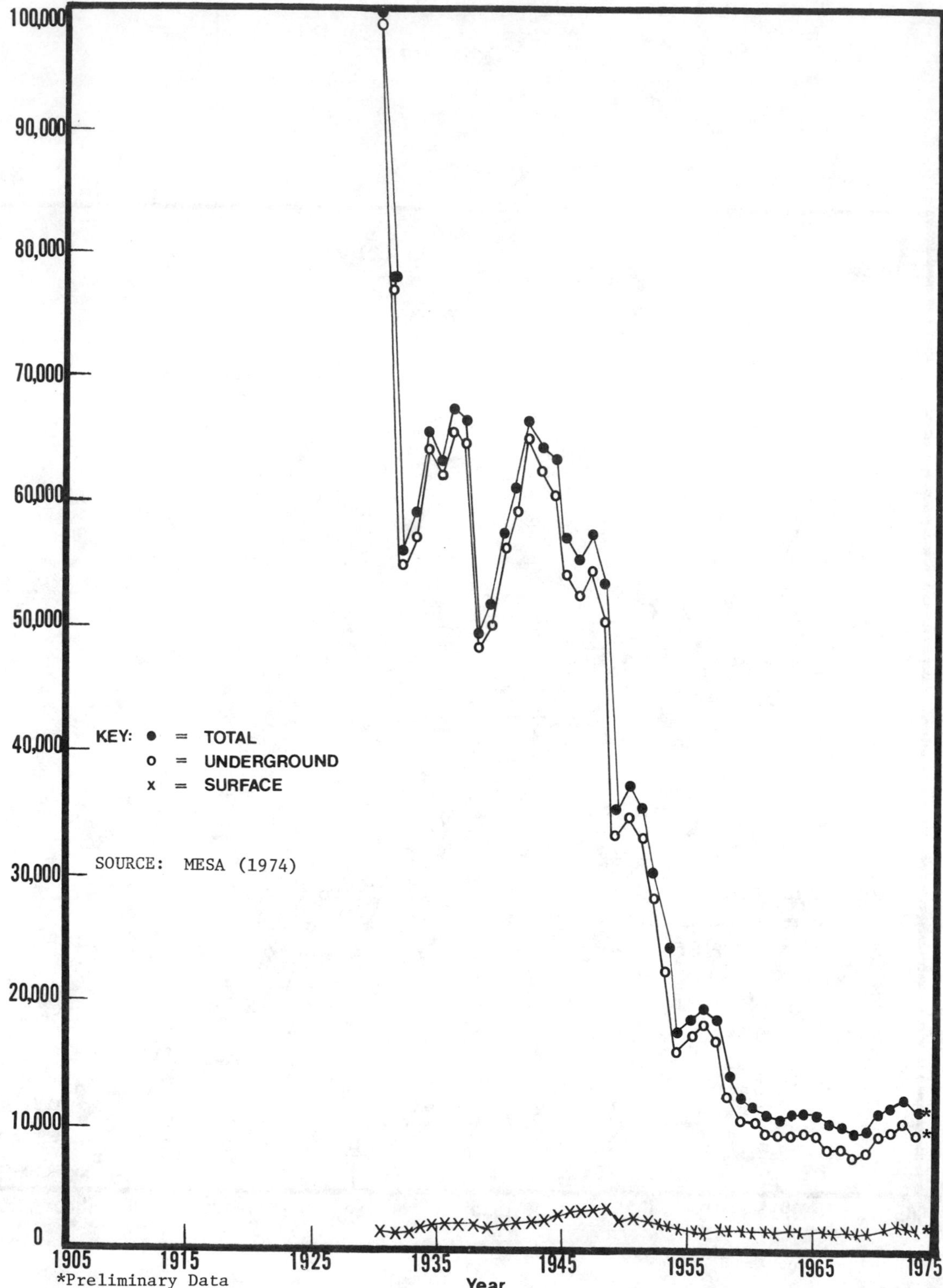

FIGURE 10: Total Nonfatal Injuries -- U.S. Coal Mines 1930-1973

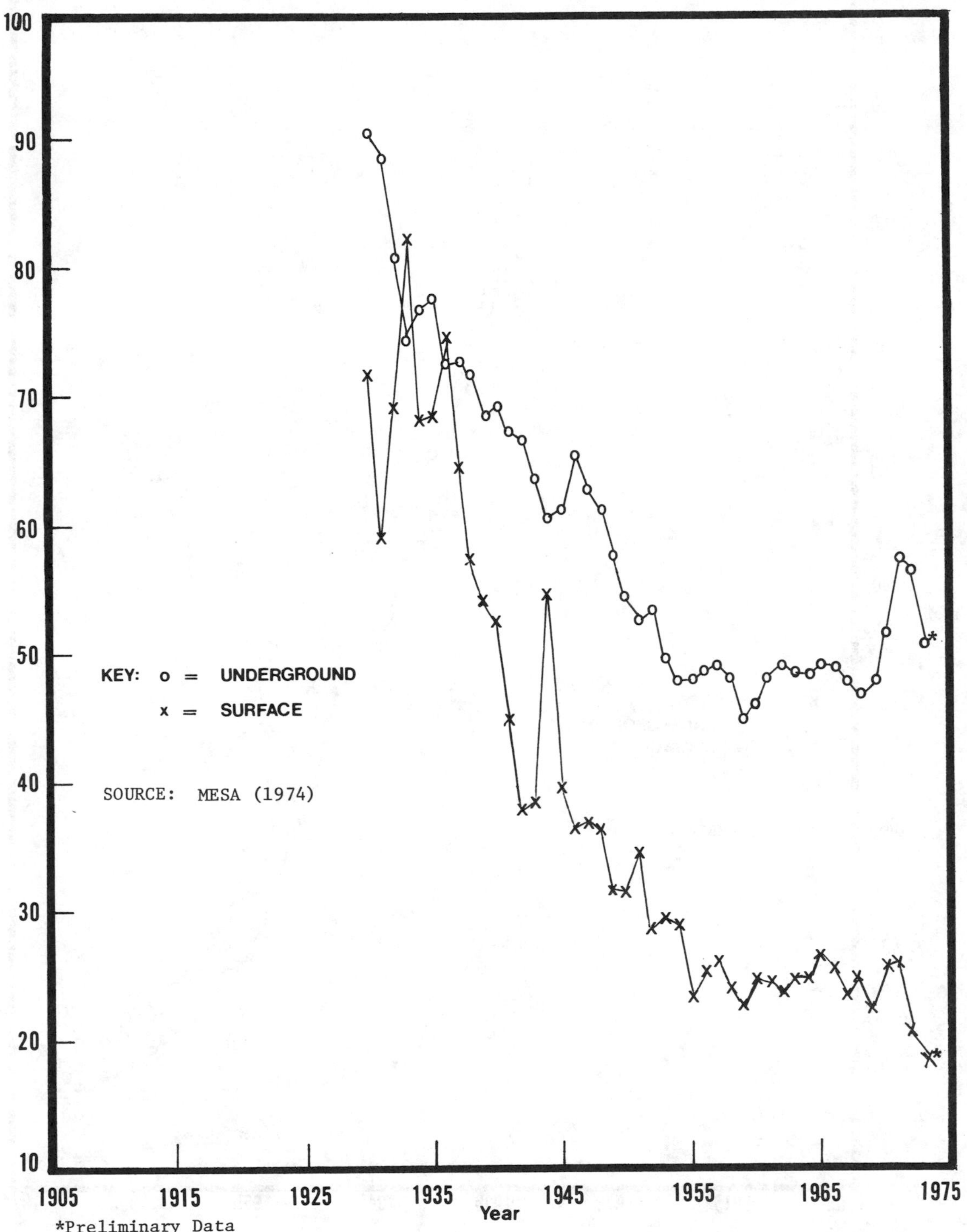

FIGURE 11: Frequency Rate of Nonfatal (Disabling) Injuries in U.S. Coal Mines 1930-1973 Per Million Man-Hours Worked

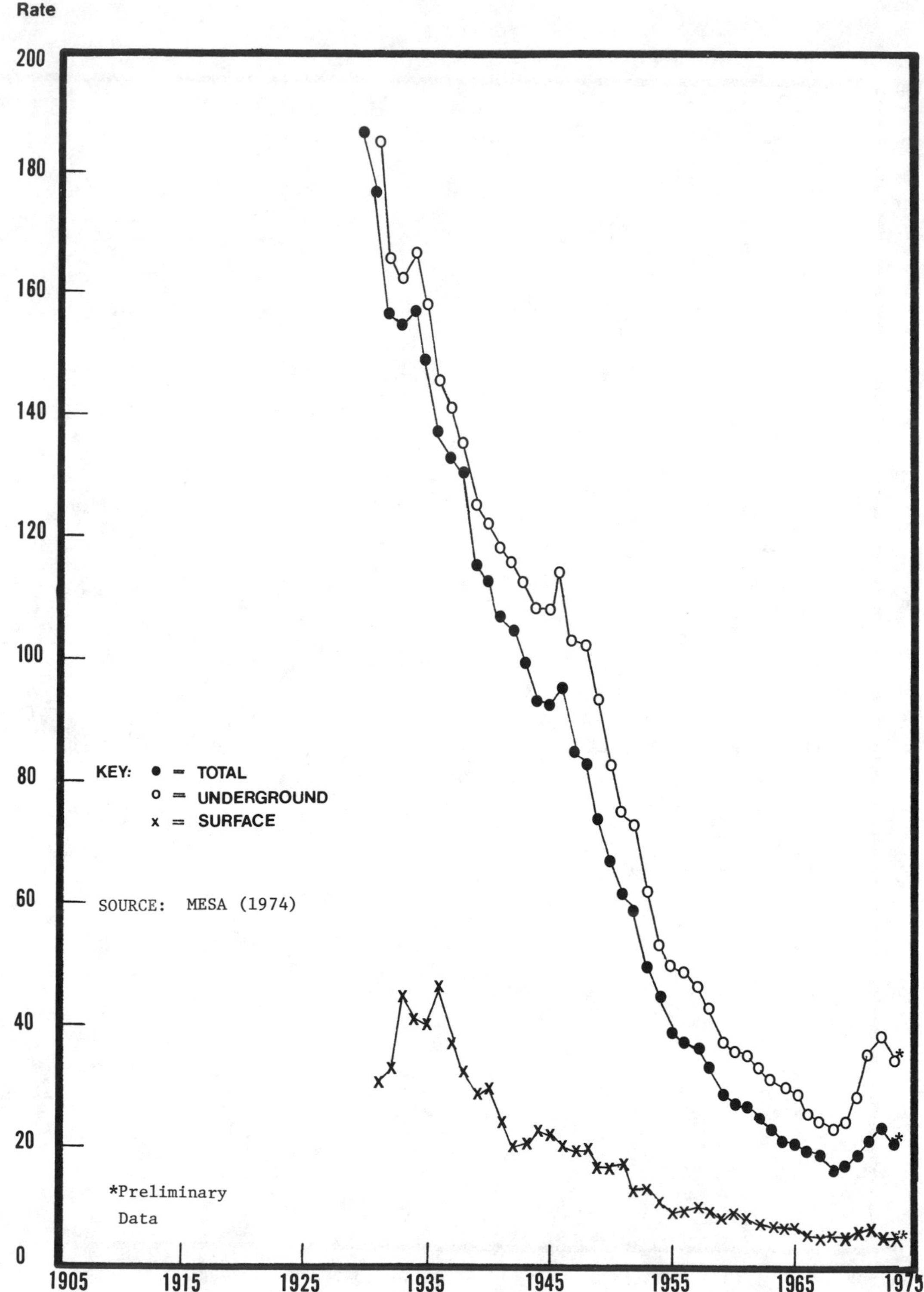

FIGURE 12: Nonfatal (Disabling) Injuries Per Million Tons of Coal Mined in the U.S.A.

COAL WORKERS' PNEUMOCONIOSIS (CWP)

The 1969 U.S. Coal Mine Health and Safety Act established dust standards for coal mines that were designed to reduce the risk of developing coal workers' pneumoconiosis (CWP), and provided compensation to miners adjudged to have the disease. The 1972 "Black Lung Law" relaxes the criteria for diagnosis. Traditionally, epidemiologic studies of CWP have relied on chest X-ray changes as the principal means of identifying CWP. The 1972 "Black Lung Law" specifies that X-ray findings are no longer necessary for a coal miner to qualify for compensation if he has worked 15 years underground and has impaired respiratory capacity. The law has had the effect of establishing a new definition of CWP.

In 1973, the Federal Government paid approximately $1 billion for CWP in benefits to coal miners and their dependents who qualified under the 1969 and 1972 Acts. The annual expenditure may rise to $8 billion by 1980 (Edwards, 1973).

Diagnosis

CWP is defined as a disorder of the respiratory system occurring in persons exposed to coal mine dust and presumably attributable to its inhalation. It represents a slowly developing response of the tissue to prolonged retention of coal mine dust, and is typified by scarring (fibrosis) and deformation. Neither the structural damage nor the clinical symptoms are unique. Pulmonary diseases such as emphysema and chronic bronchitis share clinical and functional but not radiographic features with CWP. A combination of diseases, arising from different causative agents, may coexist in the lung. It is this potential for overlap, interaction, and misdiagnosis that is responsible for much of the controversy about CWP.

The medical diagnosis of CWP rests on: (1) an appropriate history of exposure to coal mine dust (on the order of ten years or more); and (2) the presence of fairly distinctive abnormalities in a chest roentgenogram. The World Health Organization and the International Labor Organization state that a diagnosis of CWP cannot be made without the characteristic radiographic features. It should be noted that the interpretation of the chest film is not completely objective. Differences in interpretation occur among examiners and further efforts to standardize the procedure are in order. Despite this shortcoming, the roentgenogram affords the only epidemiological method for assessing dust retention in miners' lungs (Morgan, 1974; See Appendix of Source Material).

CWP is divided into simple (uncomplicated) and complicated forms or stages. The complicated form is referred to as Progressive Massive Fibrosis (PMF). A classification of CWP which is based on the roentgenographic findings is shown in Table 1. Simple CWP is less disabling and unlike PMF does not progress in the absence of further exposure. Indeed, PMF may develop after the subject has left mining. Fortunately, in about half of the cases, PMF undergoes spontaneous arrest without further deterioration. Unfortunately, little is known about the factors that predispose to PMF.

CWP is essentially confined to underground coal miners. Key, et al., (1971) in a survey of over 400 surface mines, found a mean airborne dust level of less than 1 milligram per cubic meter. Fourteen percent of the samples from auger mines and seven percent of the samples from surface mines contained more than 2 milligrams per cubic meter.* The findings suggest that there may be a risk of getting respiratory disease in some surface operations.

Prevalence, Disability

The prevalence of CWP as determined in three studies that relied on the chest X-ray as the prime criterion for identifying and categorizing CWP is shown in Table 2. A discussion of the assumptions, sampling procedures, and difficulties of correlating the results of the studies is found in the Appendix (Costello and Morgan, 1974).

How debilitating the exposure to coal mine dust alone may be remains an open question. In two separate surveys (Enterline, 1967; Higgins et al., 1968), the ventilatory capacity of underground miners was found comparable to that of suitable control groups.** In discussing this problem, Morgan (1973) has written that "in some regions of the United States coal miners do have an excess of respiratory symptoms and also have a reduced ventilatory capacity. The reduction of ventilatory capacity that can be attributed to occupation is in general minor, and probably has little effect either on working capacity or the health of the miner unless he is also a cigarette smoker."

Prevention

The Federal Coal Mine Health and Safety Act of 1969, in recognition of the technical difficulties of reducing the level of airborne dust in mines, designated a standard that was considered to represent no more than a reasonable degree of risk to underground coal miners. It was intended to reduce the rates at which new cases of CWP developed and old cases progressed.

---

*By comparison, the federal ambient air quality standard for total suspended particulates is 75 <u>micrograms</u> per cubic meter (Buechley et al., 1973).

**Manual laborers who were matched for age and socio-economic background, and lived in the same region.

Specifically, an initial respirable dust standard of 3.0 milligrams per cubic meter of air was established to become effective after June 30, 1970, to be followed by an upper level of 2.0 milligrams per cubic meter of air after December 30, 1972. These standards were based on a report of the National Coal Board (UK) in which data were developed which permit estimates of the probability of developing CWP following exposure to different levels of dust for 35 years (Figures 13 and 14 list the probabilities for developing categories 1 and 2 of simple CWP).

Following passage of the 1969 law the coal mining industry successfully reduced respirable dust levels by instituting improved techniques of ventilation and waterspraying.

While improvement in the reported dust levels has been significant, it remains to be proven that current dust-control technology will "permit each miner the opportunity to work underground his entire working life without incurring any disability from pneumoconiosis." (Interior, 1972.)

Cost

The principal means of compensating miners disabled by CWP lies in the provisions of the Federal Coal Mine Health and Safety Act of 1969 and the Black Lung Benefits Act of 1972. Prior to these Acts, compensation was largely determined and administered through state laws. With the new federal legislation, the Social Security Administration assumed this role. More recently this responsibility was transferred to the Department of Labor.

The Black Lung Benefits Act of 1972 was passed by Congress to correct inequities perceived in the earlier law, principally that relating to the coal miner who suffered some disability, presumably owing to the coal mine dust, but who had a normal roentgenogram. Under the second Act, no miner could be refused compensation solely on the basis of the normal roentgenogram. Any respiratory disability in a coal miner who had worked underground a minimum of 15 years was assumed to be job-related unless contrary evidence could be found. This assumption is subject to revision or rebuttal by an administrative judge if there is evidence that the miner does not have CWP or that his respiratory impairment is not caused by employment in coal mines (Costello, 1974).

The cost to the Social Security Administration of compensating CWP beginning in fiscal 1970 and extending through fiscal 1973 is summarized in Table 3. In four years the total annual costs of the program have risen almost seventy-fold. The cumulative cost during this period has been $1.7 billion. In 1973, approximately 99.5 percent of the expenditures for CWP were for benefits to miners and their dependents. Of the remainder, about $2.5 million was allotted for research on CWP, and $2.6 million for the sampling of coal dust.

In addition to this federal program, most of the major coal mining states compensate disability related to coal mining either under workmen's compensation laws or under separate disability amendments to workmen's compensation. In theory, any payments made directly by the states are deducted from the awards for compensation from the Federal Government. These deductions are called "offset." In practice, however, there are delays in reporting these state payments, so the total amount of offset may be less than the amount of money being spent by the states on compensation for CWP.* According to the Social Security Administration, the offset for September 1973 (latest month that data are available) was at a monthly rate of $637,000, or $7.6 million per year.

The Black Lung Benefits Act has become an expensive means for compensating a relatively small segment of the American working population. It is difficult to estimate what part of the approximately one billion dollars in compensation for CWP is paid to miners who are disabled by chronic bronchitis and emphysema, the causes of which are not necessarily related to coal mining. The assertion has been made that cigarette smoking is at least five times as important as dust inhalation in impairing pulmonary function in miners (Kibelstis et al., 1973). There is evidence to indicate that, in the absence of complicated pneumoconiosis, coal dust does not produce sufficient ventilatory impairment to cause disability (Morgan et al., 1974). The Black Lung Benefits Act has been accused of establishing a pension program for coal miners without regard to whether the miner is disabled and, if so, without regard to whether the disability is the consequence of occupational exposure (Occupational Health, 1974b). Such critical attitudes should be tempered by the realization that coal mining is an extremely hazardous occupation. In the past, society has acted almost as if the miners were an "expendable resource." (Tabershaw, 1970). A proper policy built on scientific and human considerations can be a difficult and elusive goal.

Recommendation

To initiate a study of the full social costs of CWP to include consideration of: (a) problems associated with the diagnosis of the disease, particularly in its incipient stage; (b) improvement of preventive measures; (c) distribution of the costs of the compensation program; (d) administration of the program.

*Social Security Administration, Baltimore, Maryland.

TABLE 1 ILO U/C 1971 International Classification of Radiographs of the Pneumoconiosis

Feature	Code	Definition
Small Opacities		
Rounded:		
Type		The nodules are classified according to the approximate diameter of the predominant opacities.
	p q(m) r(n)	p = rounded opacities up to about 1.5 mm in diameter.
		q(m) = rounded opacities exceeding about 1.5 mm and up to about 3 mm in diameter.
		r(n) = rounded opacities exceeding about 3 mm and up to about 10 mm in diameter.
Profusion		The category of profusion is based on assessment of the concentration (profusion) of opacities in the affected zones. The Standard Radiographs define the midcategories (1/1, 2/2, 3/3).
	0/- 0/0 0/1	Category 0 = small rounded opacities absent or less profuse than in Category 1.
	1/0 1/1 1/2	Category 1 = small rounded opacities definitely present, but few in number. The normal lung markings are usually visible.
	2/1 2/2 2/3	Category 2 = small rounded opacities numerous. The normal lung markings are usually still visible.
	3/2 3/3 3/4	Category 3 = small rounded opacities very numerous. The normal lung markings are partly or totally obscured.
Extent	RU RM RL LU LM LL	The zones in which the opacities are seen are recorded. Each lung is divided into three zones--upper, middle, and lower.

TABLE 1 Continued.

Small Opacities		
Irregular:		
Type		As the opacities are irregular, the dimensions used for rounded opacities cannot be used, but they can be roughly divided into three types.
	s t u	s = fine irregular, or linear opacities.
		t = medium irregular opacities.
		u = coarse (blotchy) irregular opacities.
Profusion		The category of profusion is based on assessment of the concentration (profusion) of opacities in the affected zones. The Standard Radiographs define the midcategories (1/1, 2/2, 3/3).
	0/- 0/0 0/1	Category 0 = small irregular opacities absent or less profuse than in Category 1.
	1/0 1/1 1/2	Category 1 = small irregular opacities definitely present, but few in number. The normal lung markings are usually visible.
	2/1 2/2 2/3	Category 2 = small irregular opacities numerous. The normal lung markings are usually partly obscured.
	3/2 3/3 3/4	Category 3 = small irregular opacities very numerous. The normal lung markings are usually totally obscured.
Extent	RU RM RL LU LM LL	The zones in which the opacities are seen are recorded. Each lung is divided into three zones--upper, middle, and lower--as for rounded opacities.
Combined profusion:	1/0 1/1 1/2 2/1 2/2 2/3 3/2 3/3 3/4	When both rounded and irregular small opacities are present, record the profusion of each separately and then record the combined profusion as though all the small opacities were of one type, i.e., either rounded or irregular. This is an optional feature of the Classification, but it is strongly recommended.

TABLE 1 Continued.

Large Opacities				
Size	A	B	C	Category A = an opacity with greatest diameter between 1 cm and 5 cm, or several such opacities the sum of whose greatest diameters does not exceed 5 cm.
				Category B = one or more opacities larger or more numerous than in Category A whose combined area does not exceed the equivalent of the right upper zone.
				Category C = one or more opacities whose combined area exceeds the equivalent of the right upper zone.
Type	wd		id	In addition to the letter A, B, or C, the abbreviation "wd" or "id" should be used to indicate whether the opacities are well defined or ill defined.
Pleural Thickening				
Costophrenic angle	R		L	Obliteration of the costophrenic angle is recorded separately from thickening over other sites. A lower limit Standard Radiograph is provided.
Chest Wall and Diaphragm				
Site	R		L	
Width	a	b	c	Grade a = up to about 5 mm thick at the widest part of any pleural shadow.
				Grade b = over about 5 mm and up to about 10 mm thick at the widest part of any pleural shadow.
				Grade c = over about 10 mm at the widest part of any pleural shadow.

TABLE 1 Continued.

Chest Wall and Diaphragm		
Extent	0 1 2	Grade 0 = not present or less than Grade 1. Grade 1 = definite pleural thickening in one or more places such that the total length does not exceed one half of the projection of one lateral wall. The Standard Radiograph defines the lower limit of Grade 1. Grade 2 = pleural thickening greater than Grade 1.
Ill-Defined Diaphragm	R L	The lower limit is one-third of the affected hemidiaphragm. A lower limit Standard Radiograph is provided.
Ill-Defined Cardiac Outline (Shagginess)	0 1 2 3	Grade 0 = absent or up to one-third of the length of the left cardiac border or equivalent. Grade 1 = above one-third and up to two-thirds of the length of the left cardiac border or equivalent. Grade 2 = above two-thirds and up to the whole legnth of the left cardiac border or equivalent. Grade 3 = more than the whole length of the left cardiac border or equivalent.
Pleural Calcification		
Site	Wall Diaphragm Other R L	
Extent	0 1 2 3	Grade 0 = no pleural calcification. Grade 1 = one or more areas of pleural calcification the sum of whose greatest diameter does not exceed about 2 cm. Grade 2 = one or more areas of pleural calcification the sum of whose greatest diameters exceeds about 2 cm but not about 10 cm.

TABLE 1 Continued

Pleural Calcification			
Extent		Grade 3	= one or more areas of pleural calcification the sum of whose greatest diameters exceeds about 10 cm.
Additional Symbols	ax cp es pq bu cv hi px ca di ho rl cn ef k tba co em od tbu	ax	= coalescence of small rounded pneumoconiotic opacities
		bu	= bullae
		ca	= cancer of lung or pleura
		cn	= calcification in small pneumoconiotic opacities
		co	= abnormality of cardiac size or shape
		cp	= cor pulmonale
		cv	= cavity
		di	= marked distortion of intrathoracic organs
		ef	= effusion
		em	= marked emphysema
		es	= eggshell calcification of hilar or mediastinal lymph nodes
		hi	= enlargement of hilar or mediastinal lymph nodes
		ho	= honeycomb lung
		k	= septal (Kerley) lines
		od	= other significant disease. This includes disease not related to dust exposure, e.g., surgical or traumatic damage to chest walls, bronchiectasis, etc.
		pq	= pleural plaque (uncalcified)
		px	= pneumothorax
		rl	= rheumatoid pneumoconiosis (Caplan's syndrome)
		tba	= tuberculosis, probably active
		tbu	= tuberculosis, activity uncertain

Source: Jacobsen and Lainhart (1972)

TABLE 2 Prevalence of CWP in U.S. Coal Mines

	Number of Miners	Bituminous mines			Anthracite mines			All mines		
		Simple CWP	PMF	Total CWP	Simple CWP	PMF	Total CWP	Simple CWP	PMF	Total CWP
USPHS Study (1963-65) (Bituminous mines only)	2,549 current	6.8%	3.0%	9.8%						
	1,191 ex-miners	9.2%	9.0%	18.2%						
Federal Coal Mine Health and Safety Data (Dec. 1973)	72,433	9.6%	1.0%	10.6%	26.7%	5.6%	32.3%	9.87%	1.03%	10.9%
Interagency Study (1973) [31 mines in 10 states--2 anthracite and 29 bituminous]	9,076	28.0%	2.0%	30.0%	45.7%	14.3%	60.0%	27.4%	2.5%	29.9%

Source: Costello, and Morgan, (1974). See Appendix of Source Material.

TABLE 3 Cost of Federal Compensation for CWP

	1970	1971	1972	1973
Administrative	3,984,000	22,054,000	21,750,000	32,024,000
Benefit Payments	10,000,000	320,000,000	408,000,000	926,000,000
	13,984,000	342,045,000	429,750,000	958,024,000

Total cost = $1,743,803,000

Source: Interior (1972)

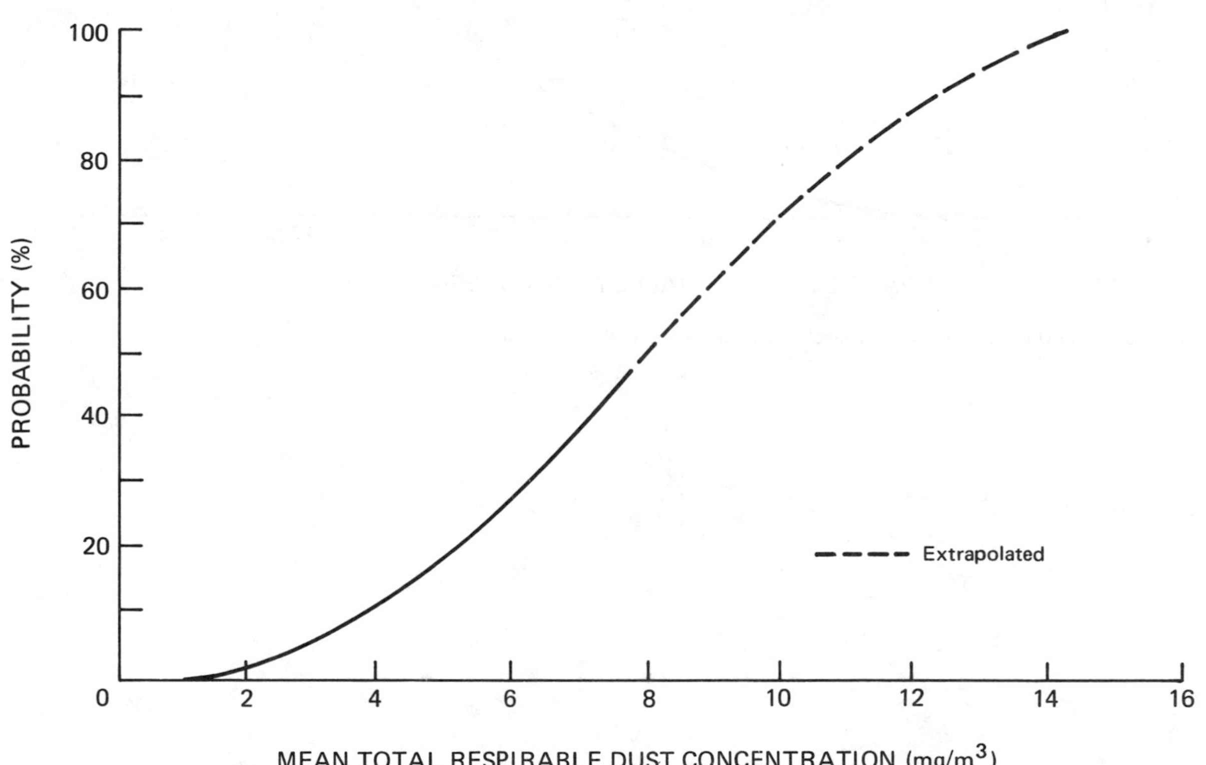

SOURCE: The U.K. National Coal Board's Pneumoconiosis Field Research.

FIGURE 13: Probability of Contracting Pneumoconiosis I L O Category '1' or Greater After 35 Years Exposure to Coal Dust

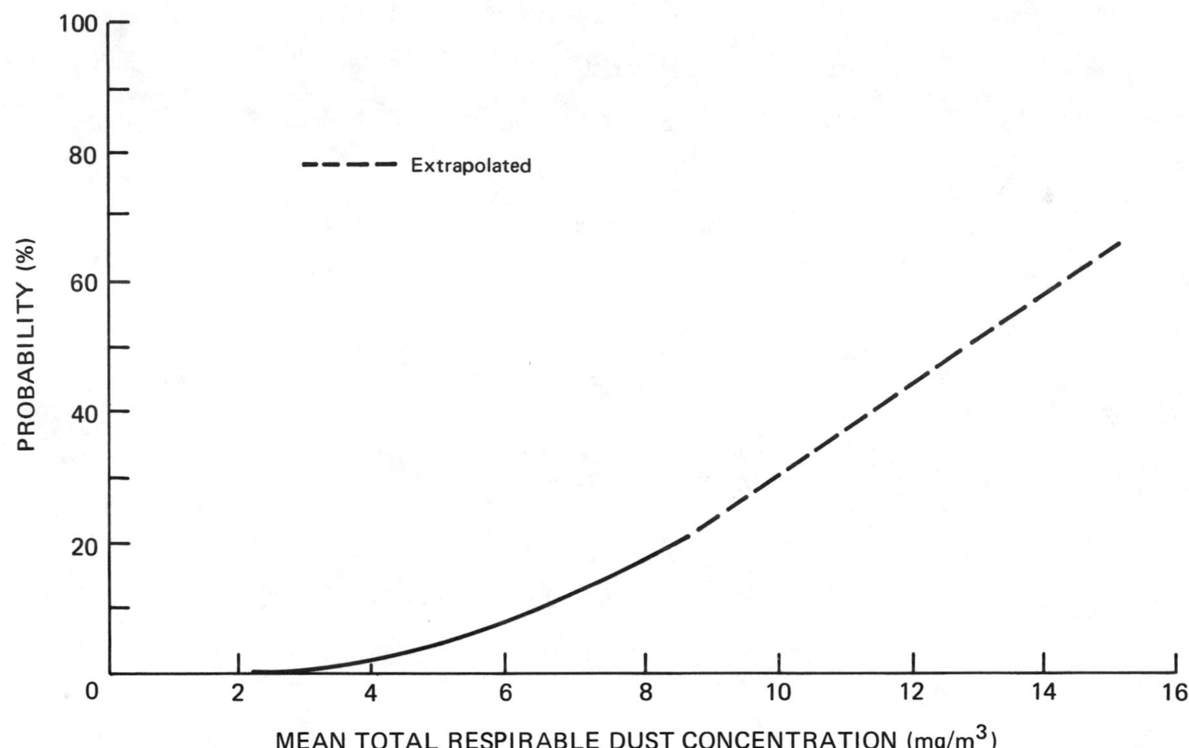

SOURCE: The U.K. National Coal Board's Pneumoconiosis Field Research.

FIGURE 14: Probability of Contracting Pneumoconiosis I L O Category '2' or Greater After 35 Years Exposure to Coal Dust

ACID MINE DRAINAGE, RECLAMATION

Debate over legislation to control or buffer the damage to the environment has engendered a number of documents on the problems of acid mine drainage and reclamation.

Acid Mine Drainage

The extential and potential impact of acid mine drainage associated with both underground and surface mining, and the attendant problems of erosion, sedimentation, and subsidence have been studied in detail (Hill, 1968; Interior, 1969). An appraisal by the Appalachian Regional Commission (1969) drew the conclusion that ". . . about 10,500 miles of streams in eight states of the Appalachian Region are affected by mine drainage . . . (and) of the total stream mileage affected, acid drainage continually pollutes nearly 5,700 miles." The Commission estimated that acid mine drainage is responsible for $3.5 million in added costs to users of water. The problem is stubbornly resistant to solution and there is likely to be a continuing need to control drainage at the mining site, to treat effluents, and to control water sources of potentially high activity. Unfortunately, a large percentage of the drainage originates in inactive mining sources which are difficult and costly to treat. The U.S. Bureau of Mines and EPA are both currently undertaking procedures and research directed toward control of this problem. As the Appalachian Commission (1969, p. 126) indicates, progress in achieving improved water quality is likely to be slow and will require better definition of priorities. Coal is widely distributed and its attendant environmental problems may be expected to vary from one region to the next. Acid mine drainage presents different challenges in Appalachia, the Midwest, and the Rocky Mountains. We join others in acknowledging the importance, and often the stubborness, of the problems of acid mine drainage. We believe the problems of human health that are linked to the mining and use of coal are of more immediate concern.

Reclamation

In its report to the Senate Interior and Insular Affairs Committee, the Council on Environmental Quality (1973) describes prospects for rehabilitation of surface mined lands in the eastern United States, which include alternative mining technologies appropriate to topographic and geologic differences. This and similar reports indicate that the costs of reclamation in the Appalachian region range from about 15 to 90 cents per ton depending on the production of coal per acre and a number of physical conditions at the mining site. By comparison, the costs of producing the coal are about $4 or $5 per ton (CEQ, 1973).

A committee of the National Academy of Sciences--National Academy of Engineering (1974a) concluded recently that the rate and ultimate success of rehabilitation of surface mined areas in the western United States will depend principally upon the moisture regime of the region. In arid regions revegetation, if attainable at all, may require decades or very much longer; in wetter areas revegetation may be possible within a few years. The cost of reclamation will depend on the thickness of the coal seam relative to the thickness of the overburden. Such costs are estimated to range from roughly $900 to $3,000 per acre, or 3¢ to 10¢ per ton, compared to the production costs which are about $1.25 to $2.50 per ton (NAS/NAE, 1974a). These costs, as with the costs associated with acid mine drainage and waste accumulations from coal mining, cleaning and processing operations, are neither trivial nor forbidding. The Panel is aware of these costs; as in the case of acid mine drainage, the human health effects appear to be more important.

CHAPTER X
USE

Scientific evidence points increasingly to the importance of SO_2 and its oxidized products as a factor adversely affecting the quality of air, human health, and elements within the ecosystem. An important byproduct of the combustion of fossil fuels, SO_2 undergoes a variety of chemical reactions in the atmosphere, the types and rates of which are not well known. Limited research is in process to understand the variety of transformations of sulfur in the atmosphere, and an impressive body of evidence indicates that human health is impaired as levels of SO_2, total suspended particles, and sulfates rise. Considerably more scientific research is required before specific recommendations can be made about the upper level of SO_2 emission control needed, and before new standards can be set.

At present we can act only upon the trends shown by the results of research and the belief that essential scientific proof may be forthcoming in the not too distant future. Nevertheless, in the interim, important decisions about control and capital investment may be made which could exacerbate the sulfur problems for human health and the environment, and result in costly, perhaps irreversible commitments. One set of such decisions we have identified is the proposed use of tall stacks and intermittent control systems. In view of the evidence implicating sulfur, great caution in adopting control strategies is needed; relaxation in air quality standards is not warranted from the evidence at hand.

The following four sections elaborate upon these problems.

SULFUR EMISSIONS, ATMOSPHERIC REACTIONS

Globally man generates about one-third of the total gaseous sulfur present in the atmosphere. Because he occupies only a small fraction of the earth's surface, amounting to about 5 percent, his contribution is dominant in populous regions. Estimates of the combined outputs of sulfur from the generation of power, and from industrial and domestic activities, suggest that coal is chiefly responsible for the emission of this sulfur.

Regional Distribution

Our knowledge of the physico-chemical processes that ensue once SO_2 is emitted into the atmosphere is incomplete. SO_2 reacts to produce a number of substances as it moves downwind from the source of emission. Evidence indicates that submicronic particles, many of which react as acids, are formed as the SO_2 is oxidized (Junge & Scheich, 1969; Lodge & Frank, 1966), but explicit proof of the mechanisms and reaction rates is still lacking. In laboratory studies designed to simulate the natural environment it has been possible to demonstrate the conversion of SO_2 to H_2SO_4. Recent sampling data from St. Louis and its environs (Charlson et al., 1974) and from Scandinavia (Bolin, 1971) support the hypothesis that a similar transformation occurs in the atmosphere. Experiments designed to test unequivocally this hypothesis are now underway.

Sulfur-containing aerosals are widespread geographically. They now extend over one-third to one-half of the continental United States, affecting principally the East and Midwest (Figure 15a, 15b). The area involved differs completely from any jurisdictional settings organized to control air pollution.

It would be erroneous to consider coal-burning for the generation of power as being principally an urban activity. Urban sites, because they are plainly visible, are likely to fix our attention. Examples may be cited of the power plants along the East River in New York, and of those bordering Chicago, but such plants represent a small fraction of the total. Instead, coal-burning is distributed widely over the country (Figure 16). The bulk of our power is generated in rural areas and is then transmitted through great electrical networks to metropolitan centers. Typically the spacing between these sources of SO_2 has an order of magnitude of about one hundred kilometers. The distance for the removal of SO_2 from the atmosphere, or for its chemical conversion to higher oxidative states, is on the order of about one thousand kilometers. As noted earlier, virtually all of the eastern third of the U.S. is now exposed to these emissions and to their end-products (Figure 17; see also Figure 19 under Acid Rain).

Chemistry

Because of the evidence to suggest that SO_2 is converted in part to acid sulfates, and the additional evidence that such aerosols are hazardous to health (see Public Health), it may be necessary to achieve and maintain strict control of SO_2 if the concentrations of these aerosols are to be kept within acceptably safe levels. How much control will be needed to achieve a particular level of sulfates is unknown. The reason is that we do not understand the quantitative relation between the amount of gaseous SO_2 emitted and the amount of particles produced. Our emphasis is on submicronic particles which are formed in the atmosphere, and not on particles produced by mechanical or abrasive events (Figure 18).

The relation between SO_2 emissions and aerosol formation cannot be assumed to be a simple linear one. This point is illustrated in the following calculations: For SO_2, 1 part per million by volume (ppm) is roughly equivalent to 2,500 micrograms per cubic meter. Were this amount to oxidize completely to ammonium sulfate $[(NH_4)_2 SO_4]$ which has roughly twice the molecular weight of SO_2, about 5,000 micrograms per cubic meter of the salt would be produced. Therefore, even a fraction of a ppm of SO_2 is potentially a large source of aerosol. The results of a recent study (Altshuler, 1973) suggest that nationally the atmospheric ratio of SO_2 to sulfate is about 5 molecules to 1 molecule (5/1). A number of factors may change this ratio from one community to another. In Los Angeles, the ratio is about 2/1 or even 1/1, an indication of the relatively large amount of SO_2 that is oxidized to sulfate (Altshuler, 1973). The ratio is potentially important since it has implications for the strategy of controlling sulfate pollution and thereby protecting the public health.*

Atmospheric aerosols are principally produced in three ways: (1) by mechanical events including natural phenomena (wind, ocean spray) plus a miscellany of human activities (plowing, grinding, and the action of wheels on gravel); (2) by direct emission of particles during the combustion of fuels--an example is fly ash; (3) by condensation in the atmosphere--the chemical transformations of pollutant gases such as SO_2 and nitrogen oxides contribute to this form of aerosol. It is difficult to generalize about the fraction of airborne particles that arise from each source. Recent evidence suggests that in a community like Los Angeles, 75 percent or more of the suspended particles on a hazy windless day are generated in the atmosphere from gaseous reactions (Whitby et al., 1972).

In general, atmospheric aerosols that are well aged and are removed from their sources can be divided into two basic groups according to size: those above and those smaller than 1-2 micrometers in diameter (Figure 18). The larger particles are generated principally by the aforementioned mechanical events. Typically, they contain silicates or other metallic oxides. They have a mass or volume mode that is on the order of about 10 to 20 micrometers in diameter and may reach a size of 100 micrometers. The second group of smaller particles, virtually all of which are submicronic, are different in

---

*SO_2 is not ordinarily considered to be a significant pollutant in Los Angeles, where several decades ago the sources for this gas were subjected to strict control. As a result, the use of high sulfur fuels was, until most recently, sharply limited. Because of the current shortage of fuels, consideration is being given to relaxing this control. While Los Angeles does not have levels of gaseous SO_2 approaching those of the industrial centers of the east and midwest, in part because of its strong oxidizing environment, its levels of sulfate are among the highest regional levels found in the United States: concentrations of 10-20 micrograms per cubic meter are not infrequent (Altshuler, 1973).

many important respects. The great majority of submicronic particles is produced by condensation. If the condensation occurs at high temperatures, as in the region of the coal-burning flame, the aerosol will be rich in trace metals and metallic oxides. If the condensation occurs at a distance from the source, as with the oxidation of SO_2 and oxides of nitrogen, the aerosol will contain sulfates and nitrates. It would be impractical, therefore, to attempt to reduce atmospheric sulfate aerosols by controlling the emission of particles from combustive sources, for only a small fraction of the significant submicronic species have been formed at this stage. Current strategies for controlling airborne particles tacitly assume that the concentration of the particles in the atmosphere can be controlled by trapping those that are emitted at the primary sources. This ignores the large formation of particles that takes place in the atmosphere by condensation. Some other means, more likely related to the capture of SO_2, is indicated.

Limitations of Air Sampling Data

At present, the bulk of information on airborne particles has been obtained with high volume samplers. The data are expressed in milligrams per cubic meter. Unfortunately, the samples collected by this means are dominated by particles several micrometers in diameter or larger. (Most of the total weight of the sample may be contained in relatively few large particles.) When particles above several micrometers in diameter are inhaled, they deposit principally in the upper airways (nose, mouth, pharynx), or in the trachea and large bronchi (Bates et al., 1966). Clearance of these particles is usually accomplished efficiently and rapidly, reducing their hazard to health. The submicronic particles are more likely to deposit in the small peripheral airways (bronchioles) and parenchyma (alveoli) of the lungs where clearance is slower and less certain and where the effects of irritation or injury assume great significance.

Conventional practice is to measure the "sulfate" concentration on the filter-collected sample without determining its specific molecular form.* This chemically unqualified designation is of dubious value, for the biologic effects of the aerosol should depend not only on its size (and solubility) but on its molecular composition. It would be of considerable value to know what types of sulfate are present, for the purpose of improving the evidence used to relate air pollution to illness, as well as for setting more cogent air quality standards and establishing more precise methods of controlling pollution. To illustrate, calcium sulfate (gypsum) or magnesium sulfate (Epsom salt), present in the sea spray from the ocean, can be expected to be quite different in toxicity from $(NH_4)_2SO_4$ or H_2SO_4, yet all of these

*Other difficulties are inherent in this method. The values for sulfate determined on the filter-sample are not necessarily identical with those present in the air, that is, chemical reactions may proceed after the sample has been collected.

compounds will appear in the typical aerometric analysis as "sulfate."

Other Emissions

In addition to the oxides of sulfur a partial list of the substances known to be emitted in the combustion of coal includes:

(1) Gases: Carbon dioxide (CO_2), carbon monoxide (CO), nitric oxide (NO), and water vapor;

(2) Particles condensed from gases near the combustion zone: Carbon and high molecular weight hydrocarbons, H_2SO_4, lead,* beryllium,* cadmium,* arsenic (possibly as As_4O_6), selenium,* antimony,* mercury,* and vanadium;*

(3) Particles that are mechanically generated: Fly ash from grates, carbon as soot.

The ambient levels of CO_2 have been rising at a number of widely separated locales. From 1959 to 1969, CO_2 increased by 6 to 8 ppm at Mauna Loa, Hawaii, a change corresponding to about 2 percent of the presumed pre-industrial concentration of 290 ppm. It has been estimated that almost half of the contemporary production of CO_2 derives from the combustion of fossil fuels and from industrial sources (Ekdahl & Keeling, 1973).

There are as yet no reports concerning the response of any natural ecosystem to changes in atmospheric CO_2. Indirect efforts to predict such effects are underway utilizing computer models (Botkin, 1973; See Appendix of Source Material). In our judgment, this problem is important and in need of additional study, but it is not as immediate as the other issues considered in this report.

Recommendations

1. Research: To give high priority in federally sponsored research to a more complete understanding of the variety of transformations of sulfur in the atmosphere, including such factors as: (a) residence time in the atmosphere; (b) rates of reaction; (c) distances over which the reactants travel.

2. Data collection: To improve the current sampling and analytical techniques for aerosols (a) establish a more complete and standardized grid of urban and nonurban sampling stations; (b) identify the mass concentration and chemical composition of the submicronic particles.

---

*Of unknown molecular species.

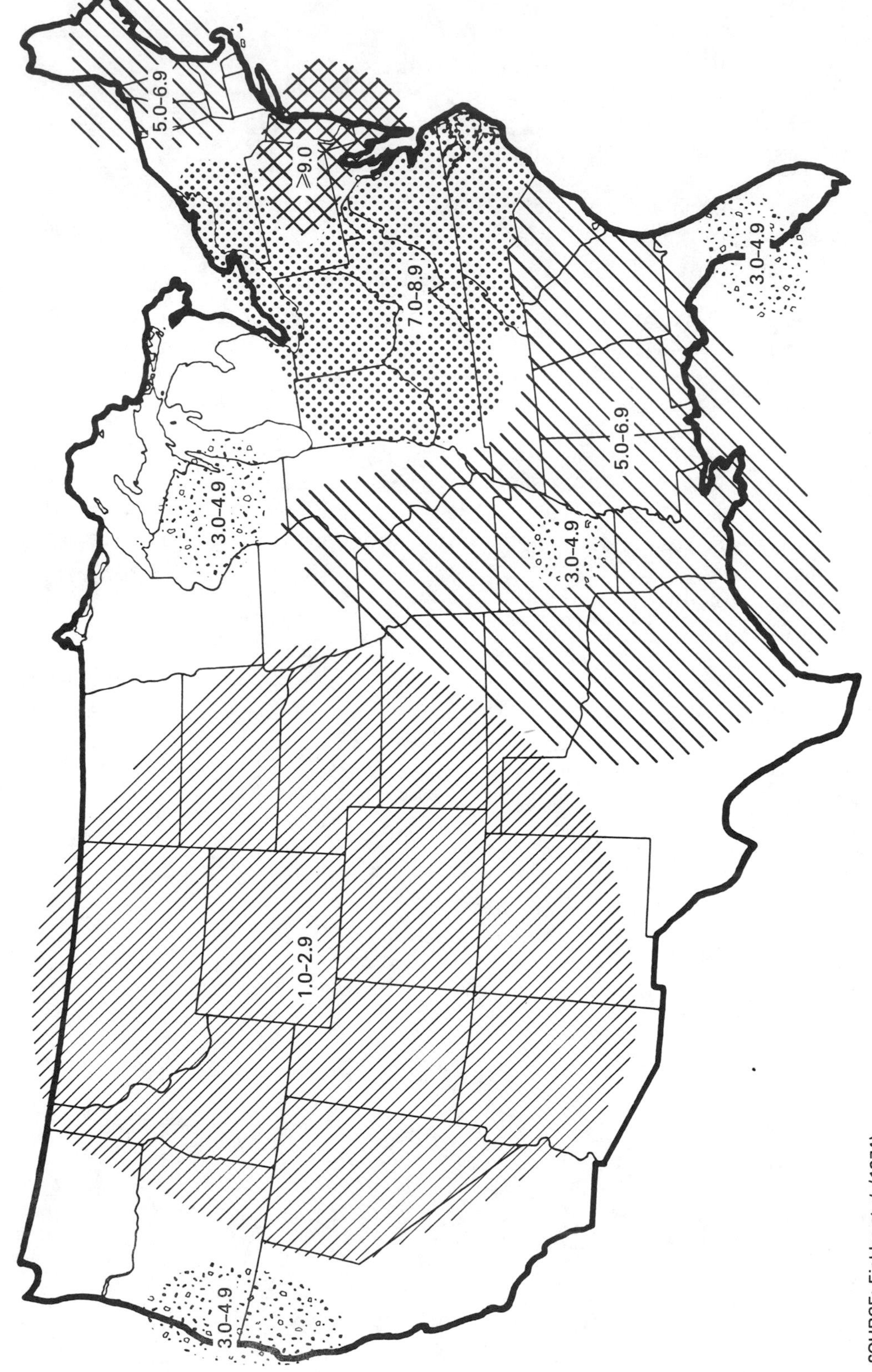

SOURCE: Finklea *et al.* (1974).

FIGURE 15A: Nonurban Average Sulfate Concentration.

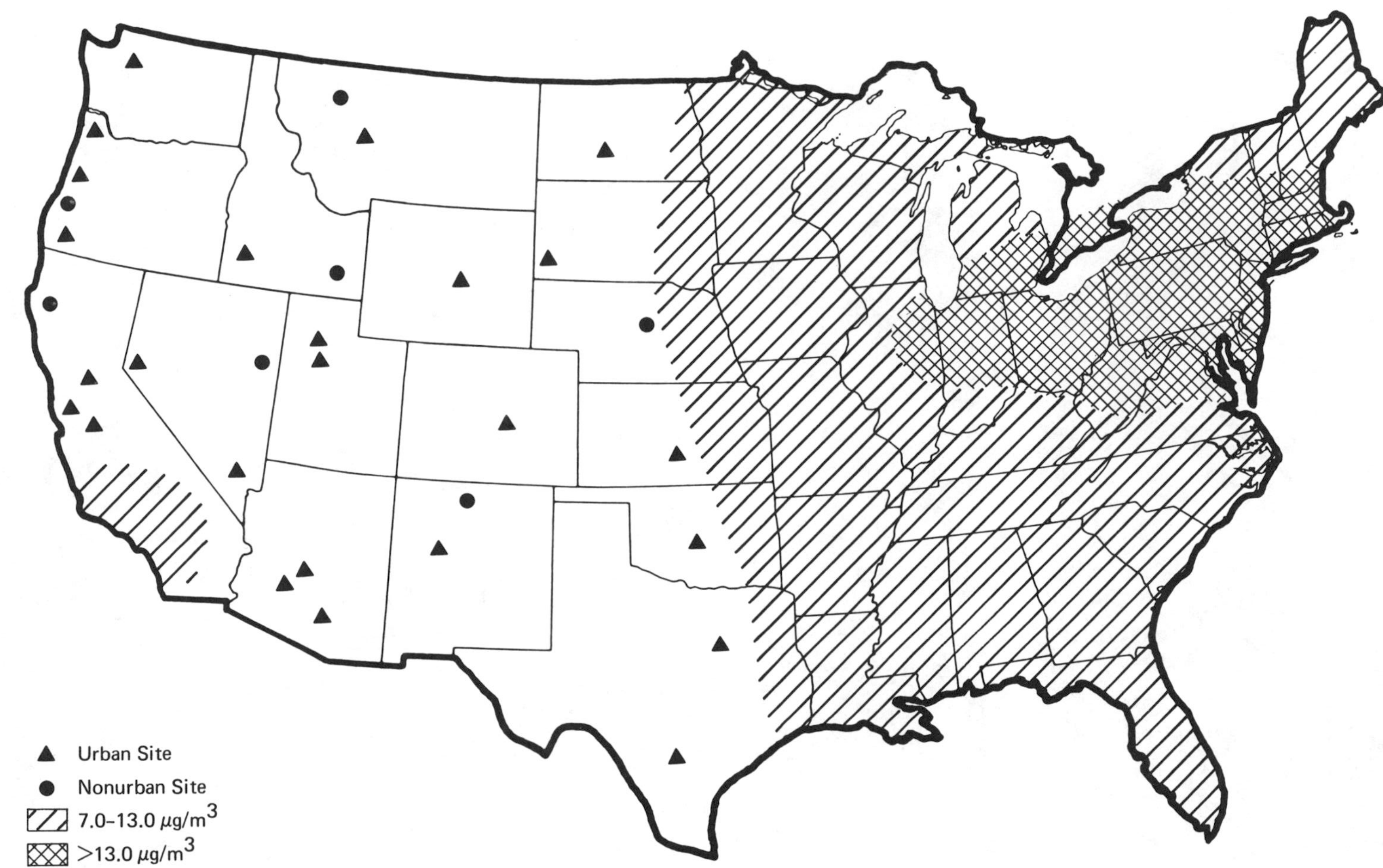

FIGURE 15B: 1970 Sulfate Concentrations, Urban. Within approximately the Eastern half of the Continental United States, levels of sulfate were higher at urban sampling sites than at non-urban sites, (see Figure 15A).

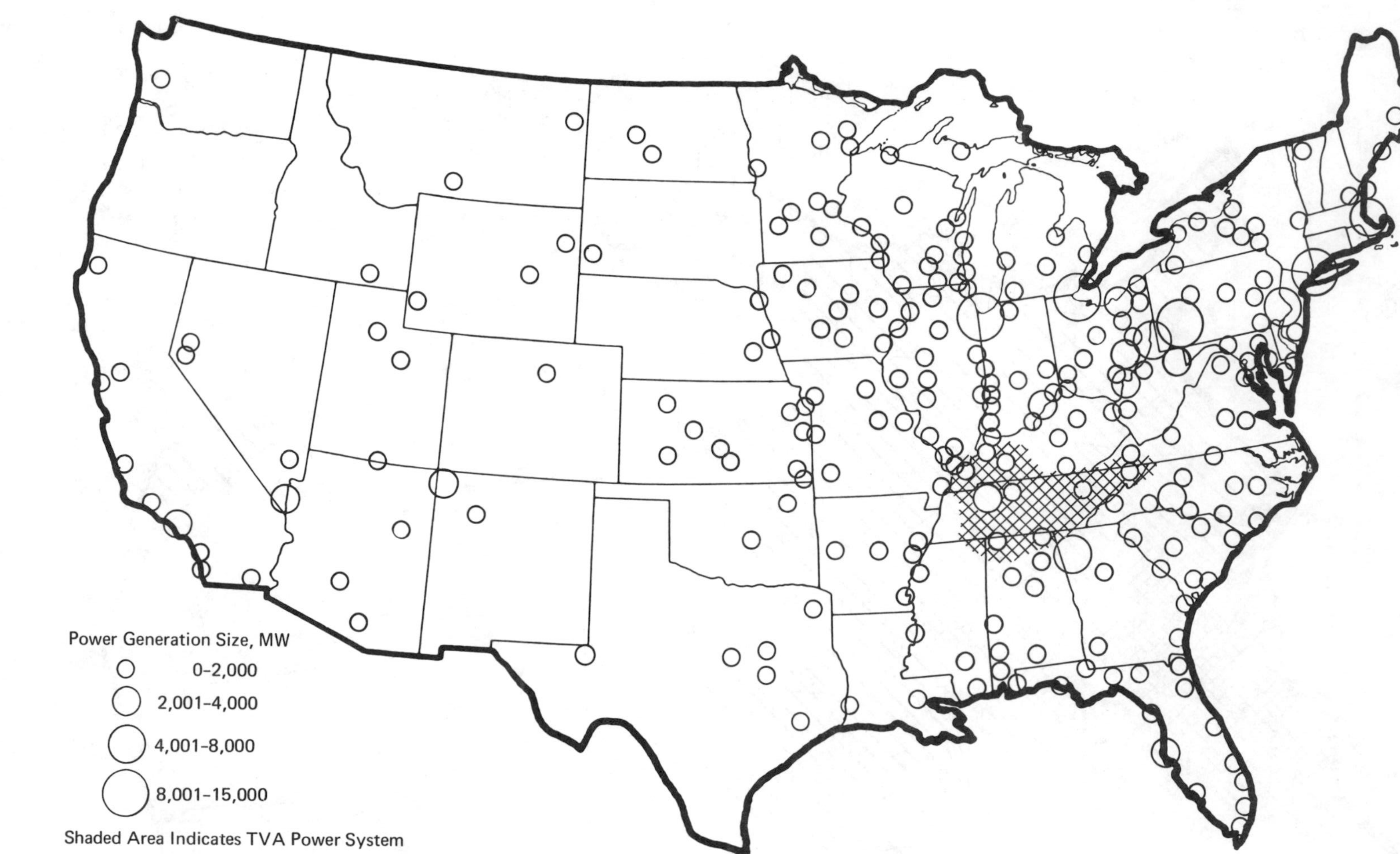

SOURCE: Finklea *et al.* (1974).

FIGURE 16: Location of Major Coal- and Oil-fired Power Units, 1971.

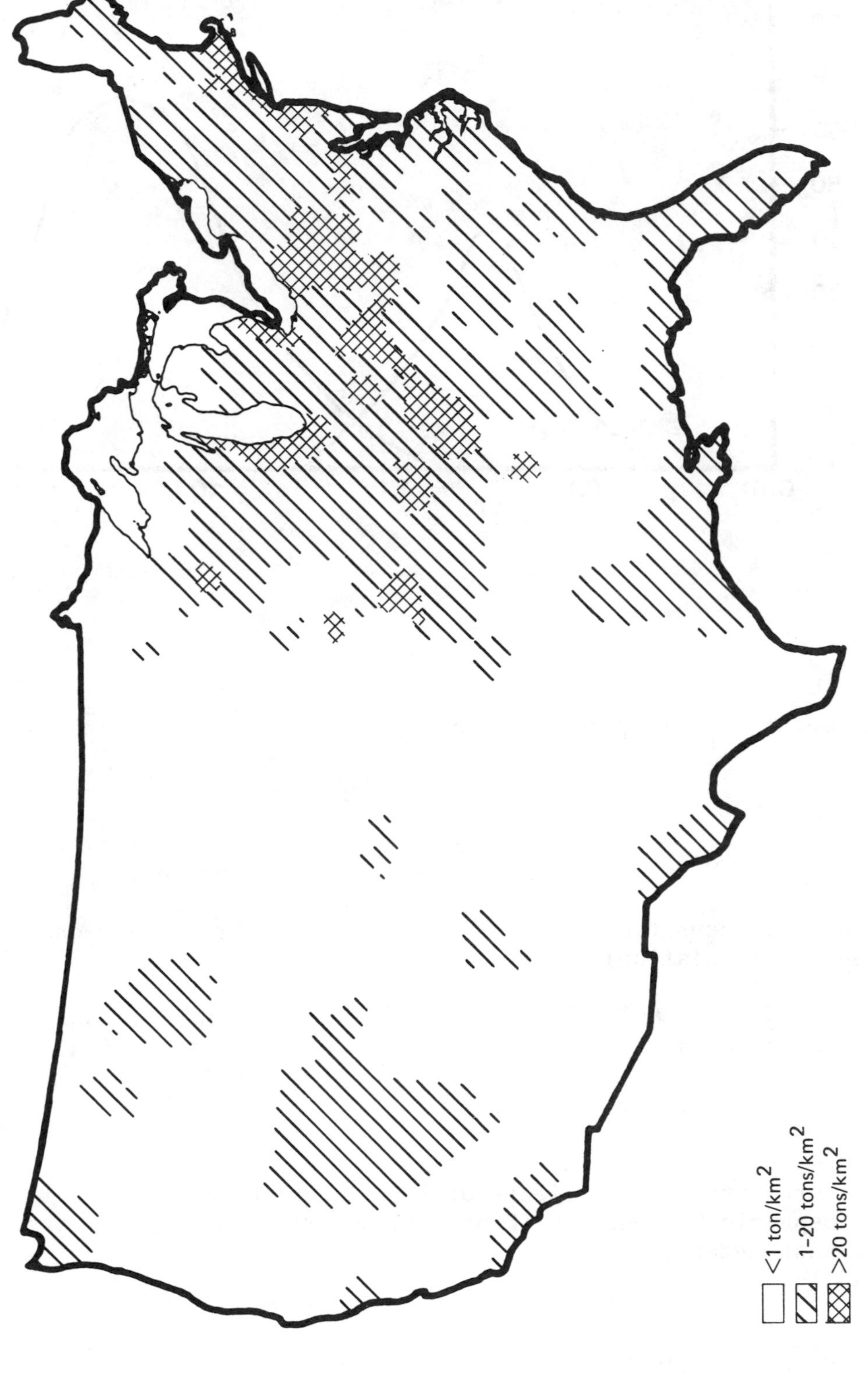

SOURCE: Finklea *et al.* (1974).

FIGURE 17: Nationwide Geographic Variation in SO_2 Emission Density.

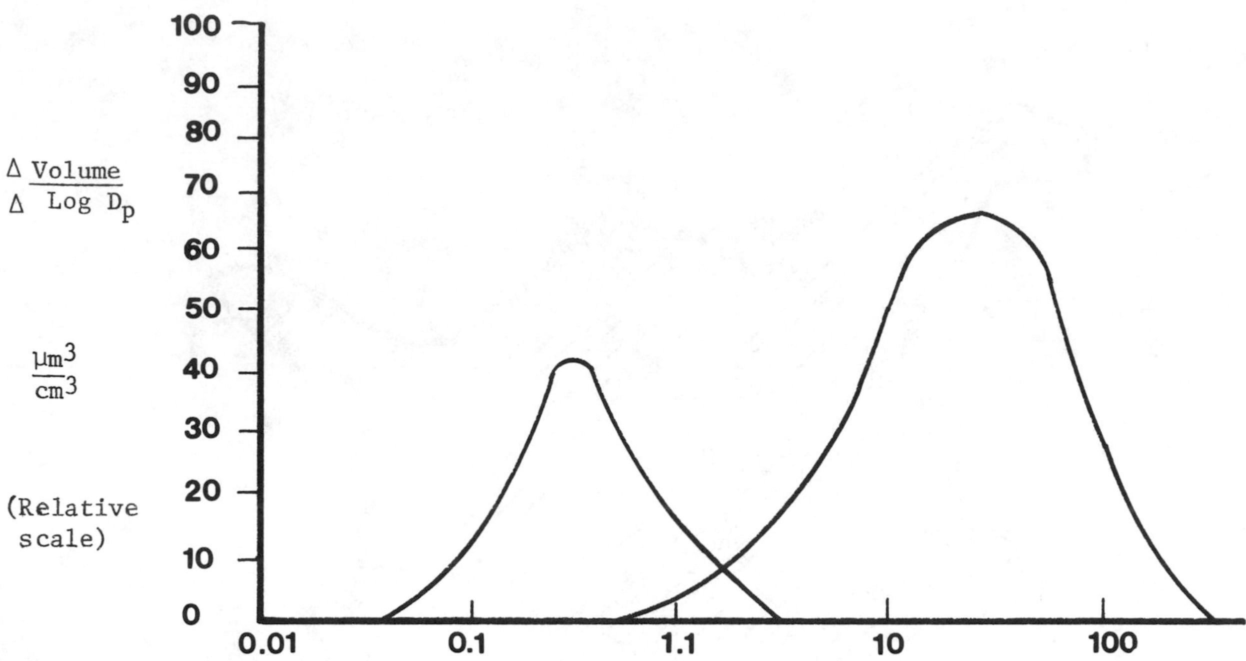

Particle diameter, D_p, micrometers (μm)

GENERAL AEROSOL PROPERTIES

0.01 - 2μm

1. Soluble
2. Produced by condensation (enlarge by coagulation)
3. Contain: $SO_4^=$, NO_3^-, NH_4^+ organic condensates, trace metals, (lead, arsenic, beryllium, cadmium, mercury, etc.)

2 - 100+μm

1. Insoluble
2. Produced mechanically
3. Contain: soil, tire dust, sea salt, fly ash, SiO_2, $CaCO_3$

Source: Measurements of the effects of Fossil Fuel Combustion on Atmospheric Composition (Professor Robert Charlson, University of Washington.)

FIGURE 18: Bimodal Distribution of Airborne Particles.

ACID RAIN

For the past two decades, scientific concern has grown steadily over the increasing acidity of the rain and snow in the Northern Hemisphere.* This concern has now spread to the public realm (Likens and Bormann, 1974; Time, 1974). Probably the first major study of "acid rain" was reported on behalf of the Swedish government before the United Nations Conference on the Human Environment. The report was entitled "Air Pollution Across National Boundaries: The Impact on the Environment of Sulfur in Air and Precipitation" (U.N., 1971). Its conclusion is the acidity of the rain in Sweden has increased substantially since the mid-1950s. The change is attributed to sulfuric acid formed from the oxidation of pollutant sulfur compounds. In the report the claim is made that about half of the sulfur deposited on Swedish soil comes from foreign sources, chiefly in central Europe and Great Britain. Munn and Rodhe (1971), in a separate study, report that sulfur deposition from precipitation in Sweden is increasing at a rate of 2-3 percent per year. This increase is roughly parallel to the rise in emissions of anthropogenic sulfur.

While long-term data on the pH of rain in this country are scarce, there is evidence to suggest that a similar trend is underway. Certainly the rainfall in large areas of the United States, especially in New England, is now strongly acidic (Figure 19).** The annual, weighted average pH of rain falling in the Hubbard Brook Experimental Forest in New Hampshire ranged from 4.03 to 4.19 during the years 1965-1971 (Likens et al., 1972), while the lowest individual pH was 3.0.# Measurements of the acidity of precipitation in the Finger Lakes region of New York State are comparable (Likens, 1972). Reports of similar findings in Massachusetts, Connecticut, Pennsylvania, West Virginia, and Maryland have appeared recently (Pearson and Fisher, 1971; Gordon, 1972a).

Clean rainwater is slightly acidic, having a pH ranging from 5.5 to 5.7. This acidity arises from the small amount of CO_2 that is in solution. SO_2 in

*Barret and Brodin, 1955; Carrol, 1962; Feth, 1967; Likens, 1972; Likens and Bormann, 1974; Oden, 1968; Wiegolaski, 1971.

**Figure 19 charts the acidity of all rain that fell in the U.S. during the two-week interval from March 15 to March 31, 1973. It shows that heavy pollution acidifies substantially all rain that falls east of the Mississippi. More important, it dramatizes the kind and quality of data that can be gathered simultaneously throughout vast areas by enlisting the enthusiastic cooperation of grammar school children (16,000 of them in this case).

#pH is the symbol for the logarithm of the reciprocal of the hydrogen ion concentration $[H^+]$ of a solution. A neutral solution has a pH of 7.0. When $[H^+]$ increases, the solution becomes acid and pH falls.

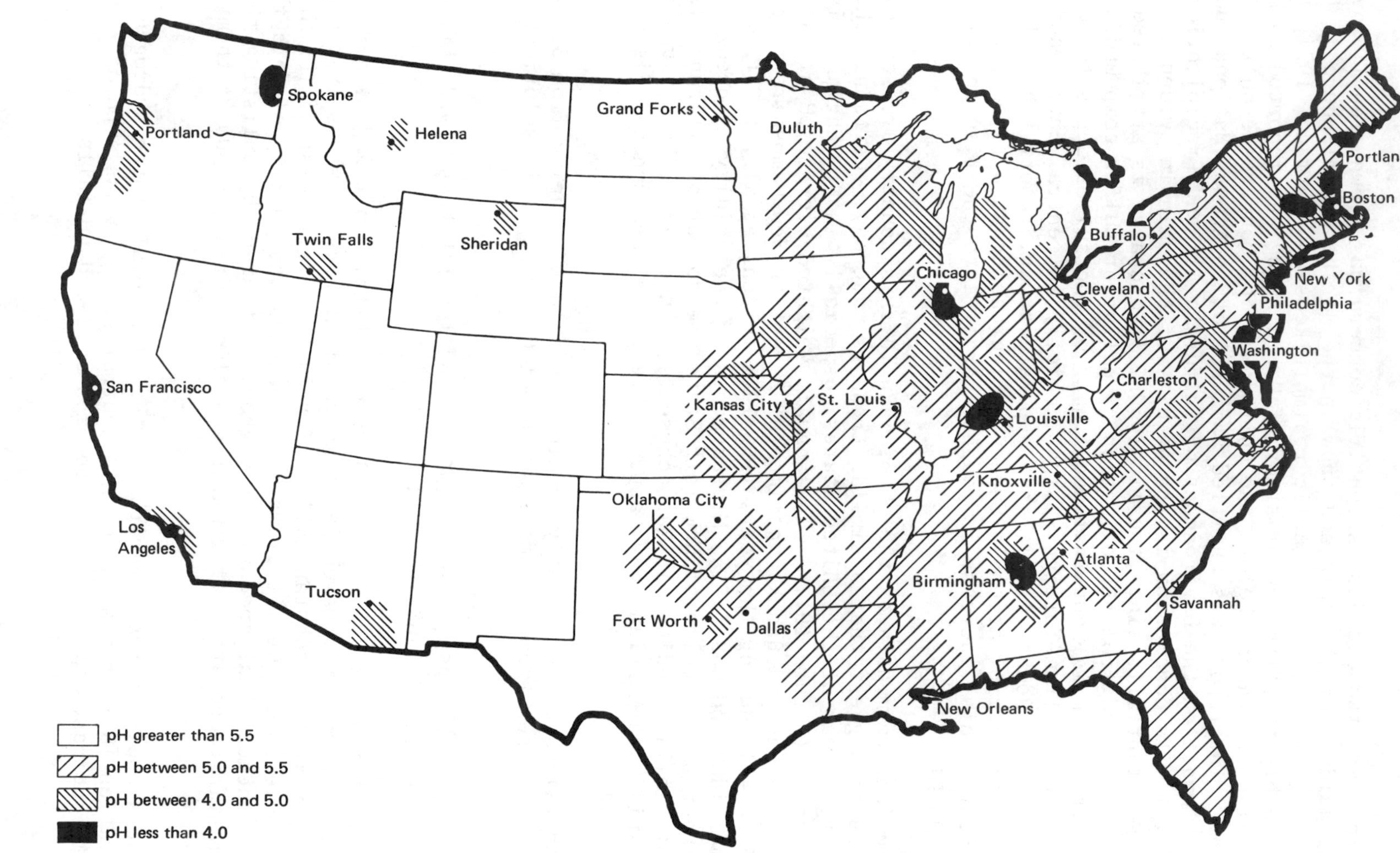

SOURCE: *Scientific American:* The amateur scientist: *230*:122–127 (June 1974).

FIGURE 19: Acidity of Rainfall in the U.S.

the atmosphere dissolves to form sulfurous acid (H_2SO_3) which dissociates in two stages to produce hydrogen ions:

$$SO_2 + H_2O \leftrightharpoons H_2SO_3$$

$$H_2SO_3 \leftrightharpoons H^+ + HSO_3^- \leftrightharpoons 2H^+ + SO_3^{=}$$

The process contributes to the hydrogen ion concentration of rainwater and to acidification. Sulfuric acid is highly soluble in water and dissociates almost completely in forming hydrogen ions. It has been hypothesized that the pH of rainwater near sources of pollution is controlled by the ratio of SO_2 to H_2SO_3, but at greater distances is controlled principally by H_2SO_4. The oxides of nitrogen and hydrofluoric acid are among the other pollutants which may acidify rain.

The environmental and ecological effects of acid rain are coming under increasing study. There is evidence that the low pH may be harmful to vegetation, aquatic communities, and possibly to soils. Over the past twenty years, reductions in forest growth have been found in both Sweden and northern New England in association with the increasing acidity of rain (U.N., 1971; Whittaker et al. in press). A corresponding decrease in the pH of tree bark has been reported in southern Poland and Scandinavia (Grodzinska, 1971; Staxang, 1969). Species of pine and spruce growing in the vicinity of a coal-fired electrical power station may exhibit a variety of abnormalities, including randomly dwarfed needles, excessive dwarfing of needles and shoots, and reductions in lateral bud formation, all of which have been attributed to acid fall-out (Gordon, 1972a and b; EPA, 1971). In laboratory experiments, it has been possible to reproduce most of these defects with inoculations of sulfuric, nitric and hydrofluoric acids (Gordon, 1972a and b).

Simulated acid rain, (i.e., H_2SO_4, pH 1.5-3.5) damages some species of flowering plants (Ferenbaugh, 1974), and impedes the development of leaf tissue on yellow birch seedlings (Wood and Bormann, in press). The growth of timothy, an herbaceous plant, is reduced by inoculation with H_2SO_4at a pH of 2.7 to 4.0.

Acid rain may also harm aquatic ecosystems, particularly where the ratio of direct rainfall to land drainage is high (Likens and Bormann, 1974). In Scandinavia and Canada, numerous lakes and rivers have shown changes in pH which in some instances reflected about a hundred-fold increase in hydrogen ion concentration (Oden and Ahl, 1970; Beamish and Harvey, 1972; Beamish, 1974). Once the pH of a body of water has fallen below 5, serious damage to fish is likely to follow (Brock, 1974). The reduction in lake trout and other species of fish in Lumsden Lake, located near a large smelting complex at Sudbury, Ontario, is attributed to the increasing acidity of the rain and snow (Beamish and Harvey, 1972; Beamish, 1974). Of 150 other lakes in the same vicinity as Lumsden Lake, more than one-fifth had pH values less than 4.5; they were termed "critically acidic." In some Swedish lakes, trout, roach, arctic char, and perch have been eliminated as the waters gradually became acidified. Below a pH of 4, the natural life-cycle of fish may be interrupted and the diversity of species is greatly reduced. Acidification of surface waters also threatens the micro-flora and fauna (Almer et al., 1974; Brock, 1973). Green algae, diatom algae, and daphnid zooplankton have been eliminated following the acidification of lakes in Sweden.

Acid rain and snow may leach out critical ions from the soil. In areas of calcareous rocks and soils, the acid is likely to be neutralized rapidly, but podzol soils which are already acidic lack this capacity for neutralization. Overrein (1972) showed in a Norwegian study that acid precipitation could remove considerable quantities of calcium, an essential plant nutrient, from the soil. Theoretically, acid rain might be expected to accelerate the weathering of rocks. It is known that the major mechanism for the chemical decomposition of rockforming minerals is through solution in acids. The Hubbard Brook ecosystem study, however, did not find any increase in the rate of weathering associated with excess acid (Johnson et al., 1972).

Recommendation

To establish a national network for continuous monitoring of pH levels (acidity) of rainfall which might be incorporated into the air sampling network suggested previously.

PUBLIC HEALTH

The following are summary remarks on the relation between air pollution and ill health. A number of comprehensive reviews of the subject either have recently appeared or are being compiled.* The obstinate problems of what constitutes acceptable air quality standards and whether there are threshold levels below which no harmful effects are experienced will not be considered.

Epidemiology

Just as the information on air sampling is ambiguous, so are much of the data on the biological effects. We do not know which pollutant or combination of pollutants is injurious to health.** While SO_2, total suspended particulates and, more recently, "sulfates," are relied on as indices of

---

*Rall, 1973; Committee on Public Works, 1973; EPA, 1974; NAS/NAE, 1974c; NERC, 1974; Finklea et al., 1974; Higgins and Ferris in preparation.

**This assertion does not apply to industrial and occupational settings which may generate elements having specific toxicity.

pollution,* the exact relation between these elements and illness is uncertain. It is conceivable that other pollutants not measured routinely may play a role. Still, it is worth emphasizing that most of the evidence imputing adverse effects on health to air pollution has emerged from communities in which SO_2 and airborne particles were found to be prominent.

Another factor contributing to the difficulty of assessment is that air pollution is only one of many stresses to which urban populations are exposed. Others that may impinge on health are crowding, poor hygiene, malnutrition and, particularly, cigarette smoking. Pollution probably interacts with and is reinforced by these additional stresses. But to isolate the role of pollution is an elusive task requiring extreme care in the design and execution of epidemiological research.

Despite such caveats, there remains a growing body of evidence to show that as atmospheric levels of the oxides of sulfur rise, the risk to health increases. This evidence comes from other industrialized nations as well as the United States. At present we have ambient air standards for SO_2 and total suspended particulates (without reference to the size of the particles), but not for "sulfates." Yet there are data to suggest that levels of sulfate particles falling well below the standard for total suspended particles are associated with rising rates of illness (EPA, 1974; Finklea et al., 1974). It is reasonable to conclude that further increases in the ambient concentration and the geographic distribution of these pollutants through the relaxation of standards of emissions and air quality would pose an unacceptable risk for the population.

Certain segments of the population appear to be particularly susceptible to the effects of pollution. Most prominent among these are the aged, persons with underlying cardiovascular or pulmonary disease (for example, bronchitis and asthma), and perhaps most ominously, infants and children.**

---

*A simplifying assumption typical of such studies is that the aerometric data from one or even a few sampling sites are representative of the pollution to which all of the subjects have been exposed. But differences in levels of activity and in the amounts of time spent out-of-doors may influence both the type (chemical) and magnitude of exposure within the same community. Usually the assumption is dictated by technical and financial constraints.

**(EPA, 1974; Douglas and Waller, 1966; Lunn, 1967; Lebowitz et al., 1973) Children offer a distinctive advantage as subjects for study since they are not likely to smoke or work in potentially hazardous industrial environments. The reason for their apparent vulnerability to air pollution is uncertain. It may reflect some form of biological immaturity, or a greater degree of physical activity out-of-doors compared with adults. Laboratory experiments show that the effects of pollutants are magnified by exercise.

Perhaps the most dramatic consequences of air pollution are seen in acute episodes kindled by changes in weather, especially inversions. They are manifested by excessive rates of mortality or morbidity. An example of an association that was found between fluctuating changes in SO_2 and mortality rates in the New York-New Jersey metropolitan area is shown in Figure 20 (Buechley et al., 1973).

By comparison, it has been difficult to assess the adverse effects of protracted exposure to low levels of pollution, and particularly to determine if such exposure alone might induce disease. Some of the most compelling evidence to suggest that chronic exposure has a cumulative effect has been adduced in the young. Two examples can be cited: A study was performed in Britain on over three thousand children born during a single week in March 1946 and followed thereafter for about 15 years (Douglas and Waller, 1966). Their respiratory illnesses were related to levels of pollution estimated from the amounts of coal consumed by the local communities. The investigators found a significant correlation between these estimated levels of pollution and the occurrence of lower respiratory infections (Table 1, 2).* The second example is from a study conducted by the Environmental Protection Agency (1974) as part of its Community Health and Environmental Surveillance System School (CHESS). These subjects ranged in age from infancy to 12 years. The children living three or more years in communities with relatively high levels of oxides of sulfur and suspended particles had more lower respiratory illnesses than did their peers who came from similar socio-economic backgrounds but lived in cleaner environments. Significantly, this association between illness and pollution was not evident before the third year (Tables 3, 4), suggesting identification of a gradual, cumulative effect from air pollution.

Experimental Toxicology

Laboratory evidence supports the thesis that submicronic and sulfate particles, some of which have been identified in urban atmospheres, can be irritating to the low airways (Amdur, 1973; McJilton et al., 1973). In realistic concentrations SO_2 alone appears to be of little consequence; it is so soluble in tissue fluids that little penetrates beyond the upper airways. But if SO_2 is combined with an aerosol that may itself be physiologically inert, an interaction between the two agents can occur which causes an exaggerated irritant response (Amdur, 1973). This exaggerated response resulting from the combination of agents is termed synergism. Recent evidence (McJilton et al., 1973) suggests that synergism is enhanced once the submicronic aerosol is converted in the atmosphere from a dry crystalline state to a droplet (accomplished by raising the relative humidity of the reaction chamber). The SO_2 dissolves in the droplet producing an acid, irritating aerosol. The result may be interpreted to suggest that either

*The lower airways include the larynx, trachea, bronchi and most peripherally, the bronchioles; and the upper airways, the nasopharynx and oropharynx.

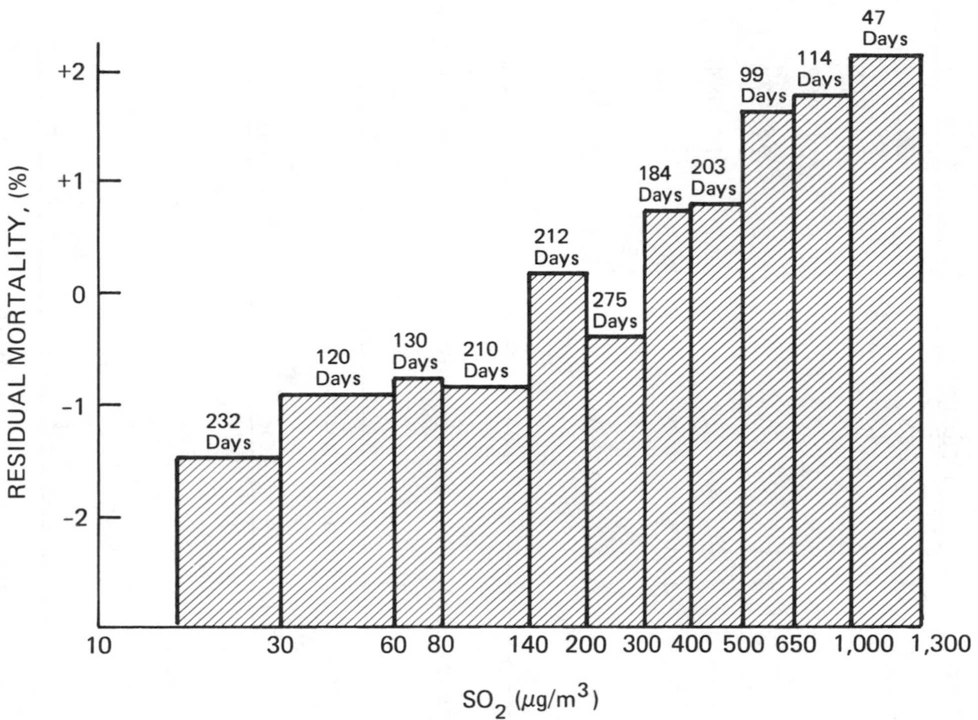

Source: Buechley et al., (1973) Arch. Environ. Health, 27, p. 137, copyright 1973, American Medical Association

FIGURE 20: Mean Residual Mortality by SO_2 Classification New Jersey Metropolitan Area 1962-1966.

TABLE 1 Lower Respiratory Tract Infections (Percent) in Infancy by Category of Air Pollution

Lower Respiratory Tract Infections		Air Pollution Group A Very low	B Low	C Moderate	D High	Significance X^2(m=3)	P
First attack in first 9 months		7.2	11.4	16.5	17.1	35.342	<0.001
At least one attack in first 2 years		19.4	24.2	30.0	34.1	39.800	<0.001
More than one attack in first 2 years		4.3	7.9	11.2	12.9	31.887	<0.001
More than one attack in first 2 years	Boys	5.7	8.1	10.9	16.2	21.574	<0.001
	Girls	2.9	7.7	12.1	9.7	18.884	<0.001
More than one attack in first 2 years	Middle Class	3.0	4.0	7.7	9.3	12.201	0.01-0.00
	Manual Working Class	5.1	10.8	13.9	15.4	22.092	<0.001

TABLE 2 Hospital Admissions (Percent) for Respiratory Tract Infections During the First Five Years of Life by Category of Air Pollution

Reason for Hospital Admission	Air Pollution Group A Very Low	B Low	C Moderate	D High
Acute upper respiratory infections	0.4	0.3	0.4	1.1
Tonsillitis and tonsillectomy	4.4	6.2	5.7	5.2
Lower respiratory infections*	1.1	2.3	2.6	3.1
(a) Bronchitis	nil	0.9	1.0	1.4
(b) Pneumonia and bronchopneumonia	1.1	1.4	1.6	1.8

* Comparing children in area "A" with those in areas "B", "C", and "D" X^2=4.06; =1;η0.05>P>0.02.

TABLE 3 Age-Sex-Adjusted 3-Year Attack Rates for any Lower Respiratory Illness by Residence Duration and Number of Episodes

Community	Number of illness episodes	Attack rate, percent		
		<3 year residence	≥ 3 year residence	Any length residence
Low	≥ 1	25.6	27.3	27.1
Intermediate I		27.8	26.5	27.0
Intermediate II		27.5	29.0	28.7
High		22.9	38.2	36.1
Low	≥ 2	19.2	15.2	15.7
Intermediate I		14.6	14.6	14.6
Intermediate II		14.2	17.2	16.9
High		12.3	23.4	21.9

Source: EPA, (1974)

Legend: The number and types of respiratory illness were determined through questionnaires filled out by the parents. The sampling data used to establish "low," "intermediate," and "high" levels of pollution appear in Table 4.

TABLE 4 Projected Air Pollutant Exposures in Four Utah Communities, 1959-1971

Pollutant and community	Concentration μg/m³ 1959-1962	1963-1966	1967-1970	1971
Sulfur dioxide				
Low	c	c	c	8
Intermediate I	27	16	18	15
Intermediate II	32	33	33	22
High	94	94	91	62
Total suspended particulate				
Low	c	109	92	78
Intermediate I	133	114	82	81
Intermediate II	<50	<50	<50	45
High	51	59	62	66
Suspended sulfates[a]				
Low	c	6	4	6
Intermediate I	8	8	5	7
Intermediate II	9	9	9	8
High	15	15	15	12
Suspended nitrates[b]				
Low	---	---	---	2.7
Intermediate I	---	---	---	3.3
Intermediate II	---	---	---	2.4
High	---	---	---	2.0

[a] Data given for 1959 to 1970 are estimated.

[b] Only fragmentary data are available for suspended nitrates from 1959 to 1970.

[c] Not available.

Source: EPA, (1974) p.2

Legend: The monthly average of SO_2, suspended particles, sulfates, and nitrates were measured during one year only. The values for the remaining year were based on estimates of emissions and dispersion.

the transformation of gaseous SO_2 to an acid sulfate particle (see Sulfur emissions, atmospheric reactions), or the mixing within the atmosphere of soluble gases such as SO_2 and droplet aerosols may be expected to increase the toxic potential of pollution.

Costs

Owing to uncertainties which have been discussed, it is difficult to estimate the total costs of illness and death arising from coal-related pollution. Yet such attempts serve a useful purpose. Even though the final figures are arguable, they do call attention to the economic burdens we must accept in not controlling pollution. Society pays these costs whether or not they can be readily categorized. It has been estimated that a 50 percent reduction in all pollution would produce a savings, in terms of reduced medical expenses and recovered income, of at least two billion dollars annually (Table 5) (Lave and Seskin, 1970). The health costs related to the pollution from a large coal-fired plant located near a metropolis may be as high as the cost of fuel required for generating the electricity (Appendix: Lave, L.B.: Short term health considerations regarding the choice of coal as a fuel).

TABLE 5 Estimated benefits in health and annual savings (billions of dollars) from 50% reduction in air pollution in major urban centers

Health Benefit	Saving/Year
1. 25% reduction of all morbidity and mortality due to respiratory disease	1.222
2. Reduced morbidity and mortality of all diseases, including extra-pulmonary	2.08

Source: Lave and Seskin (1970).

Recommendations

1. To give high priority to collaboration between the biological and physico-chemical scientists in health-related research on air pollution. The importance of such collaboration both in laboratory and epidemiological studies cannot be over-emphasized. Otherwise, considerable time and effort may continue to be wasted on studies of unrealistic design.

2. To encourage the formulation of ambient air standards that take cognizance of newly developing information about the physico-chemical properties of air pollution.

CONTROL STRATEGIES*

There is under development by the Federal Government a policy which encourages and may ultimately direct a doubling of the use of coal in the next decade. Such expanded use of coal will require the rapid implementation of strategies aimed at reducing total pollutants, and especially sulfur compounds. This action is imperative since even current levels of emission are strongly implicated in deleterious effects on man and other ecological elements.

The options available for controlling the total atmospheric loading of sulfur include: (1) reduction in demand for energy and increased efficiency of coal conversion; (2) use of low sulfur coals (less than 1 percent) to fill expanded demand as well as to substitute, at least in part, for high sulfur fuels currently in use; (3) employment of tall stacks together with an "Intermittent Control System" (ICS); and (4) removal of sulfur prior to, during or after combustion of the fuel. Option (1) has been reviewed by COMRATE Panels on Demand and on Technology in this report and will not be repeated in this section.

Option (2), the substitution of low sulfur coals as a major fuel for meeting the increased demands in electricity, would appear to be a practical solution that can be implemented in a relatively short time. Strong advocates are to be found for this option both inside and outside of government.

There are, however, a number of unsolved problems related to this option. The number of on-line boilers that can accommodate low sulfur western or eastern coal as either a total replacement or a mixing fuel is unknown. Boiler characteristics may prohibit the use of low sulfur coal without extensive retrofit.** For the new coal-electric generating facilities, the problem is one of basic design and is easily manageable.

With large reserves of low sulfur coal available in the East,# and also in the West where surface methods can be used, we are faced with the choice of where to mine. Western coals are less expensive to mine, the costs ranging from $1.25 to $2.50 per ton. Generally, western mining can be started or expanded at a smaller capital investment and with less delay than

---

*Reference for this section: Dunham et al., (1974).

**A change in design or construction of equipment to accommodate a new process.

#The quantity of eastern low-sulfur coal that is available for boiler fuel is not known, as it makes up the reserve for metallurgical coking coal.

in the East. However, western coal is typically lower in thermal value per unit weight and is far removed from load centers. To ship this energy by unit train, pipeline slurry or transmission wire would involve additional considerations: the capacity and efficiency of alternative modes of transport, the accessibility of land, water and other nonfuel resources that will be required, and the large social, aesthetic and ecological impacts that may be expected.

Deep-mined low-sulfur coal despite its high thermal value and proximity to markets, has important disadvantages. It requires large capital and higher operating costs, expensive safety systems, and exacts excessive human costs in terms of high rates of accidents and CWP. These problems have received increasing attention from responsible state and federal agencies. By contrast, the problems indigenous to the West, the rehabilitation of semi-arid land, the need for vast amounts of water in the mine-mouth conversion (electrical generation plant located at the coal mine) of coal to electricity, syngas (gas synthetically produced from coal) or syncrude, (crude oil synthetically produced from coal) and the social and economic impacts of extensive industrial development in settings that are predominantly rural, have yet to be addressed seriously on national or local levels. What may at first appear to be an easy, attractive solution of shifting the geography of energy production to supplies of coal that are more efficiently mined and less polluting, could ultimately have complex and more profound impacts than would maintaining the current distribution of mining.

Option (3), the recourse to a combination of tall stacks and an "Intermittent Control System" (ICS), has received strong support within industry and certain government agencies as a method for controlling ambient concentrations of SO_2. The method, however, would not curb total loading of the atmosphere by sulfur oxides although it could improve local and perhaps regional air quality during periods of decreased atmospheric dispersion. The quality of the air regionally may ultimately be threatened more by emissions from tall rather than short stacks. The ground may serve as an efficient sink for SO_2, thereby reducing the formation of sulfates aloft.

In normal climatic circumstances, the tall stack acts to lower ambient concentrations of SO_2 and suspended particles at ground-level. Nonetheless, it has other undesirable features; it ensures wider dispersion of the pollutants, a longer interval for the conversion of SO_2 to particles in the atmosphere and, following periods of inversion, leads to fumigations at ground level. The tall stack is an inadequate means of controlling pollution. It may be said to provide a lesser risk to nearby populations and a greater risk to populations of distant, wider areas.

If tall stacks are to be combined effectively with ICS, the following supportive measures will be needed: continuous field monitoring of the oxides of sulfur (as well as other toxic agents), continuous weather forecasting, and a capacity for rapidly reducing through-put during periods of inversion or stagnation. At present most localities cannot provide the monitoring and forecasting that are essential for early, secure decisions to curtail activities. As a consequence, those systems which are in operation in this country have been known to violate ambient air standards. It should be recognized that if air standards are enforced in conjunction with this system, periodic curtailment in the generation of electricity may be required. The result would be to increase the long-term cost of electric power to the consumer.

Option (4), the removal of sulfur prior to, during or after combustion, is a feasible alternative to control atmospheric loading. The pre-combustion removal of sulfur by a combination of mechanical and solution techniques needs further development if high sulfur coals are to be used safely. Adequate federal funding for research and development in this field should have high priority.

The scrubbing of stack gas for sulfur removal is at present an issue of national debate. We suggest that technology now available, if applied efficiently, could help to achieve clean air standards. The desulfurization systems include wet limestone or lime scrubbing, magnesium oxide scrubbing with regeneration, catalytic oxidation, wet sodium base scrubbing with regeneration, double-alkali sodium scrubbing and the citrate process. The limestone or lime wet scrubbing and the double-alkali sodium scrubbing systems yield an end-product that is virtually without value and which must be permanently impounded. However, the electrical power industry has a number of other possible systems of controlling emissions. In time, with competitive upgrading of these systems there could be a substantial lessening of our present reliance on costly low sulfur fuels for eastern and midwestern power plants.

The cost of flue gas desulfurization might reach 5 mills/kwh which would raise the price of electricity to the consumer by an average of 15-20 percent. This estimate is based on recent indications that flue gas desulfurization can cost from $45 to nearly $80 per kwh (Hesketh, 1974). Estimated operating costs for 500 megawatt and smaller facilities are in the range of 22.5-35 cents per million BTU. The smaller figure is equivalent to a charge of approximately $5 per ton added on to the cost of the coal burnt. Waste disposal costs for FGD systems could be as high as scrubbing system operating costs. Yet, in terms of the rapid rising costs of other forms of energy, the need for restricting the release of sulfur into the atmosphere, and the security provided by a vast, economically reasonable supply of domestic fuel, we believe that the benefits to be derived from FGD in terms of health and welfare will be so great that this option should receive priority consideration. (See Table 10 for dollar value alone of health benefits.)

Recommendations

1. That the Federal Government not approve the tall-stack intermittent control systems for dispersing emissions of sulfur oxides as a permanent solution. A few exceptions may be permitted in older plants in specific sites, but it should be understood that such exceptions are limited in time and are subject to reversal.

2. *That the Federal Government instead require the installation of pollution abatement equipment in both new and existing coal-fired plants, while accelerating financial support for research and development on clean fuels from existing sources.*

CONCLUSION

A century ago, the Victorian social critic John Ruskin (1819-1900) referred to the air pollution in cities like Birmingham and Manchester in an article auspiciously titled "Storm Cloud Over Europe" (1871). While Ruskin's specific intent was to show that the harmony between science and theology had been irrevocably broken, the tenor of the article reflected his concern that pollution had "blanched the Sun." He was calling attention to an undesirable ramification of the use of energy, and particularly of coal.

Ruskin's concern with the environment has greater validity today. Society should recognize that coal as a means for providing electrical power poses serious hazards to health and the environment. A realistic appraisal of these hazards becomes a convincing argument for a policy of "low-growth" in the use of coal. If we are forced to proceed beyond such a policy because of sharp reductions in the other conventional sources of energy, we should then expand the use of coal only with great caution, only where necessary, and only when accompanied by strict regulations and technological controls.

ANNEX
THE GLOBAL SULFUR CYCLE

The main reservoirs of sulfur on earth are listed in Table 1. They range widely in concentration from about 17 percent in evaporite deposits (primarily gypsum) to mere traces in continental ice and air. On a mass basis, the largest deposit of sulfur is in rocks, while seawater contains the most mobile form. With few exceptions, freshwater has less concentrated sulfur than seawater (Table 2). The variations in sulfur content among freshwaters are mostly attributable to the kinds of rock present in the drainage basin. The atmosphere, despite having a relatively low concentration of sulfur, is a primary conduit for the transfer of sulfur.

There are a number of estimates of the amount of sulfur entering the atmosphere as SO_2 as a consequence of human activity (Friend, 1973). In calculating the relative contributions of the major source of sulfur in the atmosphere, we have relied on the estimate of Kellogg et al., (1972) that the total human contribution of SO_2, expressed in terms of elemental sulfur, is 50×10^6 tons per year. These calculations are summarized in Table 3. Only the combustion of fossil fuels should add any net sulfur to the atmosphere. The overall process is cyclic, and inputs from other natural sources should be balanced by subtraction. The addition of sulfur from fossil fuels produces only a transient and geographically limited change in the sulfur budget which becomes counterbalanced by increased sedimentation of sulfur.

About 27 percent of the sulfur entering the atmosphere globally is estimated to arise from the burning of fossil fuels. Berner (1971) estimated that 27 percent of the sulfate in the world's rivers also arises from pollution. Man-made pollution as a source of atmospheric sulfur is exceeded only by biogenic hydrogen sulfide and organic sulfides. Hydrogen sulfide has been estimated only by algebraic difference, and hence is not a measurement with independent validity.

TABLE 1 Main Reservoirs of Sulfur

Source	Mass of sulfur in metric tons	Percent of sulfur content in each source
Sea	1.3×10^{15}	0.09
Freshwater	3×10^{9}	0.0011
Ice	6×10^{9}	0.00003
Atmosphere	3.6×10^{6}	0.00001
Evaporites	5×10^{15}	17
Sediments (mainly shales)	2.7×10^{15}	0.19
Metamorphic and igneous rock	7×10^{15}	0.03

Source: Holser and Kaplan (1966). The values listed are approximate; for ranges and standard deviations, see the original paper. For other values, see Friend (1973).

TABLE 2 Sulfate and Chloride Analyses of Natural Waters in Parts Per Million

Water Source	$SO_4^{=}$	Cl^-	$SO_4^{=}/Cl^-$	Remarks
Lake Temiskaming, Ontario	8.9	1.5	5.9	Precambrian shield
Lake Superior, Ontario/U.S.	4.8	1.5	3.2	Precambrian shield
St. Lawrence River, Ontario	21.5	15.7	1.37	Drains Great Lakes
Ohio River	69	19	3.6	Sedimentary basin
Rio Grande, Colorado	84	10	8.4	Arid sedimentary
Pecos River, N. Mex.	1,020	90	11.3	Arid semi-sedimentary
Colorado River, Ariz.	289	113	2.56	Arid sedimentary
Salton Sea, Calif.	4,139	9,033	0.46	Colorado River drain
Devil's Lake, N. Dak.	3,460	787	4.4	Closed basin lake
Basque Lake, Brit. Col.	195,710	1,690	109.9	Closed basin lake
Rainwater	0.58	3.57	0.16	World average (Jacobs 1937)
Riverwater	11.2	7.8	1.44	World average (Livingstone 1963)
Seawater	2,650	19,000	0.14	Average seawater (Goldberg 1957)

Source: Livingstone (1963).

TABLE 3 Components of the Global Sulfur Cycle
(Values listed are in units of million (10^6) tons per year, calculated as sulfur)

Component	Land	Sea	Total	Percent of Total
Sources to atmosphere:				
Wind-blown sea salt ($SO_4^=$)	-----	43	43	24
Biogenic H_2S, organic sulfides	-----	---	88	48
Burning fossil fuels, Smelting (SO_2)	50	---	50	27
Volcanoes, etc. (H_2S, SO_2, S°, $SO_4^=$)	0.66	---	0.66	0.36
Total			182	
Sinks from atmosphere:				
Rain over the oceans (SO_2, $SO_4^=$)	-----	72	72	39
Rain over the land (SO_2, $SO_4^=$)	85	---	85	47
Plant uptake and dry deposition (SO_2)	15	---	15	8
Plant uptake ($SO_4^=$)	10	---	10	5
Total			182	

Source: Data of Kellog et al. (1972), recalculated in terms of amounts of sulfur.
SO_2 = 50% sulfur, H_2S = 94% sulfur, $SO_4^=$ = 33% sulfur.
The agreement between sources and sinks is artificial since the quantity or biogenic H_2S was calculated in order to make the values agree.
For a comparison of other estimates, see Friend (1973).

SECTION IV

DEMAND FOR FUEL AND MINERAL RESOURCES

CHAPTER XI
INTRODUCTION TO SECTION IV

The immediate importance of mineral resources to society is in their relationship to energy--firstly in that the fossil fuel minerals and, to a lesser but growing extent, uranium provide most energy. Secondly, but less directly, all mineral extraction and use (including the fuel minerals themselves) involves major energy inputs. Section IV, therefore, examines demand for mineral resources in its relationship to demand for energy as a whole, rather than giving separate consideration to each mineral.

Policy decisions now being made to increase exploitation of these resources are of overwhelming importance for the future since present use of nonrenewable resources precludes possible future use. Energy cannot be recycled. Because commitments are already being made, it becomes a matter of urgency to determine that the supposed needs for increased exploitation are based on valid data and assumptions.

Two major conclusions of the demand panel are firstly, that the basis of demand projections which are influencing national energy research and development policy is questionable, and secondly that the potential for policy for influencing demand has been neglected relative to that for influencing supply.* Aspects of the demand question lending themselves to constructive action therefore fell naturally into two categories: 1) <u>integration and analysis of information</u> on all interdependent aspects of the materials question which relate to demand and which form the basis of demand projection--demand itself, supply, environmental impacts, technical questions, and political and social considerations, both national and international. A

---

* Evidence for this latter tendency was provided by what the Ford Foundation report (1974) called the "alarmist overreaction" illustrated in President Nixon's "Project Independence" speech of January 23, 1974 (Nixon, 1974), and the supply-oriented policies arising therefrom. It is to these policies, primarily based on the Dixy Lee Ray Report: The Nation's Energy Future (Dixy Lee Ray, 1973), that the panel refers when commenting on "current national policy" or "Project Independence." (For a discussion of the distinction between these policies and the contents of the Project Independence Report (FEA, 1974) which appeared in the final editing stages of this report, see page 269.)

corollary to this would be improvement in the actual techniques of demand forecasting; 2) the potential for influencing demand on the basis of such information.

1. INTEGRATION AND ANALYSIS OF INFORMATION

Projection of demand forms the basis for policy related to supply. The two cannot be considered independently of one another. A first-base proposition, therefore, is that estimation procedures for demand projection be examined as rigorously as those for supply. Where the information base is found wanting, every effort should be made to improve its sophistication and its usefulness to the policy makers.

The demand panel, like the supply panel, found the methods of estimation of demand inadequate in many respects. As with estimates of supply, the data are often inadequate, some of the underlying assumptions untested and the methodology as yet unsophisticated. The panel commissioned and analyzed a detailed Bibliographic Review (Overly, Schell, 1973, See Appendix) of current demand projections for energy minerals. Chapter XII examines a number of the most influential of these projections, and identifies cumulative biases which appear to make the projections too high. The danger of such a bias is that it stimulates efforts to increase supply to such an extent that the demand projections themselves become self-fulfilling.

In the panel's opinion, therefore, a first step in policy directed toward demand must be to establish efficient machinery for the integration and analysis of information on which demand projections are based, and to attempt to improve the sophistication of forecasting techniques.

Centralization of Information

A preliminary move in the formulation of policy for addressing this issue is centralization of information, on a national and international basis. Henry Kissinger proposed in his speech to the U.N. in April, 1974, that:

> "an international group of experts work with the U.N. in surveying resources and developing an early warning system for scarcities and surpluses."

His call for international cooperation stressed the interdependence of developed and developing nations, and the need to shore up confidence in the credibility of U.S. free trade principles. Priority consideration in formulating policy must be given to the repercussions on the world monetary system, the world food situation and on the poorer countries in particular, of an alarmist U.S. policy of "splendid isolation" in resources. From a national standpoint alone the potential for backlash is evident.

The environment too is a global as well as national problem. The global environmental costs of materials policy aimed at increasing supply must be factored in to the information based on which demand projections and policy relating to demand rest.

On a national basis, there is need within the United States to overhaul the federal machinery for making and using projections of demand and supply of resources. A report to Congress by the General Accounting Office (GAO, 1974) endorses a conclusion of the National Commission on Materials Policy that:

> "Almost every aspect of policy work in this area is handicapped by inadequate, inaccurate, or inaccessible information." (NCMP, 1973).

and stresses that essential data are unavailable to government except as the industry is willing to provide it. Further, the available data are processed through unsophisticated projective techniques which fail to employ modern statistical methods and, more importantly, neglect considerations which are difficult to incorporate into numerical models. The GAO report places high priority on an effort to improve and centralize data because the government:

> "already deeply involved in resource management...needs proper information on which to base its actions." (GAO, 1974).

The GAO report deplores the lack of a centralized government mechanism to coordinate policy planning. The report calls for the establishment of an interdepartmental commodities committee with monitoring, analysis, and forecasting responsibilities, to permit more accurate projections of domestic and foreign supply and demand trends. A central agency with these functions should be concerned to provide a forum for domestic interest groups through public hearings.

Enthusiasm in Washington for such an agency is tempered by fear of too much government interference in the free enterprise system* but, as the GAO (1974) and Ford Foundation (1974) reports point out, the government is already

---

*George P. Schultz, (then Secretary of the Treasury) in a March 27 letter to the GAO, said adoption of such a system "would constitute a fundamental change in the economic philosophy of this nation," and would imply that it was more desirable for the government to make essential decisions than to leave them to a "free, competitive and open market." In the Dupont Context, No. 2, 1974, (Madden, 1974), John F. Lawrence writes that there is fear that too much government control could harm the country's energy program, and that there is little enthusiasm in Washington at present for creating a Federal Corporation to take over the oil industry functions. Dr. Carl H. Madden, chief economist for the U.S. Chamber of Commerce, suggests in the same issue that a number of new institutions should be established for the purpose by multinational corporations.

heavily involved; what is needed is an improved and coordinated data base for activities and programs already being used to support the effective operation of the market system. Cogent arguments in support of centralized planning are provided by Harvard Professor of Economics Wassily Leontief in a recent article stressing the need for "a well-staffed, well-informed and intelligently-guided planning board." (Leontief, 1974).

It is plain, whichever view we take, that this issue warrants immediate consideration by the legislature.

Improvement of Forecasting Techniques and Use

Because of the long lead times involved in mineral and fuel exploitation operations, it will always be necessary to project future demand. However, the intelligent use of projections for policy cannot be based on weakly founded long-range projections which try to foretell an immutable future. Rather, policy needs short, medium, and long-range conditional forecasts which are addressed to the assessment of the direct and indirect consequences of diverse policy options. These forecasts must be continuously revised in the light of new information about changing circumstances and the intended and unintended consequences of policy. Chapter XVII discusses the state-of-the-art in demand forecasting, and gives some suggestions on ways it might be improved.

2. POTENTIAL FOR INFLUENCING DEMAND

Of equal and parallel importance with its call for improvement of information and forecasting techniques is the panel's conclusion that policy for influencing demand has been neglected relative to that for influencing supply. Even given accurate projections showing a large gap between supply and demand, there is no empirical basis for supposing that the projected figures for demand are any more recalcitrant to policy influence than those for supply. Factors such as rising per capita income, more equal distribution of income, increased conversion to electricity, and the need to use lower grade ores may augment energy use. There are, however, important offsetting factors likely to reduce energy consumption that current demand projections have not sufficiently taken into account. Some of these relate to the effect of supply constraints on reducing energy waste and increasing efficiency of use, and others result from broad social and especially economic and price trends leading to less energy-intensive production and consumption patterns. Those factors tending to increase, as well as those tending to decrease consumption would appear to be amenable to a national policy aimed at reducing demand which sought to modify the former while enhancing the effects of the latter.

In sum, by assuming that current demand forecasts are not only accurate but immutable targets, policy makers emphasizing increased supply have allowed themselves very little scope for decision making. One aim of our discussion is to stimulate enquiry into courses of action alternative to this very limited policy.

The emphasis of policy aimed to limit demand, as determined by the three major policymaking groups (government, industry, individuals), to be effective must be keyed to voluntary, self-interest motivated restraints, rather than being based on potentially unpopular fiats.

The survey results described in Chapter XV indicate the public's lack of trust that other persons will voluntarily restrict their consumption without regulations. It is therefore necessary to assure the public that in the event of shortages some mechanism will be devised to provide equitable allocation. If such safeguards were offered, there is evidence to suggest that people will be prepared to make a conservational effort. In order to foster this receptive climate of opinion, it is necessary to publicize and encourage voluntary restraint in consumption and to "sell" the alternate life-styles: for example, substituting simple near-home pleasures for far-flung resource consuming recreational activities. That such goals are not Utopian, is proved by the overall public response of the American public to the Winter 1973-1974 petroleum shortage, wherein energy usage for heating, lighting and cars was effectively diminished.

The attempt to curtail consumption will meet resistance. Obviously there will be strenuous opposition by interest groups whose sales will be affected. Owners of marinas and flying fields will clearly fight propaganda inimical to growth in the use of their nonproductive energy consumers, as will manufacturers of large eight cylinder cars, <u>unless</u> each can reorient profitably to less wasteful modes of earning a living. Such transitions will take time and should be encouraged by incentives in the interim; so that the individuals managing the businesses involved and the owners are personally motivated to make the changes to less wasteful modes.

In short, it will be necessary to develop policy to foster basic long range changes that encourage conservation and efficient use of resources. At the same time it must be recognized that a sizeable fraction of the American public is now on a sub-standard level of subsistence which must be improved, with consequent dedication of resources.

We list below some specific areas susceptible to policy influence by government, industry, and individuals.

Externalities

There is an increasing awareness of hidden costs which could be added to the producers' costs (rather than being borne by the society as a whole) meaning that they would be reflected in the price paid by the users. (Some examples are given in Chapter XIV). Institutional means should be found to transfer a major accounting of costs to the users, so that demand will adjust to the reality of what it costs to supply. The two major items here are the transfer to the particular commodity user of tax and other policy costs that artificially maintain low energy/materials prices and the transfer of many environmental costs directly to the materials/energy sector so that these may also be reflected in pricing.

Changes in Taste

Changes in tastes are a potentially powerful determinant of the rate of energy use, and there are many today who call for a return to a simpler and less energy-consuming life. It is possible that advertising, which may have contributed to certain forms of excessive energy consumption in present life-styles may be employed to achieve the reverse effect by reinforcing trends toward simpler life styles and conservation of energy materials.

Government Leadership

Although the market place is the preferred mechanism for allocation of resources, and direct government control in the form of price regulation, rationing, etc. is to be avoided as far as possible, the government has had and still has a role to play in influencing market trends.

Government at several levels (but with national uniformity) should set and enforce norms and standards in energy utilization including design standards for public buildings and federally insured housing, controls on the size and maintenance of motor vehicles, restriction in the use of feedstocks for certain types of consumer items, regulation of the intensity of use of fertilizers in agriculture, national speed limits, and so forth, to the extreme of rationing of energy. The government should also employ tax incentives, use disincentives and subsidies as a means of discouraging disposable products, reward durability and encourage innovation. Profit regulation could be employed as an extreme measure to reduce demand for mineral products; however, it is more appropriate in a free-enterprise system that appeals be made to reason and ethics to keep demand in bounds voluntarily. Finally, the tools of import-export reciprocity, regulation and incentives can be useful as a means of insuring supply-demand balance. Some estimates of the effectiveness of such measures are now becoming available, but they are probably subject to broad margins of error. Their indirect effects, on the general economy or on particular industries, or particular regions and localities, are not well understood, and in many cases they may be quite large. In a complex mixed economy, and even in those which attempt central planning, the unanticipated consequences of such actions have often dwarfed their direct effects. It is essential, therefore, that these measures receive very careful consideration.

SYNOPSIS OF RECOMMENDATIONS

The panel's recommendations, given in detail at the end of this report, arise from the two major themes outlined above. The first two concern information, and the last five, policy for influencing demand.

1. *Collect accurate data to allow meaningful analysis of demand functions.*
2. *Establish institutions and technical ability to make demand forecasts under various policy options.*
3. *Whenever possible, use the market place to influence demand rather than direct government controls.*
4. *Establish means to ameliorate particular hardships caused by public policies that alter demand.*
5. *Examine all Federal actions for effects on energy consumption.*
6. *Encourage a National conservation ethic.*
7. *Exercise restraint in all-out efforts to increase supply.*

Matrix for recommendation action

	Fed. Govt.	State Govt.	Industry	Academia	Public
1	X	X	X		
2	X	X	X	X	
3	X	X			
4	X	X			
5	X				
6	X	X	X	X	X
7	X				

CHAPTER XII
DO RECENT PROJECTIONS OF ENERGY "DEMAND" PROVIDE REASONABLE POLICY TARGETS?

The primary function of demand forecasting is to provide a basis for contingency planning. The particular constraints of theory, information and institutional needs under which forecasts are made make it disfunctional to take their estimates in a literal, quantitative sense as a basis for policy. Rather, insofar as forecasts of demand for any commodity enter into policy making, they should do so not as "givens" but as estimates (of varying accuracy) of what would happen under certain assumptions. Unless the assumptions are examined, and possible influences on them analyzed, the policymaker has generated no alternative courses of action--much less evaluated them.

If forecasts of demand for energy are to be used as a basis for national policy, we want to know, at the very least, what entity has been projected and what assumptions have been made in the process of estimation. When recent projections of energy demand to the year 2000 are examined with these points in mind, they are found to contain conceptual errors and questionable assumptions. The resulting estimates are thus likely to engender inappropriate policy decisions.

To help carry out its charge, the Demand Panel commissioned a review (Overly, Schell, 1973, see Appendix) of recent, readily available demand forecasts of minerals and fuels. Some 40 forecasts made between 1965 and early 1973 were included; however, about 30 of these were taken from previous reviews and summaries. Ten more recent forecasts and sets of forecasts, including, we feel, those being used for national policy regarding energy, were analyzed in some detail. Although there were differences in methods and details, the discussion that follows summarizes the general approach, and emphasizes the shortcomings of the majority of these forecasts. Details of individual forecast analyses are given in the review cited. (Overly, Schell, 1973; see Appendix to Section IV.)

DEMAND, CONSUMPTION OR NEED?

One reason the projections are misleading as future targets is that they fail to distinguish among "demand" and "consumption" and "need." Demand is essentially the quantity of a commodity that will be desired at a certain price (or changing price) over some period of time. "Consumption" is simply how much of a commodity is used. Need or "requirements" differ again in presupposing that extrapolated levels of consumption are normative or "required" by society. Since there is no objective way in which one can specify what living levels _should_ be at some future time, the notion of "need" or "requirements" founders in two ways--it ignores supply factors and their role in limiting demand, as well as assuming that one can specify some level of future consumption that is, objectively speaking, "needed." We shall concentrate our attention here on the distinction between "demand" and "consumption."

Most of the projections of total energy "demand" for the United States, are actually projections of "consumption," made using more or less sophisticated analyses of past use trends. They tell us what might be consumed, if _there were no supply constraints that, in turn, were reflected in price_, and if trends of all other factors that affect use don't change. But, as we know from the report of the Supply Panel (Section II), we must expect continuing supply constraints. It may be possible for energy policy to allow Americans to believe, in the short term, that such constraints do not exist. But, unless policy is geared at all costs to making cheap energy available, supply shortages will cause energy prices to rise. Demand, in the real economic sense, is most unlikely to remain unaffected by price increases, but the projections of consumption cannot take this into account.

The importance of the distinction between demand and consumption is well illustrated by the underestimates of future consumption that characterized the "demand" projections of the 1950s. The latter, when compared with the actualities of the 1960s and early 1970s, typically turned out to understate future consumption. Why? One of the principal reasons was that these projections did not take account of the fact that energy was becoming a bargain in our society--it was getting cheaper over the period and, hence a demand was created that was far greater than would have been the case at the higher prices that existed earlier. Indeed, the consistent underestimates of future consumption led estimators to believe that they should compensate on the high side. This reaction is too simplistic. The conceptual entity being projected must be one that reflects changing prices--whichever direction they are going.

Supposing national policy saw to it that energy prices reflected the true replacement costs of materials, as well as the social costs of production and use, in what proportions should the projected consumption levels be scaled down? This question concerns how responsive, or "elastic" consumption of energy would be in relation to rising prices. Unfortunately, we have no elegant answer because energy has been so plentiful that we have no large-scale econometric studies of how demand for energy varies with price. We have attempted to provide some answer to this question in Chapter XV. It seems highly unlikely that demand for energy is "inelastic" and, hence, insofar as prices will rise in the next 25 years, most of the existing estimates of future consumption which do not consider rising prices, will overshoot the mark.

Of course, demand for energy is not only a function of the price people must pay, but also of their incomes. There are not only "price elasticities," there are "income elasticities." What assumptions do the forecasts make about growth of income? In general, they assume levels of growth in our real Gross National Product that now also appear very optimistic--rates of 4.5 percent annually to the end of the century. This is higher than our average growth rate between 1951 and 1971 of 3.4 percent, and recent events have led to a scaling down of projected economic growth rates as far as the year 2000. Thus, we cannot expect that rapidly rising levels of real income will easily compensate for a shift upward in the price of energy. Trends in GNP growth will be discussed in Chapter XIII.

Population growth is another factor influencing demand. Most of the "demand" projections reviewed not only presuppose that no supply constraints are in the offing, they also assume that there will be more people around by the year 2000 than seems probable at this writing. The projections for energy consumption typically assume rates of population growth over 1 percent--sometimes as high as 1.5 or 1.7 percent. Since population growth rates have been declining quite rapidly, more recent population projections by the Census Bureau, and comparable projections to be found in Chapter XIV of this report, show that the American population will most probably be considerably smaller by the year 2000 than is assumed by most consumption forecasts.

Finally, most projections assume that the relationship between economic output (GNP) and the amount of energy used is fairly constant. Actually, however, this relationship has been variable over our history and it also varies among other highly industrial countries. (Ford Foundation, 1974.) Per unit of GNP, we have used diminishing amounts of energy over the long run. Causes of this trend appear to include greater technological efficiency, sectoral shifts in our economy, and changing tastes.

In sum, we must reevaluate the so-called projections for energy because these figures embody a concealed policy directive. Projecting consumption unrestrained by supply considerations, they say, in effect, that there cannot be supply constraints any greater than we experienced in our energy-cheap period. And, when the government makes an effort, on the supply side, to "meet" this projected consumption--when it uses the projections as targets, it is carrying out a policy whose message is as follows: sufficient energy must be found so that it will be cheap enough to enable Americans to consume at the projected level.

SOME RECENT INFLUENTIAL DEMAND PROJECTIONS

The self-fulfilling aspect of these demand projections has been seen in operation in an alarmist reaction earlier this year emphasizing all out efforts to increase domestic energy supplies.* It seems important, therefore, to analyze the projection on which these were based, according to the criteria outlined above. This is the projection of U.S. energy consumption to the year 2000 of the Dixy Lee Ray Report, The Nation's Energy Future, (Dixy Lee Ray, 1973) and other documents.

---

* The major input of information to this panel on national administration policy concerning supply and demand has been the December 1973 report, "The Nation's Energy Future" (Dixy Lee Ray, 1973). This document recognized a large gap between projected supply and demand and proposed a major research and development program aimed principally at increasing energy supplies, its stated purpose: "...to recommend the national energy research and development program needed to regain and maintain energy self-sufficiency." (Dixy Lee Ray, 1973, p. 1). Early in 1974, this program was widely referred to as "Project Independence: Background Paper," (FEO, 1974) and President Nixon, speaking before the Conference of Young Republican Leadership on February 28, 1974 stated:

> "...The major point I would make with regard to energy, however, is this: You have heard about the big Government program we are going to have. It is necessary. We are going to put $15 billion from the government into developing our energy resources over the next 5 years. We call it Project Independence for 1980."

The FEO paper, like the Dixy Lee Ray report, recognized large gaps between projected supply and demand and made recommendations for research and development aimed at markedly increasing domestic energy production, as well as decreasing energy use growth rates.

As this panel's report was undergoing final editing in November of 1974, the Federal Energy Administration (FEA) released a lengthy document entitled, "Project Independence." (FEA, 1974). Unlike the previous documents mentioned, this report does not make recommendations, but rather analyzes in some detail many aspects of the U.S. energy situation under several assumptions of world oil prices, and possible U.S. actions involving efforts at increasing domestic supplies as well as conserving fuels. Although the panel has not been able to consider this document in any detail, it appears that the FEA has made a much more thorough analysis of the demand question than did the Dixy Lee Ray study, and in fact may have anticipated some of the recommendations made here. To that extent, it is hoped the reader familiar with the FEA document will consider those of the panel's recommendations which coincide with its findings, as supporting of these findings. There are, however, many matters considered here that do not appear to have been taken up by the FEA analysis.

Finally, we would like to note that all reference in this panel's report to "current national policy," or "Project Independence," etc. (although we have attempted to replace this latter term wherever it appeared) are to previously recommended research and development programs (and other measures)

The forecast places annual U.S. gross energy consumption at about 194,000 trillion BTU's by the year 2000, nearly three times the 68,810 trillion BTU's consumption of 1970. Thus, throughout the 30-year period 1970-2000, gross energy consumption is expected to increase at an annual average rate of 3.5 percent, compared to a rate of 3.1 percent for the 24-year period 1947-1971. Although the main thrust of the Dixy Lee Ray Report concerns the period of 1980 (the desired date for "Independence") it is clear that forecasts to the year 2000 are fundamental to actions taken to reach the 1980 goals.

The discussion of energy consumption and demand in the report has two sources: a Department of the Interior Report (Dupree and West, 1972) for past data, and a Joint Committee on Atomic Energy (JCAE) summary (JCAE, 1973) for projections through the year 2000. The Dixy Lee Ray projection uses a near intermediate value of the several projections (including that of the Interior report) summarized in the JCAE publication.

Although neither the Dixy Lee Ray nor JCAE reports discuss population projections in any detail, the latter does refer to the Bureau of Census series "D" and therefore suggests a value of about 285 million persons for the year 2000. Combining this with the energy consumption forecast for the year 2000 of 49 x 10^{18} cal. (194 x 10^{15} BTU), we calculate an implied per capita consumption of 170 x 10^9 cal. (680 million BTU's), compared to a 1970 value of 108 x 10^9 cal. (429 million BTU's). Although per capita consumption is rarely one of the variables projected in most demand forecasts, we find it a useful one to consider when quantitatively discussing changes that are expected to take place between now and the year 2000.

Since few basic data are available in the Dixy Lee Ray report or the JCAE report, it is necessary to go back to the reports that are summarized by JCAE if we are to determine how the projections were obtained. The Interior report is typical and yields values of total consumption close to those used in the Dixy Lee Ray report.

The failure openly to analyze component variables, observed in the analysis of most forecasts, is apparent in the Interior projection, although the authors of this report make clear that the projection is of consumption, not demand. Although the report states that supply limitations, population growth, increased relative prices, GNP projections, and other factors were considered in that these "...data were correlated with energy consumption and any important trends extrapolated..." it is not clear how this was done, nor which were the "important trends." (Dupree and West, 1972).

(footnote continued from previous page)
made by the federal administration to reach energy self-sufficiency. The fact that the more recently established FEA has released a lengthy analysis of possible U.S. actions <u>without</u> recommendations suggests that the national program is not yet decided. We hope, therefore, that the findings and recommendations here might assist in leading to decisions for the nation's future energy (and minerals) program.

This Panel's report, in part, will deal with a revision of population projections for the United States, and also address itself to a discussion of implied projections of per capita energy demand. We wish to note again that per capita consumption is not a variable actually projected in order to calculate total demand. However, in that consumption strongly influences standard of living, or well being, we have found it a useful one in the implications of future demand calculation.

The likelihood that the population projection mentioned above may be too high is demonstrated in Chapter XIV. As will be made clear, the population of the United States in the year 2000 will probably lie between 265 million (on the assumption of a constant Net Reproduction Rate of 1 between 1970 and 2000) and 252 million assuming an NRR of 0.9. Allowing for a 15 percent increase in fertility through 1985, and only then presupposing a decline to an NRR of 1, (an unlikely scenario) the projection would rise to 285 million.

Chapters XIII, XV, and XVI argue against as large an exponential increase of per capita energy demand as implied by the Interior report or Dixy Lee Ray report. This argument is based on the high probability of a decline in the use-intensity of energy as a consequence of supply constraints, continuing sectoral change in the economy, and changing tastes. The implications for the Interior Report (and *a fortiori* for the Dixy Lee Ray report) of incorporating such fundamental considerations are examined briefly here.

The Department of Interior report states that per capita energy consumption will more than double between 1971 and 2000--89.4 - 173 billion cal. (333.3 - 686.1 million BTU's). Although per capita energy consumption increased at an average annual rate of 1.6 percent per year between 1947 and 1971 (although for shorter periods it varied considerably), the Interior projection implies average annual increments of 2.1 percent per year between 1970 and 1980, and of 2.7 percent per year between 1980 and the year 2000. Even assuming no constraint on supply, no rationale is offered for such a marked increase in per capita energy consumption.

As an illustration of how important the assumptions leading to these figures can be, Table 1 shows expected total gross energy consumption for the United States in the year 2000 under three population assumptions: 252, 265, and 285 millions the latter being our high estimate and that in JCAE, (the Interior report used a value of 279.1), and four per capita energy consumption figures: actual levels for 1965 and 1970, as given in the Interior report, the year 2000 level implicit in the Dixy Lee Ray report, and 10 percent reduction estimate suggested as possible with conservation. This table thus affords a notion of the nation's energy consumption in the year 2000 given projected population changes plus a number of per capita energy assumptions, only one of which is less than actual experience in 1970. Per capita energy consumption at 1965 or 1970 levels might still allow for increasing levels of living that could be brought about through reduction of energy wastage and higher efficiency. As can be seen from Table 1, the results of lowering the projected per capita energy assumption for 2000 from the level postulated by Interior in its 1972 projection, can be spectacular.

TABLE 1 Projected Gross Energy Consumption in the Year 2000 (in Quadrillions of BTU) for the United States Under Selected Assumptions Concerning Population and Per Capita Consumption

Assumed Per Capita Consumption (10^9cal. or 10^6 BTU)	Projected Population* (millions of persons) 252	265	285
69.1 or 274.4 (1965 value)	69.1	72.7	78.2
87.8 or 429.1 (1970 value)	82.9	87.8	93.8
171 or 680+ (Implicit in JCAE)	171	180	194
154 or 610 (JCAE with 10% conservation)	154	162	174
Total 1970 Gross Consumption was 17.3 x 10^{18}-cal. (68.8 x 10^{15}BTU)			

* The population projection of 252 million assumes a constant Net Reproduction Rate of 0.9 from 1970 to 2000, the 265 million figure assumes a constant NRR of 1 from 1970 to 2000, and the 285 million figure assumes a 15 percent increase in age specific fertility 1970-1985, after which age specific rates are assumed to be consonant with an NRR of 1. All the estimates include a constant 1970 level of migration

+ The per capita consumption of 171 x 10^9cal. (680 x 10^6 BTU) was derived from the Dixy Lee Ray projections of 49 x 10^{18}cal. (194 x 10^{15}BTU) gross consumption and the JCAE population of 285 million.

Similar marked reductions in electricity demand resulting from probable price increases have been deduced by Chapman et al., (1974).

Returning to 1965 per capita levels of gross energy consumption, and assuming a population of 265 million, would require only somewhat more total energy input in the year 2000, than was actually expended in 1970: 18.4 x 10^{18} cal. (72.8 x 10^{15} BTU) compared with 17.4 x 10^{18} (68.8 x 10^{15}). Even assuming an unlikely increase of population to 285 million by the year 2000, total energy consumption at 1965 levels would mean an increase over actual 1970 consumption of less than 15 percent. The low population (252 million) and 1965 per capita consumption rate would give essentially the same total consumption in 2000 as actually occurred in 1970.

At the 1970 per capita consumption level in the year 2000, and a population figure of 265 million, the nation's total energy consumption would be 22 x 10^{18} cal. (87.8 x 10^{15} BTU)--an increase of 25 percent over total consumption in 1970. Quite clearly, estimates such as these are of a quite

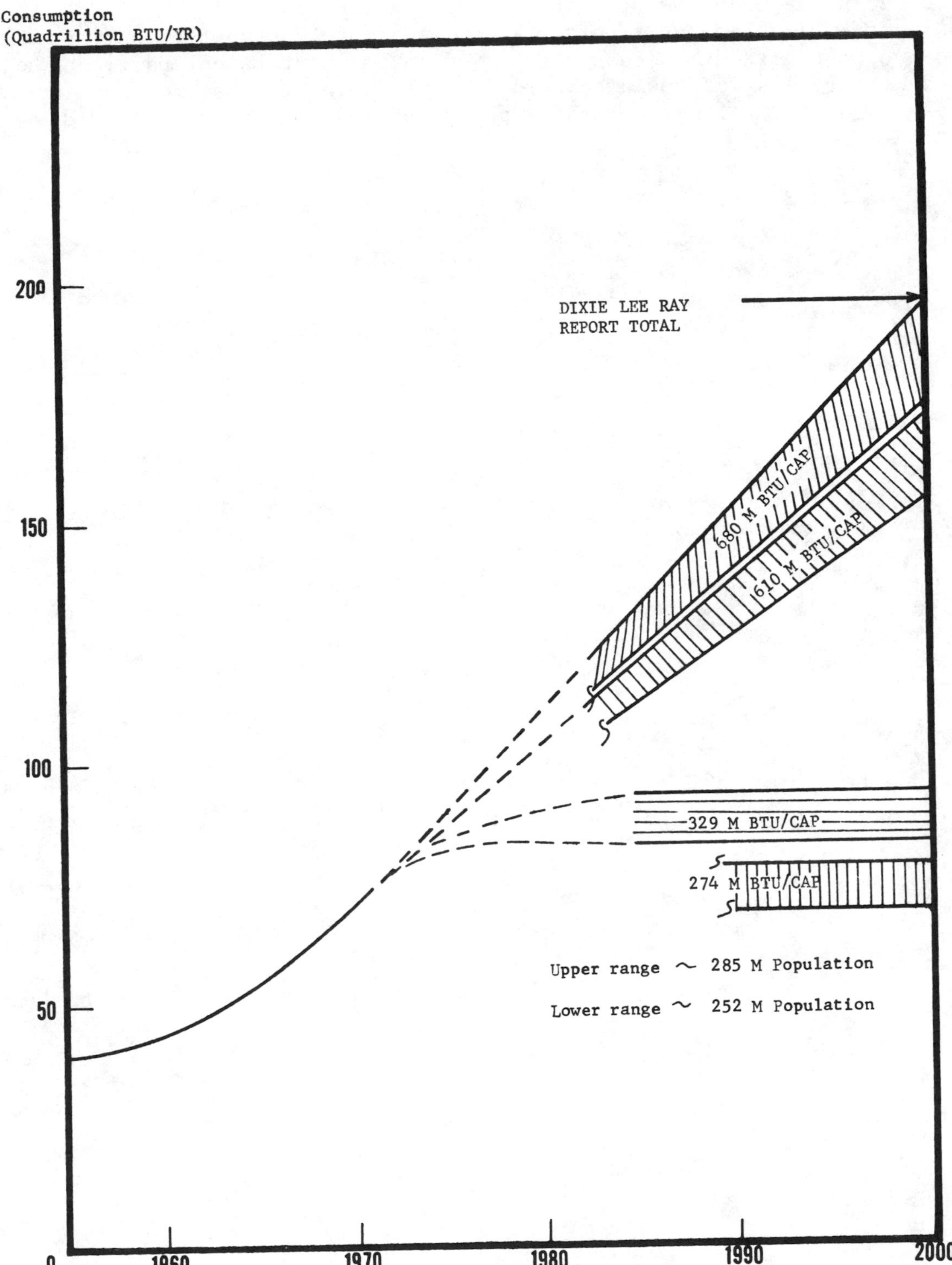

FIGURE 1: Total Gross Energy Consumption (Quadrillion BTU/Yr) Under Selected Assumptions of Population and Per Capita Consumption

different scale than the projections of 49×10^{18} cal. (194×10^{15} BTU) for 2000 by the Dixy lee Ray report. They illustrate consumption options with which we have already successfully lived in the most recent past, and combined with realistic population projections, point to the enormous importance of assessing the validity of the assumptions involved in so-called demand projections.

CHAPTER XIII
GROWTH OF GNP AND USE OF MINERALS

The ratio of materials used per unit of GNP (e.g. per dollar output, measured in constant dollars) has been declining for many raw materials in the past quarter-century, although absolute quantities of use have still been increasing. In this chapter, we will review the historical evidence on this ratio (rate of use) and then turn briefly to the question of the prospects for GNP growth itself. In brief, the line of argument will be that economic growth during the next decades very likely will not repeat the performance of the postwar era. But even to the extent growth will occur, requirements for many minerals, and especially energy, can be expected to expand more slowly than GNP. A further diminution of demand below the levels to be expected on the basis of historical experience would be likely, should the recent steep rise in relative prices for minerals be sustained.

RATES OF USE

As societies grow, it is possible to distinguish two ways in which they modify their use of minerals. First, demands for final consumption change the total composition of gross domestic product. In rich lands, the relative expansion of the services component of gross product will lower, on the whole, the rates of use for minerals. In poor lands, modernization and increased per capita income tend to raise the use intensity for minerals.

Second, man's capacity to innovate means that technological progress will serve to lower the rates of use. Throughout the world, we may expect the trend reflected in the United States' experience: smaller inputs of a mineral to accomplish essentially that which required larger inputs earlier.

What actually has been the observed relationship between materials use and the total output (GNP) to which they contributed? From 1930/34 to 1951/55, the amount of energy used per GNP dollar (1971 prices) was reduced by 20 percent (Sun Oil Company, 1972). Clearly, U.S. energy use was growing _less_ rapidly than was U.S. GNP. For the period 1951 through 1959 the U.S.

income elasticity for energy use was less than 1 (actually 0.89).* That is, relative increases in energy use were less than the relative increases in income.

At the broadest level, as U.S. per capita GNP expanded, the value of total resource use--minerals, lumber, agriculture--relative to total United States gross product, declined from 36 percent in 1870 to 12 percent in 1954. From 1920 to 1954, the minerals share alone declined by 50 percent. During the decade 1957 to 1966 industrial production increased by 57 percent whereas the use of important minerals expanded by well under 20 percent (copper, 18.6 percent; steel, 16.4 percent; zinc, 4.2 percent). During the same period, however, the use of alloy steels increased by 49 percent; synthetic rubber, 82.5 percent; and plastics, 240 percent. Since 1900, the efficiency of coal use has also improved markedly; one-eighth the amount of coal required in 1900 is now needed to generate a kilowatt-hour of electricity.

While energy use grew at a lesser rate than did total product, it still grew at a more rapid rate than did population; U.S. per capita energy consumption increased at an annual average rate of growth of just under 2 percent between 1951-55 and 1966-69. But the important point is that even _before the days of recent changes_ in the price and/or availability of energy raw materials, the forces of economic change were apparently working to _lower_ the rates with which the U.S. economy used many raw materials and energy.

To be sure, there were short-term fluctuations in use rates. There was a rapid expansion in this ratio for energy in the immediate post-World War II years after 1945, but the downward pattern was again resumed before 1950. Similarly, the years 1969-72 reflect another reversal to increased use of total energy per unit of GNP.** This was a period when energy prices moved down relative to other prices; it was also a period of intense business activity to expand consumption of energy-using appliances, it was a period of large investment in air conditioned structure--both commercial and residential.

One decisive element in the declining rate-of-use pattern in the U.S. is the changing composition of our GNP. Demand forces have moved our economy

---

*For other major materials, the same phenomena are revealed in the U.S. data. Thus over the same years, income elasticities were crude steel, 0.67; iron ore, 0.38; refined copper, 0.65; zinc, 0.58; and sulfur, 0.94. Exceptions are aluminum and fluorspar, where only rates of increase have declined markedly over the last decades.

**Thus, the recent Ford Report (1974) shows data for 1960 and 1968 which reflect the downward movement in use-intensity. But preliminary data for 1972 (p. 42 of the Ford Report) suggest higher intensity rates than in 1968. See also Sun Oil Company (1970).

toward more service output relative to a more goods output. This has, on the whole, resulted in a shift to a less energy intensive sector, especially when comparison is made between energy use over decades.*

Similar statements pertain for other rich lands in Europe and elsewhere, although there are specific exceptions which can be explained in terms of other forces than consumer demand for a changed structure of GNP. In the USSR and Canada, rich lands where new industrialization was widespread in recent decades, energy intensity has been pushed up, although moderately and perhaps temporarily. In the world's poor lands, on the other hand, changing composition of GNP reflects the new industrial emphasis on the developing nations. Rates of use of energy are still increasing; income elasticities are much higher than in the rich lands, and can be expected to continue so.

Relevant data are shown in Table 1. Over any long growth cycle from poor to rich nation status, the use-intensity of important materials may be

TABLE 1 Total Energy Related to GNP

	Intensity of Use [000 m.t. per $ billion GNP (1971) prices]		Income Elasticities
	1951-1955	1966-69	1951-55 to 1966-69
A. World	1760	1780	1.03
B. U.S.	2160	2070	0.89
C. Other Developed Lands	1940	2100	1.18
Western Europe	1600	1420	0.78
Eastern Europe	2430	2330	0.87
USSR	1890	2130	1.23
Japan	1600	1290	0.74
D. Poor Lands	872	1425	2.28
China	1280	3080	4.16

Source: Malenbaum, 1973, p. 11.

*Even a 1960 and 1968 use comparison shows energy in industry and transportation relatively smaller, with commercial and residential uses relatively higher. (Ford Foundation, 1974.)

expected to show an increase during early years followed by a flattening and decline later. When nations push their economies to modernization and industrialization, their use of materials and energy grows more rapidly than their total national product. Wealthier nations ease off in such inputs, relative to GNP growth. While the turning point cannot be precise, past experience suggests it falls in the \$2,000-\$2,500 GNP per capita range (1971 prices). Such a generalization should contribute to relevant hypotheses for materials use in future years in different parts of the world.

While poorer lands tend to have lower rates of use than do richer lands, such interregional differences depend more upon the specific characteristics of individual economies. Thus, Japan's rates of use of energy tend to be higher, at any given level of GNP per capita, than are those for Western European lands. While Table 1 suggests that poorer lands tend to have lower intensities of use than do richer lands, the GNP-materials relationships among nations is not easily generalized. Past records indicate the intensity level for each nation, but we can hypothesize only on its future _time_ pattern.

Another interesting angle on this problem can be explored by comparing the ratios in different countries. As shown in the chart from _Scientific American_, September 1971 (see next page), there is considerable variation in the ratio for different nations. The United States is fairly typical, with a value of about \$17.00 per million BTU. The Soviet Union is not as efficient, with about \$12.00 per million BTU, while South Africa manages only around \$8.00 per million BTU. So, even limited to technologies and prices currently in existence, one might, for instance, argue the possibility of the United States doubling its GNP without increasing its energy consumption (by attaining a ratio similar to that of New Zealand, but at higher absolute values).

THE PROSPECT FOR GNP GROWTH

The GNP growth rates of the two decades 1951-71 (Table 2) for the entire world, or for any major regional component, exceed significantly the average growth rates, for any twenty year span prior to this period. Today, moreover, the world encounters new problems of inflation with unemployment and of international monetary instability. It is struggling with short-term materials crises. The long-term outlook for rapid progress seems less assured than might have been projected a few years earlier. Given these special circumstances, it thus seems most unlikely that the unusual levels of 1951-71 will continue for the rest of the century. Rates of economic growth in the world as a whole, and in very important component areas, can be expected to average below what they were in the two preceding decades.

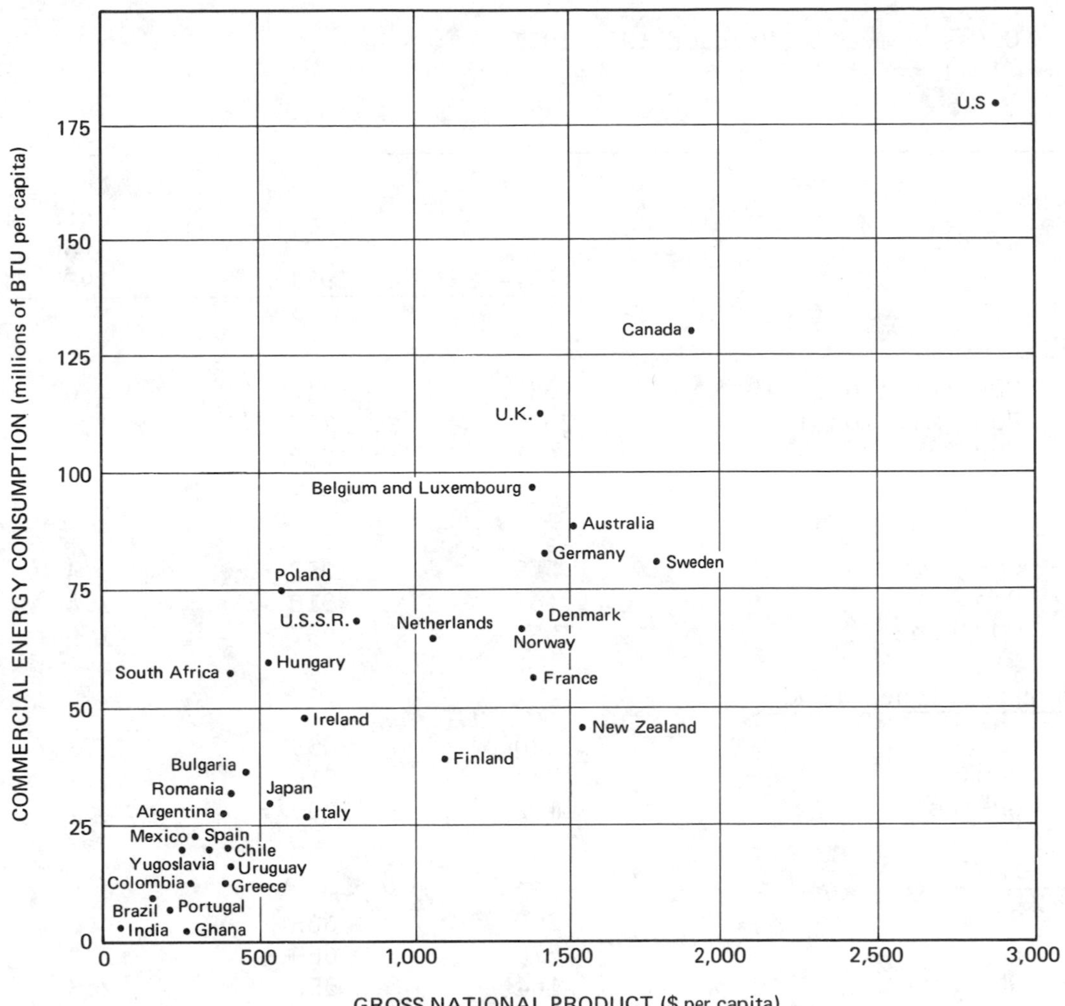

SOURCE: Cook (1971): by Courtesy of *Scientific American.*

Rough correlation between per capita consumption of energy and gross national product is seen when the two are plotted together; in general, high per capita energy consumption is a prerequisite for high output of goods and services. If the position plotted for the U.S. is considered to establish an arbitrary "line," some countries fall above or below that line. This appears to be related to a country's economic level, its emphasis on heavy industry or on services and its efficiency in converting energy into work.

Source: Cook (1971): The Flow of Energy in an Industrial Society:

FIGURE 2: The Flow of Energy in an Industrial Society.

TABLE 2 Gross Domestic Product 1951-1971
(1971 prices)

	1951-55	1971	Annual Rate of Growth (%) 1951-71
A. World			
GDP (Billions $)	1689	3810	4.6
Pop (Millions)	2677	3784	1.9
GDP/Pop ($)	631	1006	2.6
B. U.S.			
GDP	579	1050	3.4
Pop	163	210	1.4
GDP/Pop	3559	4997	1.9
C. Other Developed Lands			
GDP	820	2095	5.3
Pop	744	914	1.1
GDP/Pop	1102	2292	4.1
D. Poor Lands			
GDP	290	665	4.7
Pop	1770	2660	2.2
GDP/Pop	164	250	2.3

Source: Malenbaum (1973) p. 11.

At least in industrial countries, economic growth appears to have lost much of its magic as a national goal. Should conflicts with different goals come up in the future, growth cannot be expected to win out as easily as it has in the past. This statement applies in particular to those GNP components that are capital and materials intensive and place greater strain on natural resources (through exploitation) and environment (through processing), than do labor intensive components. The apparent physical limits to the expansion of materials production, as documented in Sections I and II on Technology and Supply, must also be considered in assessing the prospects of GNP. Because of the great extent of uncertainty we offer no quantitative forecasts of future GNP levels, but rather point only to the biases inherent in an extrapolation of past trends of both GNP and minerals use into the future, as is common in most projections analyzed.

CHAPTER XIV
NATIONAL AND INTERNATIONAL DEMOGRAPHIC TRENDS: THEIR RELATION TO THE DEMAND FOR MINERAL RESOURCES IN THE UNITED STATES UP TO THE YEAR 2000

Assuming, as most projectors do, a continuation of the positive relationship between consumption of mineral resources and population growth in the U.S., the latter is an important factor in considerations concerning possible future consumption of mineral resources. Since the amount of mineral resources on the earth is finite, is distributed around the earth in a particular way, and with varying access, the availability of, consumption of, and demand for mineral resources in the United States will depend also on population growth trends in other parts of the world.

The current socio-economic and demographic situation would indicate that U.S. and "developed countries" population growth trends will be reasonably slow during the foreseeable future.

For the U.S. in 2000, an acceptable estimate would be a population in the order of 10-25 percent larger than at present as a result of "natural growth" (births minus deaths). Assuming a continuation of the stable immigration levels of the last two decades (about 0.4 million annually), the range estimate would be increased to 15-35 percent above the 1970 population size.

Population growth trends could be influenced somewhat by regulations concerning international migration, and by attempts to modify fertility via policy measures.

The population growth trends of the developing countries are likely to exceed considerably those of the United States and the other developed countries. It is by no means unlikely that the population of the developing countries will almost double by the year 2000.

U.S. POPULATION GROWTH: CURRENT AND PROSPECTIVE

Factors Affecting Population Growth Mortality and Fertility

The mortality decline seems to be slowing down; the outlook for the next decade is a slow further mortality decline, or a practical leveling off of life expectancy, even more for men than for women.

The fertility decline of the past 15 years has continued in 1973. The crude birth rate is now around 15 and age specific rates are below replacement levels. What will happen in the next several years is questionable, but fertility rarely proceeds in drastic jumps, therefore a further moderate decline, a small increase, or a leveling off (not excluding the possibility of minor fluctuations) are likely.

Socio-economic influences that militate against a rise in age specific birthrates are the continuing inflation which, at least in the short run, could lead couples to attempt to prevent an erosion of their level of living by postponing childrearing; recent impaired employment and occupational opportunities for young people which are experienced as relatively painful because this generation was brought up in parental homes of unprecedented affluence; a probable increase in female labor force participation possibly at occupational levels that are more demanding than the typical "female" occupations of the past; a rise in the education level of the population which leads to more informed decisions concerning reproduction and greater potential costs to young parents for having children; and the greater variety of effective techniques of fertility control now available, including "back-up" methods such as abortion.

On the other hand, survey data show that few young people desire to be childless or have only one child. Marriages and births are quite possibly only being postponed. Since there is by now a large contingent of young women who have not yet had the number of births that surveys show young Americans want, childbearing among these grown-up cohorts of "baby boom" babies could augment the crude birth rate quite substantially. If, in addition, younger women started to marry earlier and have children at more youthful ages, the birth rate would receive an additional boost. Thus, changes in timing and spacing that may seem quite negligible to the individuals involved can cumulate into substantial effects on the crude birth rate. Relatively late childbearing of women currently in their early twenties coupled with possibly slightly earlier childbearing of women currently in their teens could lead to a significant increase of period fertility above replacement level in the late 1970s or early 1980s.

International Migration

The legal flow of migrants in and out of the U.S. under current conditions results in an average annual net gain of close to 400,000 people. If this flow were to be maintained, with roughly its present structural characteristics, the future size of the U.S. population in the year 2000 would be only marginally larger than without immigration (at the most about 8 percent larger). Since, however, the growth of the total U.S. population due to natural increase by the year 2000 is expected to be relatively small (i.e., 10 to 25 percent larger than in 1970) a 5 to 8 percent additional increase due to migration is in any case a meaningful part of the total population increment.

Internal Migration

Population redistribution within the country could alter patterns of demand for mineral resources. Different climates have varying demands for energy (heating, air-conditioning); the relative location of residence, place of work, place of training, place of entertainment and recreation, create varying demands on transportation (materials and fuel). During the 1960s and early 1970s the main migrational flows have been to the South and West, from metropolitan areas to the suburbs, and these have been strong enough to redistribute significantly the U.S. population. These migrational flows could have altered the structure and size of demand for mineral resources. However, the net effect may well be negligible owing to compensation effects. For example, energy requirements for heating in the North are offset at least partially by those for air-conditioning in the South. The net increase of the population in the South and West possibly meant a decline in the respective demand for resources but the trends of suburbanization and increased leisure time and the developing recreational patterns go in the opposite direction.

In the future, it is not clear that the recent trends in suburbanization will necessarily continue. They may even be reversed if families end up by being very small, married women with young children continue to participate in the labor force at increasing rates, and costs of commuting become seriously augmented. Under such circumstances, a reurbanization trend might develop, thereby saving a considerable portion of the energy costs of commuter travel as well as chauffeuring by mothers within the suburbs.

Marriage Patterns

Age at marriage (especially of legal marriages) is increasing. This usually leads to later and lesser childbearing; later childbearing means a longer average interval between generations and thus a slowing down of the rate of population growth. Divorce rates are high and increasing but, relative to other influences, this trend does not have a major impact on the amount and timing of childbearing.

Prospects for U.S. Population Growth During 1975-2000

In order to arrive at what seem to be reasonable expectations for the upper and lower limits of population growth during the next 25 years, we have engaged in varying extrapolations of current trends based on the preceding discussion of factors affecting population growth. Depending on which extrapolation is chosen, the population of the United States by the year 2000 could be in the order of 10 to 25 percent larger than it was in 1970 (with an assumed 0.4 million annual net immigrants, 15-35 percent).

TABLE 1 Prospects for U.S.Population Growth During 1975-2000

Basic Assumption of Fertility Trends For 1970-2000 & Mix	Population Size (in millions) 1970	Natural Increase 2000 Absolute Size	2000 Index (1970-100)	Assuming 1970 Levels of Migration 2000 Absolute Size	2000 Index (1970-100)
NRR=1.0 (replacement level fertility throughout the period)	205	250	122	265	129
NRR=0.9 (below replacement fertility throughout the period)	205	238	116	252	123
15% increase of fertility until 1985 then replacement fertility	205	270	132	285	139
Constant 1965-70 fertility	205	279	136	295	144

The last two rows of the table (i.e., those that assume (i) a fertility increase during the 1970s and early 1980s of about 15 percent compared to the late 1960s (a renewed baby boom), or (ii) a constant fertility of the later 1960s, compared to actual developments and to the outlook for the near future, illustrate situations that appear to be increasingly unrealistic. According to these latter projections the U.S. population in the year 2000 would be 40 percent or more larger than it was in the year 1970. Since neither of these projections are currently materializing, they can serve as a reasonably good argument for an upper limit of expected population growth. This statement has to be qualified by the fact that a baby boom (or "boomlet") could materialize with a time lag--say only by the early 1980s. If this were to occur, its impact on the size of the population in the year 2000 would be smaller than illustrated above because of the time lag and, as a result as well, lesser second generation effects.

NON U.S. POPULATION GROWTH: CURRENT AND PROSPECTIVE

In the developed countries the situation is basically similar to that of the U.S., but within this framework there are some interesting features:

1. Many countries have more or less effective pronatalist policies. These have not necessarily increased fertility considerably but might have prevented fertility from declining further. Usually countries adopt such

measures and pursue them vigorously when fertility declines below the replacement level.

2. During the last several years some countries have fertility levels not only below replacement but also have negative rates of natural increase (both Germanys).

In developing countries there is considerable regional and country-by-country variation. Several basic features can be summarized as follows:

1. Large variation in mortality is seen in the developing countries but almost everywhere it is higher* than in the developed countries and thus there is further room for decline, which is universally considered a good thing. There are visible counterforces, for instance: world food production has had difficulties keeping up with population growth, food reserves are diminishing, there seems to be evidence of unfavorable trends in the world climate; the energy crisis may lead to a long-term cut in production of fertilizers, although an increase in production is needed.

Possibly 20 million deaths from starvation are estimated in the developing countries for 1974 over and above the "natural" (normal) total of 40 million deaths; this would give a total of 60 million. The crude death rate would increase from 14 per thousand population to 21, and consequently the crude rate of natural increase would decline from about 2.5-2.6 to around 2.0 percent per annum in the developing countries.

2. In most developing countries fertility is still very high, higher than a hundred years ago in the developed countries; the average family is 5 to 6 children per woman (if childless and unmarried women are excluded, about 6 to 7 children per family). To date only in 15 countries (out of about 90 for which at least rough estimates are available) crude birth rates are already below 30. In addition these particular populations constitute a minor proportion of the population of the developing countries: 7 of these countries have populations smaller than 1 million, only 4 have populations of around 10 to 13 million inhabitants. Roughly the same number of countries have crude birth rates between 30 and 40 per thousand, and since the People's Republic of China is assumed to be in this category, together with some other populous countries, this category does carry significant weight. The majority of the developing countries in Asia, Africa, and Latin America still have crude birth rates of 40 per thousand and more.

3. High fertility in the developing countries has created a so-called young age distribution of their population. As a rule--to which there are

*The crude death rate might be lower in a particular developing country (as a result of a favorable age structure) compared to the developed countries, but the age-specific mortalities are usually higher, and the life expectancies lower in the respective developing countries.

exceptions--over 40 percent of these populations is below 15 years of age, over 25 percent is in the young childbearing ages of 15-29, and only about 5 percent of these populations is older than 60.

Summary of Demographic Situation in Developing Countries

The current population growth rates are at an unprecedented high level, i.e., higher than ever normally experienced by developed countries. Current levels and trends of fertility and mortality together with the existing age structure of these populations provide an extremely high potential for future population growth. Consequently, competition with developed countries for materials in general and for food production in particular, is likely to accelerate. Table 2 shows projected population figures for the developing countries under the assumptions of rapid and relatively slow fertility decline, together with figures for the developed countries and the world as a whole. It is clear that the presently developed countries are due to form an ever-decreasing share of the world's people. Although it must be borne in mind that some of the currently developing countries will have become "developed" by the year 2000 thereby changing the ratios.

TABLE 2

	Population Size			Index of Growth			Composition		
	1970	2000	2050	1970	2000	2050	1970	2000	2050
Rapid demographic transition in developing countries (replacement fertility by 2000)	2.5	4.5	6.5	100	180	260	70	78	82
Developed countries (replacement fertility from 1970)	1.1	1.3	1.4	100	118	127	30	22	18
World Total	3.6	5.8	7.9	100	161	220	100	100	100
Traditional demographic transition in developing countries (replacement fertility by 2040)	2.5	5.3	11.6	100	212	464	70	80	89
Developed countries (replacement fertility from 1970)	1.1	1.3	1.4	100	118	127	30	20	11
World Total	3.6	6.6	13.0	100	183	361	100	100	100

Source: (Frejka, 1973).

CHAPTER XV
RISING PRICES FOR ENERGY/MINERALS AND ADJUSTMENT TO SUPPLY CONSTRAINTS

There are a number of reasons for past and expected increases in the costs of mineral production. The first is the physical characteristics of the resources to be exploited in the future: their concentration, location, accessibility. Second, a new awareness by the holders of natural resources of their wealth and its finite nature has taken hold. The third is the diminishing tolerance by the public of environmental damage and other external effects. Quite possibly these three factors are exponentially related to exploitation, that is only a small increase in production may cause considerably increased marginal costs in the form of higher unit costs for production and resource owners rent, and increased costs to alleviate harmful environmental effects. In concert, they may overstrain an already strained system. If supply expansion meets with increasing resistance, it is well worth inquiring about the implications of curtailing demand. The recent precipitous increase in the relative prices of some minerals not only provides instructive evidence needed to answer this question, it also points to its relevance: the possibility of continuously high and rising relative prices for minerals is real and must be seriously entertained.

We start out by drawing attention to one often neglected, probable source of relative price increases for minerals: the environmental and social costs of extraction and refinement. We then go on to reviewing some data about price elasticities mainly based on the less turbulent economic history of the 1960s to be followed by an analysis of the consumer response to the recent energy crisis. No systematic evidence is yet available on the behavioral response by industry or government to the recently emerging constraints.

ENVIRONMENTAL AND SOCIAL COSTS

The assumption of steeply rising costs for minerals/energy production is fundamental to the argument of a consumer response involving reduced demand. Much of the argument for such price rises is furnished in Sections 1 and 2 by the Panels on Technology and Supply and lies outside the scope of this Section. However, the influence on prices of external (or "hidden") costs not specifically related to supply considerations should also be examined.

These hidden costs are those not considered in the price accounting of resource extraction. Although there are equally positive or "benefit" effects involved, our primary concern here is the more widespread negative or extra cost elements.

Many of these hidden costs are concerned with degradation of the environment. The situation where wastes associated with resource extraction were small enough to permit of easy disposal no longer obtains; many parts of the environment can no longer "dispose of" or "process" all the wastes they receive without detriment. The Clean Air and Clean Water Acts have been the legislative attempt to solve this problem, and the implementation of the standards they impose naturally raises the cost of operation and, incidentally, to some extent the need for energy. This cost is passed on to the resource consumers. Assuming some price elasticity of demand, such current and expected actions should have an effect in reducing demand, or moving production to areas where these concerns are not yet felt.

Direct and indirect subsidies, usually in the form of "favorable" tax policies for particular industries, are other hidden costs which society pays for maintaining an artificially low price of selected commodities although low price may not be the primary object of the policies. A specific example in the resources industry has been the depletion allowance, whose purpose is to compensate an owner whose property must be "used up" to benefit society. Without discussing the correctiveness of this compensation, the depletion allowance in effect lowers the income taxes of resource corporations relative to those of the average industry, allowing them to market at lower prices than would otherwise be necessary. If the consumers of the resources were to pay the compensation, rather than the society as a whole, the resultant higher prices of the resources would undoubtedly raise the price of final resource or energy intensive goods and activities, relative to other more labor intensive sectors, and thus lessen demand.

It is far beyond the scope of this report to treat all such hidden costs in the resource industry. However, we will attempt an investigation of a portion of such costs in the coal industry as a case study whose conclusions can be applied to resource extraction as a whole.

Coal mining is one of the more hazardous occupations in our society. It has a high accident rate relative to similar work (both fatal and nonfatal) and coal miners are subject to a high incidence of pneumoconiosis (black lung disease). These "personal environmental impacts" lead to additional production costs, normally borne by the sales of coal, in the form of lost or inefficient labor productivity, industry compensation payments, mine property damage, etc. But other costs of these occupational health and safety impacts are borne by society in general, in the form of black lung compensation benefits (of the order of $425M annually in 1972), welfare medical expenses (annual public health cost for pneumoconiosis detection and treatment is $2-3M) (Bureau of Mines coal mine health and safety research budget is $27M), and others. Finally, miners and their families carry some costs in the form of lost income, increased medical expenses, etc.

In 1969 the value of coal production was about $4.5 Billion for 540 million tonnes (600 million short tons) ($7.50/ton), and we may ask what are the relative amounts of the external costs mentioned above.

How to distribute the annual black lung benefits cost to current production is not straightforward, since the compensation is for disease contracted over a prior period of time (as well as for current cases). Moreover, as mentioned in the Environment and Health Panel's report, it may well be that the compensation is excessive. A rough estimate might be made as follows: For the last 25 years, average production has been around 450 million tonnes (500 million short tons) for a total of 11.2 billion tonnes (12.5 billion short tons) over this period. The cumulative black lung benefits payment up to 1972 (before more liberal payment requirements were established) was about $1 billion. Thus the added costs per ton of these payments, assuming all were due to operations over the last 25 years, are 8¢ per ton. The Bureau of Mines coal mine health and safety budget of $27M (5¢/ton) and the public health service sum of $3M (1/2¢/ton) might also be considered as external health costs. These external costs (and they are not all of those associated with the industry) add up to only about 2 percent of the price of coal. However, they are illustrative of the way in which prices are lower than they might otherwise be. Enactment of the Federal Coal Mine Health and Safety Act is another example of current national policy that brings formerly external costs into production accounting. As these costs are passed on to the consumers, we can expect a relative lessening in demand compared to the large increases assumed in popular projections.

There are other "new" costs that are being assigned to the coal industry and its users, such as strip mine reclamation and sulfur and particulate emission control. These costs vary considerably regionally and according to type of coal. But, as these costs are being passed on to consumers, we can also expect a relative reduction in per capita demand for these causes, compared to the large increases assumed in popular projections.

A crude calculation shows that these environmental costs at the user end may be much more striking than those concerned with mine worker's health. The Ford Foundation (1974) has given a preliminary estimate that not more than $3 Billion per year would be required for sulfur oxide controls on power plants that produce some $4 Billion in damages. Using Robinson and Robbins' (1970) data for sources of sulfur oxide emissions, and those of JCAE (1973) for energy flow in 1970, it is seen that some 95 percent of the sulfur oxides emitted by power plants is from coal combustion. About two-thirds of total coal production is used in electrical power plants, 360 million tonnes (about 400 million short tons). Thus, were these controls put into effect, with a total cost of some $3 Billion, the costs to the companies would be greater than that of the fuel. If oil shortages cause a shift to coal for electrical power generation and sulfur emissions are curtailed, it is certain that electricity prices will increase considerably.

The hidden costs depicted above are peculiar to the coal supply and use industries. The fundamental issue is, however, applicable to resource industries as a whole. For copper, some external effects are: the health and environmental effects of copper smelter operations on air and water, over-use of groundwater raising agricultural costs, and degradation of land areas. Other miscellaneous examples are smog from auto fuel consumption, possible health effects of pollution from taconite mining operations, and mercury poisoning in marine ecologies.

An example of the effect of tax structure on external costs is also in order. In the years 1969 through 1971, U.S. oil companies paid of the order of 6 to 10 percent corporate income tax on their earnings (Steinhart, 1974), compared to a level of about 42 percent for all U.S. industry. Steinhart & Steinhart estimate that this means that the average taxpayer paid extra taxes equal to about 20 percent of his yearly energy costs (including those used in the manufacture of goods he purchased) in order to underwrite this oil company "subsidy." Were he not paying this sum in the form of taxes, and fuel prices were sufficiently higher (other things being equal) to cover these taxes, it is likely he would have used less energy relative to his other needs and desires.

In terms of the impact on demand of including such costs, in production accounting, any or all of the examples illustrate the potential for future policy. It is important to note that it may, in the best interests of society, not be wise to internalize all costs of a particular segment of the society. The extent to which policy should encourage or regulate such developments must be influenced by political and social considerations outside of the question of decreasing total demand.

However, it now seems clear that the United States is definitely tending towards "full" cost accounting in its resource industries. We think that this trend will continue, that resource prices will rise relative to other segments of the economy, and that consequently there will be continued pressure towards lowering per capita demand over what it might otherwise have become.

THE ROLE OF PRICES IN MATERIALS USE

The role of price changes for energy poses complicated conceptual and measurement problems, stemming both from interrelations among the different kinds of energy with their different supply problems (solid, liquid gas, nuclear) and from the limited inputs of energy into most goods and the intermediate (vs. final) nature of these goods. A notable relative decline in real energy prices in recent years (1966 to 1973) seems to have spurred larger energy use than income change in the U.S. would in itself indicate (see Chapter XIII).

A number of recent analyses of this problem conclude that demand for total energy is price sensitive (Edmonson, 1974). Price elasticity is in the range of -0.4 to -0.5--a range (that includes plausible distributions) between long and short-run effects of price changes. This means that a price increase of 10 percent would lead to a drop in demand of about 4 to 5 percent. The specific calculation for 1985 showed that with the price level of total energy 57 percent above the prices of other goods (the base relationship is the price ratio in 1968) and if real GNP increases by 4.3 percent annually as suggested by the Council of Economic Advisors in 1971, and if population grows according to Census Projection D (made in 1970), total energy consumption will be 43 percent _higher_ in 1985 than in 1968. This contrasts with results from the same analysis of energy consumption 76 percent above the 1968 level if all assumptions are the same _except_ that real prices of total energy retain their 1968 relationships to other goods. Thus, _in this calculation_ a relative price increase of 57 percent achieves a _decrease_ of some 30 percentage points from the demand to be anticipated without the price increase.

More substantial signs of conservation, at least partly attributable to the steep price increases since 1969, have surfaced during the recent energy crisis. _Business Week_ reported on February 2, 1974 that the nation's use of electricity during the preceding three months had been running 10 percent below expectations on which capacity planning of utilities had been based. If--surprisingly to many practitioners in the field--higher prices for electricity will cut deeply into demand, much of the planned additional capacity will not be needed if peak demand changes accordingly, as is expected.

It would be fallacious to estimate a composite "elasticity figure" from the various studies that have been made about the reaction of demand to price changes. The strength of response depends not only on the severity and recency of the price change, but also on the characteristics of the particular situation, as defined by the public's awareness of the problem, its interpretation as a matter of collective vs. only individual (see the following section) and the decision-maker (business or government vs. private households).* What has been observable recently is a vigorous adjustment of demand as both private and commercial users become more aware of the price of electricity. While between 1950 and 1969 the price of energy fell sharply relative to all prices, this trend has been reversed in more recent years.

POPULAR REACTIONS TO THE ENERGY CRISIS

Two drastically discrepant types of reactions to the price increases and reduced availability of gasoline and other forms of energy can be identified under the convenient titles of the _competitive scenario_ and the _cooperative scenario_. The former would be characterized by a conspicuous absence of

---

*Elasticity estimates range from between -.1 and -.2 in the short run (meaning that a 10 percent increase in price will cut consumption by only 1 or 2 percent) to long run reactions for commercial and industrial users in the -1.5 range (corresponding to a 15 percent decrease in consumption in response to a 10 percent price rise).

discretionary conservation--an attempt to maintain or increase consumption, despite substantial costs in money, time, or convenience. To achieve a decrease in consumption, lines would have to get long and/or prices to rise drastically. In the cooperative scenario, consumers would practice conservation in response to modest changes in prices and availability. Scenario 1 would strengthen the case of the advocates of supply-oriented strategies. Scenario 2 would mean that private households are easily activated by both market and nonmarket mechanisms in favor of conservation.

The important question for policymakers, therefore, is--which is the likely reaction? The indicative evidence provided by the response to the 1973/74 energy crisis will be described in the following sequence: (a) to what extent has energy conservation been practiced? (b) are difficulties interpreted as a personal or a societal problem? and (c) are solutions sought individual/competitive or collective/cooperative?

From two different sources survey data dealing with the experiences and reactions of Americans during the recent energy crisis are available. There is first the Continuous National Survey conducted by the National Opinion Research Center (NORC) in Chicago (Murray et al., 1974a and 1974b). This Institute has interviewed a cross-section of about 170 American adults weekly, between November 23, 1973 and April 11, 1974. Second, the Survey Research Center (SRC) of the University of Michigan interviewed two samples of different size--both representative of American adults--with different questionnaires in February 1974. The larger study was a telephone reinterview of approximately 1,400 people. The smaller study was a personal reinterview of approximately 250 respondents.

Whereas only a minority had actually experienced shortages by the end of 1973, conservation at that point in time had been much more widespread (NORC). Sixteen percent of the respondents reported difficulty in obtaining electricity and only 3 persons out of the 106 heating fuel users had problems in obtaining heating oil. However, 59 percent of those having a thermostat report adjusting it downward compared with last year, 75 percent that they have reduced their lighting, and 27 percent that they run major appliances less often. Fifty-five percent of the drivers reported less car use. These were answers to a question offering respondents a set of predetermined choices for responses. The trend of these findings is supported by answers to SRC's open-ended question in early 1974, taken when difficulties in obtaining gasoline reached their peak.

In answer to the question:

> "Is there anything that you have done during the last few months to try to cut down on the amount of energy you use?"

about 70 percent of respondents report having reduced driving, 49 percent temperature in their homes, and 35 percent electricity consumption; 62 percent in the second SRC sample report adjusting their thermostats downward.

Later on, trouble with gasoline was reported by a majority of car owners. Conservation behavior increased only insignificantly in frequency.

Conservation extends almost evenly over the whole socioeconomic and age spectrum. This is also true for home temperature reduction, although the reported average reduction compared to last year is about one degree fahrenheit more drastic for families with incomes of $20,000 yearly and over that for those with incomes under $8,000 yearly (Murray and Minor, 1974b).

It would be too narrow to view this conservational response to shortages as primarily a form of price elasticity. The following evidence suggests that the average American interprets the difficulties as a matter for collective, rather than individual, concern. In response to a question inquiring about the importance attached to the energy problem, NORC produces the following estimates: most important, 26 percent; very important, 59 percent; not a problem, 8 percent. This evaluation has been fairly stable between November and April. The higher the importance rating, the more likely is conservation. For instance, cutting down on driving is reported by 37 percent of those deeming it "fairly important," 50 percent of those saying "very important," and 61 percent of those who believe it is "the most important problem in the country." The larger SRC-sample interviewed in February identified 53 percent as considering the energy problem as serious, 22 percent as not serious, the rest falling in-between. (In explaining the difference, the question formulation may have played a role as well as the timing: "serious" is a more extreme term than "important.") From these data we conclude that there is a great deal of public awareness of and concern about the energy situation, extending far beyond the impact of the objective changes alone. Further evidence indicates that the issue, besides being taken seriously, is interpreted more frequently as a problem of conservation than one that should be solved by increasing the supply. In the smaller SRC survey the following question was asked:

> "There have already been some shortages of energy and other resources in this country, and others are predicted for the future. In general, do you think most of these shortages can be avoided by new scientific discoveries or will we have to learn to consume less?"

Thirty-five percent of the respondents believe shortages can be avoided by additional supply, 44 percent think we will have to learn to consume less, and 16 percent plead for both choices. Technology may help but we still will have to economize.

Trust in others with whom one has to share limited rsources is the precondition for this cooperative solution. The larger SRC survey contains this question:

> "If there is a shortage of gasoline, or heating fuel, or electricity--do you think most people will try to cut down

on how much they use, or do you think most people will use as much as they want to?"

The vote of confidence in the wording of the second survey is considerably more demanding:

> "If there is a limited supply of something, such as gasoline or heating oil, do you think most people can be trusted to take only their fair share, or that most people would take as much as they could get, or what?"

Seventy-four percent in response to the first question expect that most people will try to reduce consumption. But only 29 percent, in response to the second question, think that most people can be trusted to take only their fair share while 58 percent believe others will take as much as they can get. Again, a thorough analysis of the difference would go beyond the scope of this paper. What emerges is a spirit of constructive pragmatism, a widely but not universally spread belief in the fellow citizens' responsible attitude toward conservation.

While there is some belief in the effectiveness of the cooperative model, additional safeguards are needed where cooperation does not suffice to overcome a critical situation. In response to the NORC question:

> "Do you think gasoline rationing throughout the nation is necessary?"

between 10 and 40 percent of the respondents at various points during the 6-month interval answered in the affirmative, with the highest proportions reached in November--when the oil embargo hit unexpectedly, but was not yet experienced--and again in February--when gasoline lines were longest. The somewhat more suggestive SRC question:

> "If enough energy is not saved by voluntary actions, what do you think should be done?", also asked in February,

results in 61 percent of the sample mentioning gasoline rationing. And in response to the NORC question:

> "What three actions would you most like federal, state, or local government to do in order to cut fuel consumption?"

eighty-one percent refer to speed limits as at least one of the choices, followed in popularity by the improvement of mass transit (51 percent), encouraging car pools (48 percent), and gasoline rationing (32 percent), with the relaxation of anti-pollution standards being mentioned by only 23 percent of respondents. The preference for rationing, then, is clearly contingent on the belief that a serious problem exists. In answer to the question:

"Why do you think gasoline rationing is not necessary?"

seventy-five percent of those respondents who think gasoline rationing is not necessary answered: "There is no real shortage." Only one-third of the critics of rationing appear to be philosophically opposed to rationing (Murray and Minor, 1974b).

In summary, then, these results suggest that most Americans, rather than trying to get more than others, want to be sure they get as much as the next fellow. Rationing and speed limits, in the absence of sufficient voluntary conservation, would be accepted in the interest of an equitable distribution of scarce resources. Should the market allocate smoothly and efficiently, i.e., without steep price increase, this would be the preferred mechanism. Should this not be the case, the market solution would not be considered equitable by most: people do not want to see those who can afford it get by without consuming less. If others can purchase more and do so, there is less left for me, and I have to pay more for it. In return for seeing others curtailed in their consumption, people appear to be motivated to accept limitations to their own consumption. But the reverse is also true: the motivation for conservation above and beyond price elasticity is contingent on the assurance that the burden of conservation is shared by all or most. Rationing, speed limits, etc., are recognized as mechanisms for achieving this objective. To ignore people's notions of equity of energy allocation and to rely entirely on the market or other allocation mechanisms of dubious public acceptance would seriously endanger the spirit of cooperation that emerges from the data presented.

Let us then attempt a generalization on the basis of incomplete data with insufficient temporal depth. The manifest signs of behavioral and attitudinal adjustment lead to the conclusion that welfare losses from conservation have so far been small. Very many Americans have come to realize that their use of energy has been profligate; significant conservation can be achieved at a relatively minor price in terms of convenience and comfort. All in all the evidence strengthens the case for management of demand and points to the considerable prospect of discretionary conservation in the private sector.

Yet we must go beyond stating that conservation is a possible behavior pattern, and attempt to specify the conditions under which it is likely to be practiced. It must be remembered that concern and awareness of a problem often represent passing and temporary phenomena. To the extent publicity and saliency have contributed to the impressive mobilization, will the behavioral changes persist after the energy problem has vanished from the headlines? While the energy and materials problem may diminish in immediate saliency, it is not likely that it will be solved in the near future. In this case, the spontaneous, respectably strong adjustment observed during the past few months may well indicate a pattern for the future. Since the most important conservation decisions involve consumer durables and are made rarely, adjustment takes time. As additional appliances come up for replacement, houses for sale, and as the supply of consumer durables reacts to the new preferences, more and more of the energy-expensive brands and models can be expected to give way to the more economical ones. Furthermore, reinforcement of conservation behavior through the diffusion processes, social comparison, and finally social normation is time-consuming. The slow change from the model 3-4 child family of the fifties to the 2-child family of the early seventies comes to mind as an example. And finally, the importance of a collective sense of trust and purpose must be reiterated. Individual conservation is vastly more likely if and when it is perceived as conforming to a commonly shared set of values, goals, and behavior pattern. More than other institutions, the government is being charged by the people with the responsibility of articulating these values, setting the goals, and seeing to it that conservation is practiced by all.

CHAPTER XVI
U.S. CONSUMPTION PATTERNS AND FUTURE DEMAND FOR ENERGY

Do emerging personal consumption patterns among Americans suggest that demand for energy is likely to continue increasing at recent rates until the end of the century? Although it is impossible for us to provide a definitive answer to this question, it is nonetheless worth pointing out that not all trends in American Society favor increasing consumption of material goods. In effect, the American way of life may not involve an immutable commitment to consuming ever greater amounts of energy and minerals per capita; hence conservation, reduction of wastage and a slower (or zero) rate of growth of consumption of energy and minerals will not necessarily involve serious socio-cultural shocks. In this chapter we shall discuss briefly some major social and economic trends that (a) suggest a shift of personal consumption to the service sector of our economy (a sector that we will here assume to be less energy intensive than manufacturing) and (b) imply, because of shortages of time, an upper limit to the desire for consumption.

INCREASED CONSUMPTION OF SERVICES RELATIVE TO GOODS

This section explores two reasons for believing that Americans may substantially increase their consumption of labor intensive services relative to goods over the next quarter of a century. One reason relates to the price of these services versus goods in our economy, and the other concerns changing tastes.

In speaking of the service sector of the economy, we are not including energy-intensive parts of the nongoods producing sector. That is, following Victor Fuchs in his book The Service Economy we are excluding transportation, communications, and public utilities, leaving as the service sector wholesale and retail trade, finance, insurance, real estate, general government, and the traditional services such as professional, personal, business, and repair services. This sector is admittedly quite heterogeneous; however Fuchs points out that most of the industries included are manned by white collar workers, are labor intensive, deal with the consumer fairly directly, and produce an intangible product. Nevertheless, we must note that although it is generally agreed that the service sector as defined here is less energy-intensive than manufacturing, transportation, communication, and public utilities; research is badly needed concerning how much less energy-intensive services are. This is especially true since there is a tendency for services to become more highly capitalized and, hence, perhaps more energy consuming.

The price of services relative to goods may very well go down over the next 25 years both because of augmented productivity in the service area, and because the price of goods will be more affected by increasing energy and materials costs than the price of services.

The service sector now employs well over half of the American labor force. In contrast to the rapid growth of employment in services, the overall contribution of this sector to output has been relatively static, owing very substantially to lack of technological innovation, traditional management practices, and the absence of economies of scale. As Fuchs has shown, whether real output is measured in constant dollars, or current dollars, taking the period 1929 to 1965, the service sector's share of total GNP changed very little since 1929. The share in constant dollars was almost exactly the same in 1965 as in 1929--48.3 as against 48.4. This contrasts sharply with the share of employment in the service sector which rose from 40 percent to 55 percent in the same period.

Thus, according to Fuchs the rise in employment in services has not been a response to rapidly rising demand for services--due to rising incomes and high income elasticity, lower prices of services, or changing tastes--but because of "a dramatic difference in sector rates of change in output per man" between industry and the service sector. As a consequence, the price of services relative to goods has increased greatly in American society, giving rise to the oft-mentioned "scarcity" of services.

A continuation of such trends seems very unlikely in view of existing changes in some important aspects of the service sector. Leaving aside prospective technological breakthroughs, the application of existing technology to major industries in the service sector is already taking place, aided by structural reorganization making these industries more amenable to the application of advanced technologies and economies of scale. Because many of these industries are increasingly being financed out of tax dollars, there is a built-in political pressure for higher productivity and "accountability."

The price effect of increased productivity in the service sector could be augmented by the fact that the cost of producing goods may be more adversely affected by energy and materials shortages than the cost of producing services. Thus, the price advantage that material goods have enjoyed compared to services in the past may be seriously undercut both by a rise in the productivity of services and an increase in the materials/energy cost of goods.

Regardless of price changes, will people want to consume more services? There are some reasons for expecting a shift in taste to the service sector:

the aging of the population, the increased labor force participation of women, and changing symbols of consumption. We may consider each briefly.

Assuming a continuation of low fertility in keeping with the medium population projection of our report--268 million by the year 2000, the American people would by that date be older than in the current period. We would have a median age of 34 in contrast to a median age of 28 in 1970. Thirty-three percent of the population would be in the economically dependent category, and of these dependents 33 percent would be aged 65 and over. By contrast, in the United States in 1970, 38 percent of the population was in the dependent category, but only 26 percent of this dependency was made up of aged persons. Moreover, the population would be on its way to even greater aging as it reached a stationary state early in the twenty-first century.

On balance, it would seem that this shift in the age-structure of the dependency burden would entail increased demand for services. This may result in part because the health and welfare needs of the aged are greater than those of the young. For example, as standards of health care rise, they apply more to the aged than the young since the young require less attention to maintain health. An additional factor of importance leading to the expansion of services consumed by older as against younger dependents is that older persons are decreasingly living in multi-generational family units and increasingly in retirement homes and the like. Children are typically housed and cared for in families.

Another trend in American society that suggests an increased demand for services is the greatly augmented labor force participation by American women. Whereas between 1910 and 1940 approximately 25 percent of women of working age were in the labor force, by 1970 this figure had increased to over 40 percent. Moreover, the largest increases have been among married women with young children.

A consequence of such changing rates and patterns of female work participation is a major loss within the household of the services of wives and mothers. Even taking into account that some of these services are now replaced in the market by the very women who would have offered them on a non-market basis in the past, and that most women with families bear a "double burden" and attempt to keep up with their domestic tasks, we find ourselves constantly remarking on absolute gaps in service that have not yet been filled by normal adjustive market mechanisms: makeshift baby-sitting arrangements, latchkey children and teenagers, lonely and virtually unattended old people, and the heavy workload and responsibility borne by wives and mothers in the labor force.

Although it seems unlikely that the public sector will compensate for this loss, it does seem probable that more women will demand services in the market. If increasing proportions of women become earners, and particularly if women break out of the traditionally low paid women's occupations, there seems to be a major potential for increase in services that will help women retain a desired kinship status without sacrificing their jobs or virtually all of their leisure time. For example, it seems inconceivable that the trend in labor force participation among women with young children will continue without giving rise to a larger network of private professional nurseries and child-care centers.

Finally, we might postulate a major shift of symbolic and invidious consumption from goods to services. It has long been recognized that much of American purchasing behavior is conditioned by a desire to use goods, in varying manners and to varying degrees, as status symbols. Now however, increasing educational attainment is widening the range of alternative consumption symbols. Moreover, with goods widely available in the society, people desire consumption symbols that express their unique claim to superiority--typically symbols that require the consumer to exert time, effort and talent in order to acquire them. Thus status differentiation is increasingly requiring the consumption of services rather than goods--services relating to health, grooming, athletic, artistic and linguistic skills, gourmet tastes, various kinds of particular information areas, and the like. Pockets of enterprise in the service sector have already responded enthusiastically--the "beauty" industry being perhaps one of the best examples. But, it is possible to conceive of a boom in the competitive consumption of services at the private level--a boom which could greatly divert consumption from the energy-intensive goods that we have been using so lavishly to symbolize status.

TIME CONSTRAINTS ON CONSUMPTION

For all human beings time is an ultimate constraint. However, the more societies move toward a cash nexus, and the more prosperous on a per capita basis they become, the more are individuals subjectively impelled to put a market value on all of their time. They come to assess their nonmarket activities in terms of their opportunity costs--the market "worth" of their time, and to place high value on harnessing forms of energy that net them "extra" time--mechanized labor-saving devices and the like. Nonetheless, there is a limit to our ability to harness nonhuman energy without personal efficiency losses. Our labor "saving" devices require maintenance, repair, and storage. Moreover, frequently they tempt us to augment the scale or complexity of our mode of life, because their use is more efficient if attempted on a larger scale. In this manner, it is not difficult for individuals to suffer a rapid displacement of goals--they spend increasing amounts of time involved in the compulsions of a system of activities that they originally embarked on for purposes of saving time.

Equally, the enjoyment of material possession--for example, boats, skis, or second homes, as well as the consumption of services, require personal inputs of time. Because of the high value that individuals place on their time, they are tempted to try to do many enjoyable things simultaneously. As Stephen Linder has pointed out in his imaginative book The Harried Leisure Class, this effort can get to be more like work than fun.

Highly industrialized societies throughout the world have in the past decade been reacting against this so-called enslavement to market activity, mechanization, harried rushing, and material possessions. This revulsion would seem to be a stage in a learning process whereby human beings come to terms with the wholly new definitions of time that modernization has created. It would appear that the process of mature industrialization is impelling individuals to metabolize the advantages of modernization into an organized hierarchy of life-style priorities, and that this process is engendering a set of self-imposed limits to consumption. To many people, it just seems more pleasant to lead a less complex existence than to avail themselves of all possible "advantages."

Although this trend is difficult to measure quantitatively, there are significant indications of changes in tastes that are "simplifying" in their content. Blue jeans, once the badge of hippiedom, today make up a significant proportion of the daytime wardrobes of even the "squarest" teenage girls. The steadfast loyalty of modern women to sportswear (interchangeable "separates" rather than dresses and complicated "outfits" suited to limited times, places, and seasons) has effectively closed down many a large garment business in the United States, as well as virtually wrecking *haute couture* in France.

In sum, it seems as if, with experience, people in highly industrialized countries are learning to place their consumption in a ranked order along with other ways of spending their time. This learning process seems to be setting an upper limit to their taste for expensive material consumption and makes it highly precarious to believe that they will wish to go on consuming indefinitely at an increasing rate.

CHAPTER XVII
DESIGN AND USE OF PROJECTIONS FOR POLICY

FORECASTS AS INSTRUMENTS FOR POLICY

As discussed above, many existing forecasts are of weak numerical reliability. More importantly, they do not make explicit the implied conditions or the scenario to which they apply, hence they are of limited usefulness to guide policy. For policy use, projections should be conditional telling how the rates of use would vary under different policies and circumstances.

The value of existing forecasts is their consensus that at present rates of increase the "use" will exceed the supply of energy. Of course, in reality the rate of consumption cannot exceed the rate of supply, so that this contradiction in their projections means that, if crises and wrenching dislocations are to be avoided, there is need to broaden and deepen the sources of supply and, at the same time, to determine modes of adjustment in rates of use which are consistent with social objectives of equity, efficiency, quality of life, and environmental integrity. Existing forecasts, then, indicate the likelihood of the need to restrain use just as well as a need to increase supply.

Since precise, fully articulated, accurate conditional forecasts are not scientifically possible, forecasts which are useful for policy must be constructed to answer particular questions in terms of their variables, their stated relations, and their time scale. In simple terms, the need is not for an ultimate projection, but for a set of conditional projections differing among the considerations undertaken and the time-span considered.

LONG-TERM FORECASTING--THE STATE OF THE ART

At present there are two strains of modeling in long-range forecasting both highly aggregated so that they may be called macro-forecasts by analogy to macro-economics. The first of these might be termed econometric modeling and accounts for the vast majority of forecasts. The other, of which there are fewer instances, might be termed physical simulation modeling. "System Dynamics" models, such as those sponsored by the Club of Rome, are instances of this approach.

The physical simulation models have received considerable public notice, and have been much criticized by the scholarly community, on the grounds of omission of important relations and weak factual bases. However, these models embody some interesting advances in modeling techniques and introduce a greater concern for physical and environmental interdependencies than do econometric models. Some of their more useful features may in time be reconciled with the more established econometric approach, both in terms of the numerical analysis of dynamic processes and of the consideration of physical relations.

Econometric modeling is a widely practiced activity in many areas of forecasting and policy analysis. Energy studies have remained an underdeveloped province within this technical approach. Until recently, for instance, there were no estimates of the price-elasticity of the demand for gasoline, and the few recent estimates are crude and tentative. Further, there appears to be little exploration of income elasticities, cross-elasticities or technical rates of substitution, of the characteristics of derived demand, or of identification of the set of functional variables.

An ideal forecasting instrument would consist of a large number of carefully calibrated numerical relations representing well-understood individual and institutional behavior, technical factors and advances, physical and biological relations, and the responses to policy alternatives. A model such as this would ideally be linked to a comparable model for supply or production, which would tell us what is possible and how it is possible. Any forecast of demand (use) will depend on assumptions about supply, whether they are made explicit or not. Conversely, supply will to a considerable degree depend upon demand. If increasing scarcity produces higher prices and public action, supply will expand in many ways, including the exploitation of resources previously too expensive and the development of new sources through investment in science.

But the vision of such an elaborate forecasting system is patently beyond the capacity of our present theory of facts. Given our present level of knowledge, understanding of variables and their interdependence, and unreliability of data, it must be judged that any attempt to construct such an elaborate model would fail. A micro-model which tried to include a great many variables might be no better and perhaps worse than present macro-forecasting models, however crude.

For all of that, the development of such models is well worth undertaking, recognizing that it will be a slow and expensive process, involving many failures as the cost of some successes. In this respect, the development of such models would be comparable to other forms of capital investment in the energy field, being slow to bring into line, expensive, and involving intelligent risk-taking. However, well before such models could be used with confidence for numerical prediction, they should be extremely useful to help in the understanding of the complex interdependencies of the energy system. In brief, then, it is not realistic to expect dependable long-run modeling for many years, but it seems advisable to begin substantial investment in this type of modeling, preferably initiated simultaneously with different scholars in multiple institutions, recognizing that the payoffs will be gradual and developmental.

THE DILEMMA OF FORECASTS: THE NEED FOR POLICY FORECASTS

The market, as an automatic self-adjusting mechanism, functions best in situations when the rates of adjustment to change are relatively fast, or when there is high predictability or reliable forecasts of the longer term future. Neither obtain in the case of energy; hence the need for public policy and intervention. There are long lags in response. On the supply side, bringing in additional capacity within existing technologies or bringing in new technologies are very slow processes. Less often recognized, there are long lags on the demand side as well, involving the turnover or modification of vast stocks of capital in housing, transportation, and industrial installations. Coupled with these slow rates of adaptive response, there is the patent difficulty of reliable long-range forecasting which has been discussed. The dilemma is that we need to see ahead, but that we do not know how.

The way to proceed in the face of this dilemma is not, as is sometimes done, to pretend that the problem does not exist, either by relying on automatic self-correcting market processes or by pretending that projections can be made of a transparency and accuracy which are clearly beyond present scientific capacity. To apply the best intelligence within the constraints of reality, it is necessary to understand better the uses and possibilities of demand projections.

The immediate need is for the development of capacity to do short- and medium-range policy projections. Such projections must recognize that any possible future is conditional upon a host of circumstances, ranging from the structure of nature to historical accident, and including the policies chosen. At the same time, since precise, fully articulated, accurate conditional forecasts are not feasible, there is need in policy formulation for a series of partial forecasts constructed to answer particular questions. The variables, the specified relations, and the time-scale will depend upon the purpose. In the most sophisticated ones, the probability distributions of costs and benefits will be considered for choosing policies.

It must be recognized that there is today little competence and informational infrastructure for these policies with respect to energy questions. But there are a large number of technical approaches developed for other policy fields which can be transferred. The need, then, is to begin investment in research and application, in development of technical personnel, and in institutional development of public agencies for the intelligent use of these projections.

These short- and medium-term forecasts, like long-range forecasts, must be frequently revised, in the light of new data or new understanding of relations. Too often outdated forecasts become fossilized, and are quoted without reexamination. Whereas long-run forecasts serve to point to probable distant outcomes, short- and medium-term forecasts are used to modulate policy responses, responding to recent information and to the feedback from prior policies. Thus, long-run forecasts serve as an uncertain navigator, while short- and medium-term ones function as the steersman.

SIDE EFFECTS AND PUBLIC PARTICIPATION

It must be stressed that conditional forecasts must be particularly concerned with possible side effects of policy or of changing circumstances, as well as with the direct effects. For instance, a doubling of gasoline prices may lead to a given reduction in energy consumption. But it may have important secondary consequences, such as a substantial sudden loss in value of suburban property, or the going under of many family farms that depend on long-distance commuting to jobs in manufacturing for supplementary income. Such shocks and stresses may be more important than the eventual form of the adjustment, not only because of national concern with equity, but also because, as was demonstrated during the 1973-1974 energy crisis, those who feel themselves disporportionately affected may express their grievances in forms highly disruptive to the society.

The low capacity of technical planning to anticipate these distributive effects, and the fact that many problems in equity are not subject to technical solution, lead to two conclusions. First, that the process and findings of technical analysis must be made as intelligible and widely available as possible, so that diverse sectors of the public and the economy may understand and judge prospective impacts on themselves and make themselves heard in the democratic process. Second, that since the issues and consequences are novel, efforts should be made to facilitate such participation by sectors of the public and the economy in the interest both of supplementing the limited capacity of technical analysis and of providing a forum for the necessary transactions and compromises among these sectors. This will require such instruments as hearings, advisory commissions, and coordinative councils of interest groups, and may require new resources of political and legislative inventiveness. The alternative may lead to acrimony, divisive conflict and disruption, and functional vetoes and possible threats to the civil order.

CHAPTER XVIII
CONCLUSIONS AND RECOMMENDATIONS

CONCLUSIONS

As discussed at some length above, we conclude that the demand projections forming the basis for current emphasis on large increases in energy supply are likely to be much too high for conditions that might be expected during the next 25 years. This stems from the committee's findings, recently supported by others (Chapman and Mount, 1974) that these popular demand projections are exaggerated by assumptions concerning the continued cheapness of energy and minerals that seem unrealistic for the foreseeable future. Because of the many shortcomings of current forecasting data and techniques, we are not able to quantify how much too high these forecasts may be.

The Demand Panel's overriding conclusion is that all-out efforts to increase supplies of fuel (and minerals) in order to close the gap between popular demand and supply projections are misguided; and will place an unnecessary hardship on the nation. A policy that forces the nation to undertake extremely rapid development of known energy sources in order to match demand projections that have been based on price trends that appear unrealistic for the future can also generate the demand (for example by keeping prices artificially low through subsidies) to use the supply. In this sense the demand projections can become self-fulfilling. We believe that policy should not abet this outcome without grave consideration of the various costs involved. In view of the findings of other panels that supplies for the long-run are limited; that technology can provide only limited solutions, involving long lead times for their implementation; that growth in use of materials and energy is causing crucial environmental problems; this panel believes that policy should support social trends tending toward a lesser rate of consumption of energy and minerals, incorporate the long-term replacement costs of externalities involved in extraction and use, and actively augment measures designed to curtail wastage insofar as these cannot be efficiently handled by the price mechanism. Such a policy would be an essential anticipatory measure to prevent the disruptive adjustment processes likely to result from short-fall of supply.

RECOMMENDATIONS

Research and Information

1. Means should be taken to improve and require systematic collection and analysis of information, and to implement a more effective organization of government to make use of this information in order to anticipate contradictions between rates of use and supply. These functions should not be limited to agencies within the executive branch of the Federal Government, but efforts such as the current survey commissioned by the Congressional Office of Technology Assessment, should be supported and augmented as a basis for further effort in this area.

2. Greater technical ability should be developed to make better use of data and information in conditional and policy forecasting, particularly short- and medium-range. Important aspects would include consideration of alternative circumstances, policy options, and their direct and indirect social and economic effects. This will require the development of institutional capacity to generate and use policy forecasts, and substantial investment in applied social science research in appropriate institutions to develop the needed techniques.

Policy Recommendations

1. To the degree possible, policies to modify demand should be designed to function through the market mechanism rather than through direct governmental controls and allocations. This reliance on the market requires policies which improve its functioning in terms of supply prices which fully reflect costs and in terms of improving the information on which decisions are made. More specifically: (a) Legislation and regulation should be designed so that supply prices fully reflect production costs. This might include taxes or side payments which bring the costs now being imposed on others without compensation within the producer's accounting. It might also include exploitation taxes which in effect would bring the discount rate of the market, on which producers base their decisions, more closely into line with the social rate of discount in balancing the interests of present versus future levels of consumption. Such measures would reduce rates of consumption by making users pay the full social costs of production. (b) It should be required that energy-consuming products, such as buildings, automobiles, or appliances, carry when sold, clear information as to their rates of consumption, efficiency, and other pertinent information relating to operating costs, so that consumers may make decisions based on total costs rather than on first costs. Informed consumers would waste less.

2. *Compensatory programs and policies, and the necessary legislative and administrative basis, should be prepared and held in readiness to assist those particularly hurt by sudden transitions, particularly those arising from public policy. Those affected might include geographic districts, industrial and commercial sectors, or workers who are laid off or experience substantial loss of real income because of higher commuting, heating, and food costs. The logic of this recommendation is similar to that of disaster relief programs.*

3. *All elements of Federal legislation and regulation should be reexamined to assess their implicit effects on energy consumption. For instance, agricultural policies which restrict planted acreage promote the use of machinery and fertilizers; the income-tax deductibility of home-owners' property tax and morgage interest payment contrasted with the non-deductibility of rents, encourage low-density urban patterns which greatly increase the use of the private automobile. Such policies, not directly aimed at energy or resources, should take those indirect consequences into consideration.*

4. *Through education, publicity, and example, those habits and attitudes sometimes called a "conservation ethic" should be encouraged. Although there is a diversity of opinion as to the likelihood of major changes, experience in areas such as public health and traffic safety suggests that some effects are possible.*

5. *There is a strong possibility that there will occur from time to time "crises" in which a sudden sharp short-fall of supply results in severe dislocations without there being time for the smooth functioning of the market processes of adjustment. In such cases direct regulation, control, allocation, and rationing may be necessary. Insofar as such crises cannot be sufficiently anticipated by earlier recommendations, there should be developed and held in readiness a contingency set of programs and policies, together with the necessary legal and administrative mechanism. These should be continually reviewed and updated in the light of information gained (see Research & Information Recommendations 1.2). Particularly, there is need to have in readiness a flexible response for those particularly hurt (see Policy Recommendation 2).*

Such compensatory and assistance programs are needed not only for equity reasons, but quite importantly in order to avoid the rending of the social fabric of which there was a foretaste in the violence and disaffection which accompanied the rather brief 1973-74 energy crisis.

Recommendation on Supply Policy

Exercise restraint in all-out efforts to increase supply to levels that could exceed requirements. Past experience suggests that increased consumption can be induced when excess supplies are available. This would cause increased degradation of the environment and wasted use of capital, manpower, and other national assets--assets that the nation may well have greater needs for in the future.

REPORT OF PANEL ON DEMAND FOR FUEL AND MINERAL RESOURCES: A MINORITY VIEW

Wilfred Malenbaum, University of Pennsylvania

Why a Minority Dissent?

This panel--on "demand"--is concerned with minerals' use (benefit, gain); the other panels--on "environment," "resource assessment," "technology"--emphasize supply (cost, pain). The importance of close interaction and integration of the demand side with the supply side is explicitly recognized in the Charge to the Panel. COMRATE sought a balanced analysis of the U.S. minerals' position in contrast to the usual analysis with primary emphasis upon the forces of supply.

This Minority View holds that the Report* fails to offer a scientifically valid analysis of minerals demand. The Report uses a conceptual framework not suitable for assessing the benefits and gains from the use of primary inputs. Without this approach, the Report can make no contribution on demand. Its demand arguments are romantic, not scientific, its conclusions on demand are not derived from valid hypotheses or statistical relationships. The Report offers its "overriding" conclusions on a new demand role in national policy and action. This is an important objective; it is disserved by the technical inadequacy of the Report's treatment of demand.

In broader vein, the present Report deprives NAS and COMRATE of the opportunity to consider integrated and balanced views of the U.S. energy/minerals position, and of the policies and programs that might best serve energy/minerals in the U.S. and the world.

Demand for Energy/Minerals

Essentially all energy/minerals demand arises because these resources are basic inputs in the final goods and services a nation desires. These inputs are used up in product, in power, in transport. Changes in the demand for these raw materials thus tend to be closely associated with changes in national product. What about (real) price changes in these inputs? Such price changes tend to have limited influence on the amount of energy/minerals

*Hereafter "Report" means this Panel's Report.

used because in practically all cases the cost of these inputs is small compared to the cost of the final goods and services people want to buy.

Income elasticities for energy/minerals are straightforward in concept and in measurement. Price elasticities are neither, a consequence essentially of the derived nature of energy/minerals demand. Recent estimates for total energy in the U.S. suggest long-term elasticities of about unity (1.0) for income and of about (-) (0.25) for price. Increases of 50 percent in the relative price of energy over a decade or so might thus be accompanied by a 12.5 percent decline in energy use. Real national income growth of 50 percent over that decade, would in itself tend to generate a corresponding increase (50 percent) in energy use. With both price and income increases, energy demand would expand by more than 30 percent.

These round numbers are illustrative, but the point on the predominant role of income change in use of energy/minerals is valid: it reflects the strength of a fundamental technical input relationship. The relatively smaller importance of price changes is also valid: it arises from the derived nature of the demand for energy/minerals.

The historical record of energy* consumption in the U.S. provides ready evidence of the role of income in this growth. For the half century to 1970 for example, annual data reveal the high correlation of these two measures. The regression coefficient has high statistical reliability, whether total consumption levels or changes in these levels are analyzed. That consumption not associated with income level (or change) is readily studied through the intensity-of-use concept. This is the amount of energy used per unit of real GNP. Obviously energy use (or change in use) is the product of GNP (or change in GNP) and the appropriate intensity.**

Intensity of use shifts because of substitution of inputs, especially as a result of technological change in producing different kinds of energy and in different ways of using energy and as a result of price change, and because of changing composition of people's demands--that is, of the changing structure of GNP.

---

*The discussion will hereafter focus on energy, although the argument essentially pertains to both energy and minerals. The Report is primarily concerned with energy.

**Use of population as an explanatory variable has little support in the actual record (nor, as indicated below, in the logic of consumption change). Thus no significant regression coefficients for a population variable are obtained in the statistical analysis, whether the measure of population change itself is correlated with change in energy consumption or whether population and income variables are used together in the analysis of energy consumption.

Historical experience and theoretical considerations support the existence of fairly general patterns of intensity change. For energy (and for most minerals) the U.S. and other rich lands show declining intensities: such nations tend to use less energy per unit of GNP as GNP per capita increases. Study of GNP change and intensity change thus permit identification of the causal factors in changes of total energy use. Both components--GNP and intensity--are readily measured for past periods. Both have recognized bodies of disciplinary doctrine.

They are thus powerful tools for any process of projecting future demand. The change in the level of GNP focuses on the amount of energy/minerals that must be used up. Changing intensity levels reflect society's decisions on relative gains and costs from energy use, given alternative inputs, and products, methods, prices. Together these two components can illuminate energy use.

Demand Analysis in the Report

These tools suit theory and fact; they provide the framework for most demand projections. But the Report takes a completely different course. Its analytic framework treats changes in energy use as the product of changes in population and per capita energy consumption. Again there is the convenience of two components that exhaust the total (as do income and intensity). Here however neither component has theoretical (nor even logical) attributes for analysis of energy/minerals demand. Thus 70 percent of the world's population uses less than 14 percent of the world's energy. Between 1951-55 and 1966-69, U.S. population increased by some 45 million persons and our energy use by close to 800 million metric tons (coal equiv.) In the same period Mainland China's population went up by nearly 250 million persons and energy consumption by less than 270 million tons. All such data show that population numbers and energy use (or changes in them) are not related--until the income variable is introduced. For income is the key energy use determinant.

Nor is the population-income relationship straightforward. Population growth is often held to have negative consequences for income growth, particularly in poor lands. Indeed this possibility is a major reason for the existence of COMRATE! For poor nations a positive population-energy relationship is assured when population and per capita income have a positive relationship. The Report uses only a positive population-energy association, and hence precisely this positive income relationship. This is a strange assumption for a COMRATE panel to adopt implicitly, without discussion. The other demand component, per capita energy consumption, is a purely descriptive variable. It varies, but its values have meaning only in context with other variables. Thus for the U.S. high consumption in 1950 may be low in 1970; or high consumption in New Zealand in 1970 would be low in the U.S. Critical is what energy is used for, and why. The characterization, "energy-saturated," has an emotional, not a scientific thrust. At the least, per capita energy consumption must be considered with per capita income (again the energy-income relationship) but even this would not make per capita energy consumption a useful analytic variable. By itself it has no significance for an economy, society, people: its use is romantic, not analytic.

The Report's persistence in a structure without methodological validity imposes a heavy toll. Herewith a few among many possible illustrations:

1. The Report misinterprets factual matters.

To my knowledge the Report stands alone in the components it uses to study energy demand. And it assumes other analyses follow the same course. Thus it discusses such key studies as those by Interior and AEC as though they were not read, certainly not understood. These two project energy estimates on the bases of GNP and intensity-type considerations. Their specific assumptions on these components are legitimately open to question, especially in the light of later experience. It is true they do divide energy/materials projections by projected population and thus derive a per capita consumption figure, for descriptive purposes. But this is not a component of their demand analysis. Such use, again, seems confined to the Report.

The Report's methodology impedes its handling of price elasticity studies. The Edmonson manuscript could be helpful to the Report's themes, but it is misread. That study (suggesting a price elasticity that is relatively high for energy: -.4 to -.5) anticipates large increases in per capita energy consumption by 1985, its terminal date, despite marked relative price increases in energy over the intervening period. The fundamental matter here, and more so in price elasticity studies by other scholars, is again the dominant role of income in energy consumption. By essentially passing over the fundamental relationship of income and energy, the Report simply misinterprets relevant work. Such treatment of factual material is unscientific as well as misleading.

2. It endorses a population-energy fallacy (more population, more energy).

As already indicated, the basic interdependences here are complex. They can be used in the simplistic (more people more energy) form of the Report only on the assumption that population growth is always associated with greater per capita income--"limits"--considerations notwithstanding. On the other side, even where both population growth and per capita income are expanding, population does a poor job as a guide for growth in energy consumption. For energy tends to grow with income, which is assumed to be growing more rapidly than population.

The Report exploits this population-energy misconception in a way which mocks the COMRATE exercise. Thus the Report illustrates the nature of multiplication in a table (p. 319) which gives the product of alternative population projections to 2000 with alternative per capita energy consumption levels. In particular, energy per capita at the 1965 level (which the Report obviously likes for the future) and energy per capita derived from an AEC projection for 2000 (a level above 2-1/2 times the 1965 level and which the Report obviously dislikes) is each multiplied by a population estimate for

2000 which reflects rather rapid population growth for the U.S. The table is meant to dramatize how large is the product when the two high components are multiplied in contrast to products from smaller components. But results from multiplication aside, the Report has no concern about using a wide range of per capita energy consumption levels with the same population figure. The possibility that increments in population will mean diminishing increments of economic and social returns happens to be a basic concern of COMRATE.

Demand Outlook in the Report

The Report presents no projection of its own for future energy consumption. But it makes amply clear what it considers appropriate orders of magnitude. Using its own components, it presents (via the multiplication table discussed above) the products of "revised population projections"--a range with some current acceptance for the year 2000--and "credible" per capita energy consumption levels. There is no definition of "credible" except that it encompasses actual per capita levels of the recent past--1970, or better 1965--with which "we have already successfully lived." Or a "credible" per capita level yields "essentially the same total consumption in 2000 as actually occurred in 1970." This would require an inverse relationship between changes in population and in per capita use, but the Report also accepts as appropriate for 2000 "an increase of 25 percent over total consumption in 1970." While this flexibility might thus save (for a while) the simplistic population-energy relationship the Report has adopted, there is no discussion of the why of any per capita level. A 25 percent increase is "of quite a different scale" from what follows with Interior and AEC per capita energy projections: these are "highly unrealistic," without "credibility."

It is important to ask what the Report does not: how do "credible" projections relate to GNP? The Report does not advocate a stationary state or even constant per capita income. Rather it accepts "the possibility of the U.S. doubling its GNP (by 2000) without increasing its (total) energy consumption." Given the Report's population range for 2000, this would mean at least a 60 percent increase in per capita GNP. It would also mean an intensity-of-use of energy that was half current levels. These realities concealed in the Report's "credible" outlook warrant some observation. Doubling of GNP means an annual rate averaging under 3 percent; there is little basis for an average below the 3.5-4.0 percent for this long period, however much of the Report believes "economic growth appears to have lost much of its magic as a national goal." Per capita income well over 60 percent higher than current levels is not readily reconciled with 1970 levels of per capita energy consumption. That reconciliation, as the 50 percent reduction in intensity, demands head-on confrontation with demand patterns, in theory and practice.

Instead the Report offers wordy and needlessly extended accounts of social forces on the U.S. scene that will encourage more rapid expansion of service sectors of the economy (pp. 297-300). It offers some evidence (?) of popular receptivity to conservation practices (pp. 291-296). The Report fails to see that at best these are part of intensity-of-use changes which have always had much less force than GNP changes in energy consumption. To counter factual and analytic reality, it needs not assertions but alternative demonstrations. The Report has no tools capable of such argument. Indeed, after stating that intensity patterns are not appropriately compared among nations, it offers, without discussion, that U.S. energy consumption should be "like New Zealand's."

The crux of the matter is that the Report has simply bought the conclusions of the supply panels: "limitations of many nonrenewable resources", "the unlikelihood of high rates of substitution and other technology-dependent miracles".... Given these, why worry about supply and demand interactions in an expanding U.S.? Demand will somehow adjust to "credible" levels consistent with the limitations on the supply side. This provides the theme for the Report's Conclusion, as we note below.

Forecasting Design

The Report's main objectives were to evaluate current projections, identify areas in need of improvement and, from its own analysis of demand, make "constructive suggestions for improving forecasting design." Performance on the first three objectives has already been discussed. On the fourth, it offers a relatively remote and abstract brief chapter (Chapter XVII) to indicate what good long-term projecting, forecasting should be. Insofar as this describes real operations, it calls for more rigor in the application of the presently fashionable device of alternative scenarios. But in no way whatsoever does the argument of this chapter bear upon what the Report actually does. Indeed the Report's reliance upon romantic, "credible" energy consumption figures for 2000--with improbable income conditions with no supporting argument--belittles what the Report identifies as one of its major objectives. Its "scenario of how these (low levels) may be approached," as "summarized" in a simplistic Figure 1 (p. 273) makes clear how little attention the Report gave even its own vague and general recommendations on this key problem.

The Report's Conclusion

The "overriding" conclusion of the Report is that current emphasis on expanding energy supply to meet existing demand forecasts, could place an "unnecessary hardship on the nation." These "demand projections are exaggerated" and far exceed what the nation will in fact use. But even this "overriding" acknowledgement of the power of supply limitations and of the acceptability to the nation of the Report's "credible" levels of demand is hedged. First, the "expense and efforts currently being taken and planned for energy self-sufficiency should be curtailed" because the supplies it could produce "will not be required." Then, apparently supply can be increased at high cost and "the nation can be induced to consume them."

This unhappy ambivalence is understandable: the Report offers no substance on the nature of demand and is therefore in no position to interact with supply considerations. It seems to favor less energy self-sufficiency without even posing the question of the benefits of the policy it opposes. The Report thus deprives COMRATE of an opportunity to render basic contributions to U.S. minerals policy. Without any way of appraising gains from energy/minerals use, of assessing the efforts and costs man is prepared to incur to achieve these gains, COMRATE has been confined to the production of another, albeit fully-articulated, footnote to Club of Rome themes.

Theory and history do offer important hypotheses on what makes for population growth, for national product growth, for materials use in product. The seeming conflict between a fixed stock and an expanding use has always been resolved. Critical is not the number of man, the amount of material, but the imagination and skill with which man adopts the things he wants, the methods by which they are produced, the number of people that share them. The world needs social and economic progress. Only man with his limitless conceptual horizons, is able to balance these needs on a finite planet. Man is the most important world resource; the quality of man is at the core of the problems of balance in materials supply and demand.

By failing to pose the relevant resource considerations, the Report prevented COMRATE and NAS from addressing the basic policy problems in the resources area. The critical national policies and programs for resources remain yet to be explored.

BIBLIOGRAPHY OF REFERENCES

SECTION I: REPORT OF PANEL ON MATERIALS CONSERVATION THROUGH TECHNOLOGY*

Clausner, H.R., 1970, Problems and Issues of a National Materials Policy: Science Policy Research Division, Legislative Reference Service, Library of Congress, December 1970, p. 173.

Dresher, W.H., 1974, Metallurgical Engineering in the United States - A Status Report: Journal of Metals, 26 p. 31.

Holland, R.C., 1973, Public Policy Issues in the Financing of New Energy Capacity: remarks of Robert C. Holland, Member of Board of Governors of the Federal Reserve System, before the Financial Conference of the National Coal Association, Chicago, October 31, 1973.

Little, Arthur E., Inc., and the Industrial Research Institute, Inc., 1973, Barriers to Innovation in Industry, September 1973.

National Academy of Sciences, 1973, Materials and Man's Needs, report of the Committee on the Survey of Materials Science and Engineering.

SECTION II: REPORT OF PANEL ON ESTIMATION OF MINERAL RESERVES AND RESOURCES

A. The Fossil Fuels

Albers, J.P., M.D. Carter, A.L. Clark, Anny B. Coury, and S.P. Schweinfurth, 1973, Summary Petroleum and Selected Mineral Statistics for 120 Countries, Including Offshore Areas: U.S. Geological Survey Prof. Paper 817.

*For tables and full list of references, see Appendix to Section I, a Table of Contents for which appears on page 429.

Section II (continued)

Anonymous, 1972a, Reserves of Crude Oil, Natural Gas Liquids, and Natural Gas in the United States and Canada, and United States Productive Capacity as of December 31, 1971: American Gas Association, American Petroleum Institute, and Canadian Petroleum Association, 26.

Anonymous, 1972b, Worldwide Crude Petroleum and Worldwide Gas Production: Oil and Gas Journal, 70, no. 9, p. 103.

Anonymous, 1973, Large Scale Action Nearer for Orinoco, Oil and Gas Journal, 71: August 13, 1973, p. 44-45.

Anonymous, 1973a, Reserves of Crude Oil, Natural Gas Liquids, and Natural Gas in the United States and Canada, and United States Productive Capacity as of December 31, 1972: American Gas Association, American Petroleum Institute, and Canadian Petroleum Association, 27.

Anonymous, 1973b, Worldwide Crude Production and Gas Production: Oil and Gas Journal, 71, no. 8, p. 131.

Anonymous, 1973c, Worldwide Oil at a Glance: Oil and Gas Journal, 71, no. 53, p. 86-87.

Anonymous, 1973d, Worldwide Production: Oil and Gas Journal, 71, no. 53, p. 108, 110, 115, 116, 118, 120-124, 126-128, 130, 135.

Anonymous, 1974a, Worldwide Crude Production and Gas Production: Oil and Gas Journal, 72, no. 9, p. 138.

Anonymous, 1974b, Mined Oil--Another Potential Petroleum Resource: Engineering and Mining Journal, April 1973, p. 102-103.

Averitt, Paul, 1973, Coal: in D.A. Brobst and W.P. Pratt, eds., United States Mineral Resources, U.S. Geological Survey Prof. Paper 820, p. 133-142.

Cashion, W.B., 1973, Bitumen-Bearing Rocks: in D.A. Brobst and W.P. Pratt, eds., United States Mineral Resources, U.S. Geological Survey Prof. Paper 820, p. 99-103.

Culberson, W.C., and Janet K. Pitman, 1973, Oil shale: in D.A. Brobst and W.P. Pratt, eds., United States Mineral Resources, U.S. Geological Survey Prof. Paper 820, p. 497-503.

Davis, W.K., 1974, U.S. Energy Prospects, An Engineering Viewpoint: National Academy of Engineering.

Duncan, D.C., 1967, Geologic Setting of Oil-Shale Deposits and World Prospects: Seventh World Petroleum Congress, Proc. 3, p. 659-667.

Section II (continued)

Duncan, D.C., and V.E. Swanson, 1965, Organic-Rich Shale of the United States and World Land Areas: U.S. Geological Survey, Circular 523.

Emery, K.O., 1969, Continental Rises and Oil Potential: Oil and Gas Journal, 67, no. 19, p. 231-243.

Emery, K.O., 1970, Continental Margins of the World: in F.M. Delany, ed., The Geology of the East Atlantic Continental Margin: ICSU/SCOR Working Party 31 Symposium, Cambridge, Rept. 70/13, p. 3-29.

Emery, K.O., in press, Unconventional Areas of Fossil Fuel Resources in the Ocean: National Academy of Sciences-National Academy of Engineering, Proc. National Materials Policy meeting in Washington, D.C., October 25-26, 1973.

Federoff, D.R., 1973, Coal--Pennsylvania Anthracite: in Minerals Yearbook 1971, v. 1--Metals, Minerals, and Fuels: U.S. Bureau of Mines, Washington, D.C., U.S. Govt. Printing Office, p. 377-404.

Fedynskiy, V.V., and L.E. Levin, 1970, Tektonika i Neftegazonosnost Okrainnykh i Vnutrennykh Morey SSSR (The Tectonics and Oil and Gas Potential of the Marginal and Inland Seas of the USSR) trans. by P.T. Broneer: Lamont-Doherty Geological Observatory of Columbia University, p. 301.

Halbouty, M.T., A.A. Meyerhoff, R.E. King, R.H. Dott, Sr., H.D. Klemme, and Theodore Shabad, 1970, World's Giant Oil and Gas Fields, Geologic Factors Affecting their Formation, and Basin Classifications: in M.T. Halbouty, ed., Geology of Giant Petroleum Fields: American Association Petroleum Geologists Memorandum 14, p. 502-555.

Hendricks, T.A., 1965, Resources of Oil, Gas, and Natural-Gas Liquids in the United States and the World: U.S. Geological Survey Circular 522.

Herkenhoff, E.C., 1972, When are We Going to Mine Oil?: Engineering and Mining Journal, June 1972, p. 132-138.

Hubbert, M.K., 1956, Nuclear Energy and the Fossil Fuels: New York, American Petroleum Institute, Drilling and Production Practice, p. 7-25.

Hubbert, M.K., 1959, Techniques of Production with Application to the Petroleum Industry: Houston, Texas Shell Development Company Exploration and Production Reserves Division, Pub. 204.

Hubbert, M.K., 1962, Energy Resources: A Report to the Committee on Natural Resources: National Academy of Sciences-National Research Council Pub. 1000-D.

Section II (continued)

Hubbert, M.K., 1966, M. King Hubbert's Reply to J.M. Ryan: Journal Petroleum Technology, February 1966, p. 284-286.

Hubbert, M.K., 1967, Degree of Advancement of Petroleum Exploration in United States: Bulletin American Association Petroleum Geologists, 51, p. 2207-2227.

Hubbert, M.K., 1969, Energy Resources: in Preston Cloud, chairman, Resources and Man, National Academy of Sciences-National Research Council, San Francisco, W.H. Freeman & Co., p. 157-242.

Hubbert, M.K., 1971, The Energy Resources of the Earth: Scientific American, 224, no. 3, p. 60-70.

Hubbert, M.K., 1974a, Statement for Joint Hearings on May 10, 1974 of the Senate Commerce and Government Operations Committees to Consider Legislation Designed to Monitor and Alleviate Product and Material Shortages: multilithed.

Hubbert, M.K., 1974b, U.S. Energy Resources, a Review as of 1972: A background paper prepared at the request of Henry M. Jackson, Chairman, Committee on Interior and Insular Affairs, United States Senate, 93rd Congress, 2nd Session.

Hubbert, M.K., 1974c, Ratio Between Recoverable Oil per Unit Volume of Sediments for Future Exploratory Drilling to that of the Past for the Conterminous United States: memorandum of June 8, 1974, to COMRATE on the occasion of panel meeting on estimates of undiscovered recoverable petroleum resources, June 5, 1974 (See Appendix of source material.)

Humphreys, R.D., 1973, Economic and Engineering Aspects of Tar Sands Project: reprint of paper presented to the 75th Annual General meeting, Canadian Institute of Mining and Metallurgy, Montreal, April 1973.

Kalinko, M.K., 1969, Nefte-gazonosnosti akvatorii mira: Izdat. "Nedra," Moscow.

Koelling, G.W., 1974, World Natural Gas, 1972: Division of Fossil Fuels, U.S. Bureau of Mines.

McCaslin, John, 1972, Worldwide Offshore Oil Output Nears 9 Million B/D: Oil and Gas Journal, 70, no. 18, p. 196-198.

McCaslin, J.C., 1973a, International Petroleum Encyclopedia: Tulsa, Oklahoma, The Petroleum Pub. Co.

McCaslin, J.C., 1973b, World Offshore-Oil Production Soars: Oil and Gas Journal, 71, no. 18, p. 126, 130, 135.

Section II (continued)

McCaslin, J.C., 1974a, International Petroleum Encyclopedia: Tulsa, Oklahoma, The Petroleum Pub. Co. (copies of pertinent tables supplied prior to publication).

McCaslin, J.C., 1974b, Offshore Oil Production Soars: Oil and Gas Journal, 72, no. 18, p. 136, 139, 140, 142.

McCulloh, T.H., 1973 Oil and Gas: in D.A. Brobst and W.P. Pratt, eds., United States Mineral Resources: U.S. Geological Survey Prof. Paper 820, p. 477-496.

McKelvey, V.E., 1972, Mineral Resource Estimates and Public Policy: American Scientist, 60, no. 1, p. 32-40.

McKelvey, V.E., 1974, Revised U.S. Oil and Gas Resources Estimates: U.S. Geological Survey News Release, March 26, 1974.

McKelvey, V.E., and F.F.H. Wang, 1970, World Subsea Mineral Resources, Preliminary Maps: U.S. Geological Survey Misc. Geol. Investig. Map I-632.

National Petroleum Council, 1972, U.S. Energy Outlook, Interim Report: Initial appraisal by the Oil Shale Task Group, 1971-1985, Washington, D.C.

Padula, V.T., 1969, Oil Shale of Permian Irati Formation, Brazil: Bulletin American Association Petroleum Geologists, 53, p. 591-602.

Phizackerley, P.H., and L.O. Scott, 1967, Major Tar Sand Deposits of the World: Seventh World Petroleum Congress, Proc., 3, p. 551-581.

Pow, J.R., G.H. Fairbanks, and W.J. Zamora, 1963, Descriptions and Reserve Estimates of the Oil Sands of Alberta: in K.A. Clark v. Athabasca oil sands: Research Council of Alberta, Edmonton, Information Ser. no. 45, p. 1-14.

Rossinier, E.A., Chairman, 1973, Potential Supply of Natural Gas in the United States (as of December 31, 1972): Colorado School of Mines Foundation, Inc.

Snyder, R.E., 1974, 60 Billion-barrel Tertiary Recovery Potential Claimed: World Oil, June 1974, p. 71-73.

Southard, L.G., 1973, International Petroleum Annual, 1971: U.S. Bureau of Mines, March 1973.

Southard, L.G., 1974, International Petroleum Annual, 1972: U.S. Bureau of Mines, March 1973.

Section II (continued)

Theobald, P.K., S.P. Schweinfurth, and D.C. Duncan, 1972, Energy Resources of the United States: U.S. Geological Survey Circular 650.

Weeks, L.G., 1948, Highlights on 1947 Developments in Foreign Petroleum Fields: Bulletin American Association Petroleum Geologists, 32, p. 1093-1160.

Weeks, L.G., 1958, Fuel Reserves of the Future: Bulletin American Association Petroleum Geologists, 42, p. 431-438.

Weeks, L.G., 1960, The Next Hundred Years Energy Demand and Sources of Supply: Geotimes, 5, no. 1, p. 18-21, 51-55.

Weeks, L.G., 1973, Subsea Petroleum Resources: in Report of the Secretary-General, Committee on the Peaceful Uses of the Sea-Bed and the Ocean Floor beyond the Limits of National Jurisdiction, United Nations General Assembly.

Weeks, L.G., 1974, Subsea Petroleum Resources in Relation to Proposed National-International Jurisdiction Boundaries: American Association Petroleum Geologists Annual Meeting, San Antonio, Texas, April 1974, Abstracts, p. 96-97.

Westerstrom, L.W., 1973, Coal--Bituminous and Lignite: in Minerals Yearbook 1971, v. 1--Metals, Minerals and Fuels: U.S. Bureau of Mines, Washington, D.C., U.S. Govt. Printing Office, p. 325-376.

Williams, D.L., and R.P. von Herzen, 1974, Heat Loss from the Earth: New estimate: Geology, 2, no. 7, p. 327-328.

B. Resources of Copper

Amax Exploration, Inc., 1973, Unpublished graphs and list of 315 copper deposits (as of 1970) contributed by P. Parker, courtesy of S.K. Hamilton, COMRATE Copper Workshop, Estes Park, Colorado, Oct. 12-13, 1973.

Bailly, Paul A., 1966, Mineral Exploration and Mine Developing Problems related to Use and Management of Other Resources and to U.S. Public Land Laws, especially the Mining Law of 1872: Public Land Law Conference, University of Idaho, Oct. 10-11, 1966, p. 51-99.

Bailly, Paul A., 1972, Mineral Exploration Philosophy: American Mining Congress Meeting, Las Vegas, Nevada, Oct. 12, 1971, (mimeo). Reprinted in its entirety in Mining Congress Journal, April 1972.

Bennett, Harold J. et al., 1973, An Economic Appraisal of the Supply of Copper from Primary Domestic Sources: U.S. Bureau of Mines, Information Circular 8598.

Section II (continued)

Boudet, E., 1970, Perspectives de la Production de Cuivre du Monde Occidental, Annales des Mines, May 1970, p. 41-58.

Brinck, Johan, 1971, MIMIC, The Prediction of Mineral Resources and Long-Term Price Trends in the Non-Ferrous Metal Industry is No Longer Utopian: Eurospectra, June 1971, p. 46-56.

Coster, W.A., 1973, The Appraisal Systems used by the European Communities: (mimeo), COMRATE Copper Workshop, Estes Park, Colorado, Oct. 12-13, 1973. This contains a good bibliography on these appraisal systems.

Cox, Dennis P. et al., 1973, Copper in United States Mineral Resources, U.S. Geological Survey, Professional Paper 820, p. 163-190.

Cox, Dennis P., D.A. Singer, and L.J. Drew, 1973, A Geologic Approach to Copper Resource Appraisal: U.S. Geological Survey, (mimeo), COMRATE Copper Workshop, Estes Park, Colorado, Oct. 12-13, 1973.

Cranstone, D.A. et al., Are Ore Discovery Costs Increasing? Canadian Mining Journal, April 1973, p. 53-64.

Dubs, M., Verbal communication to COMRATE at Workshop of April 10, 1974, Washington, D. C.

Horn, D.R., B.M. Horn, and M.N. Delach, 1973, Copper and Nickel Content of Ocean Ferromanganese Deposits and their Relation to Properties of the Substrate: in Maury Morgenstein, ed., Papers on the Origin and Distribution of Manganese Nodules in the Pacific and Prospects for Exploration, Honolulu, Hawaii, July, 1973.

IDOE, 1972a, Ferromanganese Deposits of the North Pacific: by D.R. Horn, B.M. Horn, M.N. Delach, Lamont-Doherty Geological Observatory, Columbia University, Palisades, New York. Technical Report No. 1, NSF/GX 33616.

IDOE, 1972b, Worldwide Distribution of Ferromanganese Nodules and Element Concentrations in Selected Pacific Ocean Nodules: by J.Z. Frazer and G. Arrhenius, Scripps Institution of Oceanography, University of California at San Diego, La Jolla, California, Technical Report No. 2, NSF/GX 34659.

IDOE, 1972c, Ferromanganese Deposits on the Ocean Floor: by David R. Horn, Arden House, Harriman, New York and Lamont-Doherty Geological Observatory, Columbia University, Palisades, New York.

IDOE, 1973a, Metal Content of Ferromanganese Deposits of the Oceans: by D.R. Horn, B.M. Horn, M.N. Delach, Lamont-Doherty Geological Observatory, Columbia University, Palisades, New York. Technical Report No. 3, NSF/GX 33616.

Section II (continued)

IDOE, 1973b, Inter-University Program of Research on Ferromanganese Deposits of the Ocean Floor, Phase I Report.

IDOE, 1973c, Factors which Control the Distribution of Ferromanganese Nodules and Proposed Research Vessel's Track, North Pacific, Phase II: by D.R. Horn, M.N. Delach, Lamont-Doherty Geological Observatory, Columbia University, Palisades, New York, Technical Report No. 8, NSF/GX 33616.

Kellogg, H.H., 1973a, Energy Consumption in Flotation Beneficiation: mimeo, distributed to COMRATE members, July 1973.

Kellogg, H.H., 1973b, Energy Consumption for Mining of Metallic Ores: mimeo, distributed to COMRATE members, August 1973.

Lasky, S.G., 1950, How Tonnage and Grade Relations Help Predict Ore Reserves: Engineering and Mining Journal, 151, No. 4, April 1950, p. 81-85.

Lawrence, Floyd G., 1974, The World Fight for Materials: Industry Week, January 7, 1974, p. 27-35.

Lovering, T.S., 1968, Non-Fuel Mineral Resources in the Next Century: Texas Quarterly, XI, No. 2, p. 127-147.

MacGregor, Ian K., 1971, Statement on Behalf of the American Mining Congress, before the Senate Committee on Interior and Insular Affairs: Amax, September 22, 1971.

McKelvey, V.E. and F.F.H. Wang, 1969, World Subsea Mineral Resources, U.S. Geological Survey, Washington, D. C.

Matheron, G., 1959, Etude: Remarques sur la Loi de Lasky, Chronique des Mines d'Outre-Mer et Recherche Miniere, No. 282, December 1959, p. 463-465.

Mero, John L., 1964, The Mineral Resources of the Sea, Netherlands, Elsevier Publishing Company, p. 127, 225-230, 262-267.

Meyer, Charles, 1973, Oral presentation, COMRATE Copper Workshop, Estes Park, Colorado, October 12-13, 1973.

Pelissonnier, H., 1968, Sur la Distribution du Cuivre Economiquement Exploitable sur la Terre, in L.H. Ahrens, ed., Origin and Distribution of the Elements, New York, Pergamon Press, p. 1139-1149.

Pelissonnier, H. et al., 1972, Les Dimensions des Gisements de Cuivre du Monde, R.R.G.M. Memoir No. 57.

Peters, William C., 1970, Some Geological Comparisons--Recent Porphyry Copper Discoveries: Mining Congress Journal, October 1970, p. 28-32.

Section II (continued)

Schatz, C.E., 1971, Observations on Sampling and Occurrence of Manganese Nodules: Third Annual Offshore Technical Conference, Houston, Vol. 1, p. 389-396.

U.S. Bureau of Mines, 1973, Commodity Data Summaries: Mining and Minerals Policy 1973, Vol. 2.

U.S. Bureau of Mines, 1974, Commodity Data Summaries: Mining and Minerals Policy 1974, Vol. 2.

Wimpfen, S.P. and H.J. Bennett, 1973, Copper Resource Appraisal, U.S. Bureau of Mines: mimeo, COMRATE Copper Workshop, Estes Park, Colorado, October 12-13, 1973. This includes a brief description of the U.S.B.M.'s Minerals Availability System.

SECTION III: REPORT OF THE PANEL ON THE IMPLICATIONS OF MINERALS PRODUCTION FOR HEALTH AND THE ENVIRONMENT

Almer, B., W. Dickson, C. Ekstrom, E. Hornstrom, and V. Miller, 1974, Effects of Acidification on Swedish Lakes: Ambio, 3, p. 29-36.

Altshuler, A.P., 1973, Environ. Sci. Technol. 7: 709.

Amdur, M.O., 1973, Animal Studies: Proceedings of the Conference on Health Effects of Air Pollution, Washington, D. C., U.S. Government Printing Office, p. 175-205.

Appalachian Regional Commission, 1969, Acid Mine Drainage in Appalachia: Prepared with the cooperation of the National Research Council, Committee on Water Quality Management, Panel on Mine Drainage Pollution Control.

Barret, E. and G. Brodin, 1955, Acidity of Scandinavian Precipitation: Tellus, 7, p. 251-257.

Barry, Theodore, and Associates, 1972, Fatality Analysis Data Base Development: prepared for the U.S. Bureau of Mines, June 30, 1972.

Bates, D.V., B.R. Fish, T.F. Hatch, T.T. Mercer, P.E. Morrow, 1966, Deposition and Retention Models for Internal Dosimetry of the Human Respiratory Tract: Health Physics, 12, p. 173.

Beamish, R.J., 1974, Lost Fish Populations From Unexploited Remote Lakes in Ontario, Canada as a Consequence of Atmospheric Fallout of Acid: Water Research, 8, p. 85-95

Section III (continued)

Beamish, R.J., and H.H. Harvey, 1972, Acidification of the La Cloche Mountain Lakes in Ontario, and Resulting Fish Mortalities: J. Fish Res. Bd. Canada, 29, p. 1131-1143.

Berner, R.A., 1971, Worldwide Sulfur Pollution of Rivers: J. Geophysical Research, 76, p. 6597-6600.

Bolin, B., ed., 1971, Report of the Swedish Preparatory Committee for the United Nations Conference on the Human Environment: Stockholm, Sweden, Norstedt and Söner.

Botkin, D.B., 1973, Natural Ecosystems and the Anthropogenic Production of Carbon Dioxide: Unpublished report to COMRATE's Environmental Panel. (See Appendix of Source Material.)

Brock, T.D., 1973, Lower pH Limit for the Existence of Blue-Green Algae: Evolutionary and Ecological Implications: Science, 179, p. 480-483.

Brock, T.D., 1974, Environmental Impact of Energy Generation: The Sulfur Cycle: A report to COMRATE's Environmental Panel. (See Appendix of Source Material.)

Buechley, R.W., W.B. Riggan, V. Hasselblad, and J.B. VanBruggen, 1973, SO_2 Levels and Perturbations in Mortality: a Study in the New York-New Jersey Metropolis: Arch. Environmental Health, 27, p. 134-137, Copyright 1973, American Medical Association.

Carrol, D., 1962, Rainwater as a Chemical Agent of Geological Processes - A Review: U.S. Geological Survey Water Supply paper 1535, 6.

Charlson, R.J., A.H. Vanderpol, D.S. Covert, A.P. Waggoner, N.C. Ahliquist, 1974, Sulfuric Acid-Ammonium Sulfate Aerosol: Optical Detection in the St. Louis Region: Science, 184, p. 156.

Committee on Interior and Insular Affairs, 1972, Energy "Demand" Studies: An Analysis and Appraisal, U.S. House of Representatives, Washington, D.C., Government Printing Office.

Committee on Public Works, 1973, Proceedings of the Conference on Health Effects of Air Pollutants: U.S. Senate, Washington, D.C., U.S. Government Printing Office, November 1973.

Congressional Record, 1974, Three Area A-Plants Ordered to Shut Down, Philadelphia Bulletin, September 22, p. 1; and Nuclear Plants Springing Leaks: Experts Puzzled on Repair Technique, Boston Sunday Globe, September 22: two articles and statement by Mr. Gravel printed at the request of Mr. Gravel, in the Congressional Record, October 1, 1974.

Costello, J., 1974, Personal communication from Joseph Costello, M.S., of ALFORD, Morgantown, West Virginia. The information was provided by the Social Security Administration in Baltimore, Maryland.

Section III (continued)

Costello, J. and W.K.C. Morgan, 1974, Coal Workers' Pneumoconiosis: Its Economic Impact and Prevalence. (See Appendix of Source Material.)

Council on Environmental Quality, 1973, Report to the Senate Committee on Interior and Insular Affairs, Washington, D. C., Government Printing Office.

Douglas, J.W.B. and R.E. Waller, 1966, Air pollution and respiratory infection in children: British Journal of Preventive Soc. Medicine, 20, p. 1-8.

Dunham, J.T., Carl Rampecek, and T.A. Henrie, 1974, High-sulfur Coal for Generating Electricity: Science, 184, April, 1974, p. 346-351.

Dupree, W., Jr. and J. West, 1972, United States Energy Through the Year 2000: U.S. Department of the Interior, Washington, D. C., Government Printing Office, p. 22.

Edwards, P.E., 1973, The Washington Post, May 29, 1973.

Ekdahl, C.A., and C.O. Keeling, 1973, Quantitative Deductions from the Record at Mauna Loa Observatory and at the South Pole, in Carbon and Biosphere, eds. G.M. Woodwell and E. Pecan: A.E.C., Springfield, Virginia, p. 51-85.

Enterline, P.E., 1967, The Effects of Occupation on Chronic Respiratory Disease: Arch. Environmental Health, 14, p. 189.

Environmental Protection Agency, 1971, Gordon, C.C., et al., Air Pollution Abatement Activity; Mount Storm, West Virginia - Gorman, Maryland - Keyser, West Virginia: EPA Publication No. APTO-0656.

Environmental Protection Agency, 1974, Health Consequences of Sulfur Oxides: A Report from CHESS, 1970-1971: EPA 650/1-74-004, May, 1974. (See chapter 2, section 2.3; chapter 3, section 3.3, chapter 5, section 5.6; and chapter 6, section 6.1.)

Ferenbaugh, R.W., 1974, Effects of Simulated Acid Rain on Vegetation: PhD. Thesis, University of Montana.

Feth, J.H., 1967, Chemical Characteristics of Bulk Precipitation in the Mojave Desert Region of California: U.S. Geological Survey Paper 575C, p. C222-C227.

Finklea, J.F., D.B. Turner, G.G. Akland, R.I. Larsen, V. Hasselblad, and S.D. Shearer, 1974, Briefing Notes - A Status Report on Sulfur Oxides: U.S. Environmental Protection Agency, Office of Research and Development, National Environmental Research Center, Research Triangle Park, North Carolina, April, 1974.

Section III (continued)

Friend, J.P., 1973, The Global Sulfur Cycle, _in_ S.I. Rasool, ed., Chemistry of the Lower Atmosphere: Plenum Press, p. 177-201.

Gordon, C.C., 1972a, Mount Storm Study: Report to Environmental Protection Agency (see EPA, 1971): Contract 68-02-0229.

Gordon, C.C., 1972b, Plantations Versus Power Plants: _American Christmas Tree Journal_, _16_, p. 5-10.

Grodzinska, K., 1971, Acidification of Tree Bark as a Measure of Air Pollution in Southern Poland: _Bull. L'Acad. Polanaise Sciences_, Ser. Sci. Biol. Cl. II, _19_, p. 189-195.

Hesketh, H.E., 1974, Sulfur Dioxide Scrubbing Technology: Presentation before the Colorado Air Pollution Control Commission, for EPA Region VIII, Denver Office, November 19, 1974.

Higgins, I.T.T. and B.G. Ferris, Jr., Epidemiology of Sulfur Oxides and Particles (being prepared for presentation to National Academy of Sciences Panel on Biological Effects of Airborne Particles).

Higgins, I.T.T., M.W. Higgins, M.D. Lockshin, and N. Canale, 1968, Chronic Respiratory Disease in Mining Communities in Marion County, West Virginia: _Brit. J. Industr. Med._, _25_, p. 165.

Hill, R.D., 1968, Mine Drainage Treatment. State of the Art and Research Needs: U.S. Department of Interior, Federal Water Pollution Control Administration, Cincinnati, Ohio.

Holser, W.T., and I.R. Kaplan, 1966, Isotope Geochemistry of Sedimentary Sulfates: _Chemical Geology_, _1_, p. 93-135.

Interior, 1969, Stream Pollution by Coal Mine Drainage in Appalachia: U.S. Department of the Interior, Federal Water Pollution Control Administration, Cincinnati, Ohio.

Interior, 1972, Report of the Secretary of the Interior in conjunction with Public Law 91-173, Section 201, on the Administration of the Federal Coal Mine Health and Safety Act of 1969 During Calendar Year 1972: U.S. Department of the Interior, Washington, D.C., Government Printing Office.

Jacobsen, George and W.S. Lainhart, 1972, ILO U/C 1971 International Classification of Radiographs of the Pneumoconiosis: _Medical Radiography and Photography_, _48_, no. 3, p. 65-110.

Section III (continued)

Johnson, N.M., R.C. Reynolds, and G.E. Likens, 1972, Atmospheric Sulfur: Its Effect on the Chemical Weathering of New England: Science, 177, p. 514-516.

Junge, C. and G. Scheich, 1969, Studien zur Bestimmung des Sauregehaltes von Aerosolteilchen. Atmos. Environ., 3, p. 423.

Kellogg, W.W. et al., 1972, The Sulfur Cycle: Science, 175, p. 587-596.

Key, M.M. et al., 1971, Pulmonary Reaction to Coal Dust: Academic Press.

Kibelstis, J.A. et al., 1973, Prevalence of Bronchitis and Airway Obstruction in Bituminous Miners: Amer. Rev. Resp. Dis., p. 108-886.

Lave, L.B. and E.P. Seskin, 1970, Air Pollution and Human Health: Science, 169, p. 723-733.

Lebowitz, M., P. Bendheim, G. Cristea, D. Markovitz, J. Miseaszek, M. Staniec, and D. Van Wyck, 1973, The Effect of Air Pollution and Weather on Lung Function in Exercising Children and Adolescents: Amer. Rev. Resp. Dis., 109, p. 262-273.

Likens, G.E., 1972, The Chemistry of Precipitation in the Central Finger Lakes Region: Technical Report No. 50, Cornell University Water Resources and Marine Science Center.

Likens, G.E. and F.H. Bormann, 1974, Acid Rain; A Serious Regional Environmental Problem: Science, 184, p. 1176-1179.

Likens, G.E., F.H. Bormann, N.M. Johnson, 1972, Acid Rain, Environment, 14, No. 2, 33-40.

Livingstone, D.A., 1963, Chemical Composition of Rivers and Lakes: U.S. Geological Survey Professional Paper 440-G.

Lodge, J.P., Jr. and E.R. Frank, 1966, J. Recherche Atmospherique, II, p. 139.

Lunn, J.E., J. Knowelden, and A.J. Handyside, 1967, Pattern of Respiratory Illness in Sheffield Infant School-Children: Brit. J. Prev. Sco. Med., 21, p. 7-16.

McJilton, C., R. Frank and R. Charlson, 1973, Role of Relative Humidity in the Synergistic Effect of a Sulfur Dioxide-Aerosol Mixture on the Lung: Science, 182, p. 503-504.

Mining Enforcement and Safety Administration, 1974, Technical Support Department, April 1974.

Section III (continued)

Morgan, W.K.C., 1973, JAMA, 225, p. 4.

Morgan, W.K.C., 1974, Letter to COMRATE's Environmental Panel from Dr. W.K.C. Morgan, University of West Virginia Medical School, dated 22 July, 1974. (See Appendix of Source Material.)

Morgan, W.K.C., et al., 1974, Ventilatory Capacity and Lung Volumes of the U.S. Coal Miners: Arch. Environ. Health, 28, No. 182.

Munn, R.E. and H. Rodhe, 1971, On the Meteorological Interpretation of the Chemical Composition of Monthly Precipitation Samples: Tellus, 23, No. 1, p. 1-12.

National Academy of Sciences/National Academy of Engineering, 1974a, Rehabilitation Potential of Western Coal Lands: Report to the Ford Foundation Energy Policy Project, Cambridge, Massachusetts, Ballinger Publishing Company, April 1974.

National Academy of Sciences/National Academy of Engineering, 1974b, Energy Prospects: An Engineering Viewpoint: prepared by Task Force on Energy, National Academy of Engineering, August 1974.

National Academy of Sciences/National Academy of Engineering, 1974c, Air Quality and Automobile Emission Controls: Report by the Coordinating Committee on Air Quality Studies prepared for the Committee on Public Works, United States Senate, September 1974.

National Environmental Research Center, 1974, A Status Report on Sulfur Oxides for the Administrator of the Federal Energy Office, April 26, 1974.

New York Times, 1974, Acid Rain Found up Sharply in East; Smoke Curb Cited: The New York Times, June 13, 1974, p. 1.

Occupational Health and Safety Letter, 1974b, August 22, 1974.

Oden, S., 1968, The Acidification of Air and Precipitation and its Consequences on the Natural Environment: S.W. Nat. SA Res. Council Ecology Committee, Bull. No. 1, p. 86, Arlington, Virginia, Tr-117, Translation Constants Limited.

Oden, S. and T. Ahl, 1970, Sartryck Ur Ymer Arsbok, p. 103-122.

Overrein, L.N., 1972, Sulphur Pollution Patterns Observed; Leaching of Calcium in Forest Soil Determined: Ambio, 1, p. 145-147.

Section III (continued)

Pearson, F.J., and D.W. Fisher, 1971, Chemical Composition of Atmospheric Precipitation in the Northeastern U.S.: U.S. Geological Survey Water Supply Paper, 1535-P.

Rall, D.P., 1973, A Review of the Health Effects of Sulfur Oxides: National Institutes of Health, October 9, 1973.

Ray, Dixy Lee, 1973, The Nation's Energy Future, A Report to Richard M. Nixon, President of the United States, submitted by Dr. Dixy Lee Ray, Chairman, U.S. Atomic Energy Commission: Document Wash, -1281-, December 1, 1973.

Staxang, B., 1969, Acidification of Bark of Some Deciduous Trees: Oikos 20, p. 224-230.

Tabershaw, I.R., 1970, The Health of the Coal Miner--an Expendable Resource? J. Occup. Med., 12, p. 452.

Time, 1974, The Acid Threat: Time Magazine, July 1, 1974, p. 67-68.

U.N., 1971, Air Pollution Across National Boundaries; The Impact on the Environment of Sulfur in Air and Precipitation: Royal Ministry for Foreign Affairs, Royal Ministry of Agriculture, U.N. Conference on the Human Environment, Stockholm, Sweden.

U.S.B.M., 1972, Production of Coal--Bituminous and Lignite: Mineral Industry Surveys: U.S. Department of the Interior, Bureau of Mines, Washington, D.C.

U.S.B.M., 1973, Coal Mine Injuries and Worktime, Annual Summary 1972, U.S. Department of the Interior, Bureau of Mines, Health and Safety Analysis Center, Denver, March, 1973.

U.S.B.M., 1974a, U.S. Bureau of Mines News Release, March 13, 1974.

U.S.B.M., 1974b, Demonstrated Coal Reserve Base of the U.S. for 1974: Mineral Industry Surveys: U.S. Department of the Interior, Bureau of Mines, Washington, D.C.

U.S.G.S., 1967, Coal Resources of the United States: U.S. Geological Survey Bulletin 1275, January 1, 1967.

U.S.G.S., 1970, Stripping Coal Resources of the United States: U.S. Geological Survey Bulletin 322, January 1, 1970.

Washington Post, 1974, White House Eyes Clean Air Changes, July 28, 1974, p. A-2.

Whitby, K.T., R.B. Husar, and B.Y.H. Liu, 1972, The Aerosol Size Distribution of Los Angeles Smog: J. Colloid and Interface Sci., 39, p. 177.

Section III (continued)

Whittaker, R.H., F.H. Bormann, G.E. Likens, T.G. Siccama, (in press) Ecol. Monogr.

Wielgolaski, F.E., 1971, Acidification in Precipitation; Causes and Effects: Oslo, Norway.

Wood, T. and F.H. Bormann, Environ. Pollut., (in press).

SECTION IV: REPORT OF PANEL ON DEMAND FOR FUEL AND MINERAL RESOURCES

Chapman, T., T. Tyrell, and T. Mount, 1974: Science, 178, p. 703-8.

Cook, Earl, 1971, The Flow of Energy in an Industrial Society, Scientific American, September 1971. Copyright by Scientific American, Inc., all rights reserved.

Dupree, W.G., Jr., and T.A. West, 1972, U.S. Energy Through the Year 2000: U.S. Department of the Interior, Washington, D.C., Government Printing Office.

Edmonson, N., 1974, Real Prices and the Consumption of Mineral Energy in the United States: Ms.

Federal Energy Administration, 1974, Project Independence Report, Washington, D.C., U.S. Government Printing Office, November 1974.

Federal Energy Office, 1974, "Project Independence: Background Paper" prepared for the Washington Energy Conference, February 11-12, 1974: in Materials Shortages: Selected Readings on Energy Self-Sufficiency and the Controlled Materials Plan, prepared for the Permanent Subcommittee on Investigations of the Committee on Government Operations, United States Senate, Washington, D.C., U.S. Government Printing Office, August 1974.

Ford Foundation Energy Policy Project, 1974, Exploring Energy Choices: A Preliminary Report, Washington, D.C.

Frejka, T.L., 1973, The Future of Population Growth: Alternative Paths to Equilibrium: New York, Wiley-Interscience.

Fuchs, Victor, 1968, The Service Economy: New York and London, Columbia University Press.

G.A.O., 1974, U.S. Actions Needed to Cope with Commodity Shortages: Multiagency Report to the Congress by the Comptroller General of the United States, General Accounting Office, Washington, D.C.

Section IV (continued)

J.C.A.E., 1973, Understanding the "National Energy Dilemma": Staff Report of the Joint Committee on Atomic Energy, Washington, D.C., Government Printing Office, August 17, 1974.

Leontief, W., For a National Economic Planning Board: New York Times, March 1974.

Madden, C.H., 1974, Dupont Context, 2, p. 5.

Malenbaum, W., 1973, Mineral Requirements in the United States and Abroad in the Year 2000: prepared for the National Commission on Materials Policy, March 1973.

Murray, T.R., M.J. Minor, N.M. Bradburn, R.F. Cotterman, M. Frankel and A.E. Pisarski, 1974a, Evaluation of Public Response to the Energy Crisis: Science, 184, 19 April 1974.

Murray, J.R., and M.J. Minor, 1974b, Energy Related Data: Continuous National Survey, National Opinion Research Center, Chicago: MSS, April 18, 1974.

National Commission on Materials Policy, 1973, Materials Requirements Abroad in the Year 2000: Final Report, Washington, D.C., June 1973.

Nixon, R., 1974, The Energy Crisis, The President's Message to the Congress Outlining Legislative Proposals and Executive Actions to Deal with the Crisis: Weekly Compilation of Presidential Documents, 10: No. 4, January 28, 1974, p. 69-101.

Overly, Schell Associates, 1973, Bibliographic Review of Selected "Demand" Forecasts of Minerals and Commodities, with Commentary on Methodology: prepared for Committee on Mineral Resources and the Environment, August 1973.

Ray, Dixy Lee, 1973, The Nation's Energy Future: A Report to Richard Nixon, President of the United States: Submitted by Dixy Lee Ray, Chairman of the U.S. Atomic Energy Commission: Washington, D.C., Government Printing Office.

Robinson, E. and R.C. Robbins, 1970, Gaseous Sulfur Pollutants from Urban and Natural Sources: Journal of Air Pollution Control Assn., 20, No. 4, p. 223-5.

Steinhart, C.E. and J.S. Steinhart, 1974, Energy: Sources, Use and Role in Human Affairs: North Scituate, Massachusetts, Duxburg Press.

Sun Oil Co., 1972, Petroleum and the Capital Crunch, Staff Paper, December 1972.

ANNEX I
GLOSSARY OF RESOURCE TERMS AND ABBREVIATIONS

RESOURCE TERMS

Resource

A concentration of naturally occurring solid, liquid, or gaseous materials in or on the earth's crust in such form that economic extraction of a commodity is currently or potentially feasible.

Identified Resources

Specific bodies of mineral-bearing material whose location, quality, and quantity are known from geologic evidence supported by engineering measurements with respect to the demonstrated category.

Undiscovered Resources

Unspecified bodies of mineral-bearing material surmised to exist on the basis of broad geologic knowledge and theory.

Reserve

That portion of the identified resource from which a usable mineral and energy commodity can be economically and legally extracted at the time of determination. The term ore is used for reserves of some minerals.

The following definitions for measured, indicated, and inferred are applicable to both the Reserve and Identified-Subeconomic resource components.*

Measured

Material for which estimates of quality and quantity have been computed partly from sample analyses and measurements and partly from reasonable geologic projections.

Indicated

Material for which estimates of the quality and quantity have been computed partly from sample analyses and measurements and partly from reasonable geologic projections.

Demonstrated

A collective term for the sum of materials in both measured and indicated resources.

Inferred

Material in unexplored extensions of Demonstrated resources for which estimates of the quality and size are based on geologic evidence and projection.

Identified-Subeconomic Resources

Materials that are not Reserves, but may become so as a result of changes in economic and legal conditions.

Paramarginal

The portion of Subeconomic Resources that (a) borders on being economically producible or (b) is not commercially available solely because of legal or political circumstances.

*The terms proved, probable, and possible (used by the industry for economic evaluations of ore in specific deposits or districts) commonly have been used loosely and interchangeably with the terms measured, indicated, and inferred (used by the Department of the Interior mainly for regional or national estimates).

Submarginal

The portion of Subeconomic Resources which would require a substantially higher price (more than 1.5 times the price at the time of determination) or a major cost reducing advance in technology.

Hypothetical Resources

Undiscovered materials that may reasonably be expected to exist in a known mining district under known geologic conditions. Exploration that confirms their existence and reveals quantity and quality will permit their reclassification as a Reserve or Identified-Subeconomic resource.

Speculative Resources

Undiscovered materials that may occur either in known types of deposits in a favorable geologic setting where no discoveries have been made, or in as yet unknown types of deposits that remain to be recognized. Exploration that confirms their existence and reveals quantity and quality will permit their reclassification as Reserves or Identified-Subeconomic resources.

ABBREVIATIONS, METRIC UNITS, AND CONVERSION FACTORS USED IN THIS REPORT

Meter (m.) - the basic unit
1 m. = 3.28 feet

Kilometer (km.) = 1,000m
1 km. = 3.28 feet
= 0.62 mile

Area

Square Meter (m^2)
1 m^2 = 10.76 $ft.^2$
Square Kilometer (km^2) = 1,000,000 m^2
1 km^2 = 0.386 $mi.^2$

Volume

Cubic meter (m^3)
1 m^3 = 35.3 $ft.^3$

Mass & Weight

Gram (g.) - the basic unit
Kilogram (kg.) = 1,000 g.
    1 kg. = 2.2 lbs.
Tonne (metric ton) = 1,000 kg.
    1 tonne = 1.1 ton
    = 0.98 long ton
    = 7.5 bbl. oil (average)

Area Density

Kilogram per square meter (kg/m^2)
    1 kg/m^2 = 0.21 $lb/ft.^2$
tonne per square kilometer ($tonne/km^2$)
    1 $tonne/km^2$ = 2.85 $tons/mi^2$

Grade

Weight percent (wt.% or w/o)
    100 wt.% = 6.8 bbl/ton = 276.4 gal/ton
    = 10,000 ppm.

Energy*

Calorie (cal.)
    1 cal. = 0.00397 BTU
    = 0.00116 watt-hour
    Kilogram calorie (kcal) - the "food calorie" = 1,000 cal.
    1 kcal. = 3.97 BTU
    = 1.1 watt-hour
    = 0.00116 kwh.

Specific Energy

Calorie per gram (cal/g)
    1 cal/g. = 1.8 BTU/lb.

*Note: Calories are non-standard units. The standard (mks) metric unit is the joule: 1 joule = 0.2386 cal; 1 watt = 1 joule per second.

ANNEX II
CONFEREES, GUESTS AND OBSERVERS

COMRATE MEETING - OCTOBER 16, 1972

Guests:

Ralph Llewellyn
Environmental Studies Board, NAS

James R. Balsley
U.S. Geological Survey

Olaf Kays
U.S. Geological Survey

Government Observers:

John R. Babey
Department of the Interior

F. Gilman Blake
Office of Science & Technology

Alan P. Carlin
Environmental Protection Agency

A Gordon Everett
Environmental Protection Agency

Bevan French
National Science Foundation

Harold Kirkemo
U.S. Geological Survey

Donald E. Lawyer
Department of Defense

COMRATE MEETING - OCTOBER 16, 1972

Guests (Continued):

John D. Morgan, Jr.
U.S. Bureau of Mines

COMRATE MEETING - DECEMBER 11, 1972

Guests:

Bryce Breitenstein
University of Washington

Roland C. Clement
Audubon Society

Philip L. Johnson
National Science Foundation

Francois J. Lampietti
I R & T

Vincent McKelvey
U.S. Geological Survey

Frederick T. Moore
World Bank

Elburt F. Osborn
U.S. Bureau of Mines

Addison Richmond
Department of State

COMRATE MEETING - DECEMBER 11, 1972

Guests (Continued):

William Salmon
Department of State

Government Observers:

Hans H. Adler
U.S. Atomic Energy Commission

John R. Babey
U.S. Department of the Interior

William E. Benson
National Science Foundation

Alan P. Carlin
Environmental Protection Agency

Richard A. Carrigan
National Science Foundation

Joan Davenport
Environmental Protection Agency

Bevan French
National Science Foundation

P. W. Guild
U.S. Geological Survey

Harold Kirkemo
U.S. Geological Survey

Donald L. Lawyer
Department of Defense

John D. Morgan, Jr.
U.S. Bureau of Mines

James Owens
U.S. Department of Commerce

Richard Ray
National Science Foundation

Edward W. Tooker
U.S. Geological Survey

COMRATE MEETING - FEBRUARY 10, 1973

Government Observers:

William Benson
National Science Foundation

F. Gilman Blake
Office of Science & Technology

Harry Callaway
Department of Commerce

Alan P. Carlin
Environmental Protection Agency

Richard Carrigan
National Science Foundation

Allen Clark
U.S. Geological Survey

Gordon Everett
Environmental Protection Agency

Harold Kirkemo
U.S. Geological Survey

John D. Morgan, Jr.
U.S. Bureau of Mines

COMRATE MEETING - JUNE 11-12, 1973

Guests:

Nathaniel Arbiter
The Anaconda Company

Bryce Breitenstein
University of Washington

Richard A. Carpenter
Commission on Natural Resources

Robert English
University of California

Denis Hayes
Illinois State Fuel Energy Office

COMRATE MEETING - JUNE 11-12, 1973

Guests (Continued):

Franklin Huddle
Library of Congress

Ian Kaplan
University of California

Wilfred Malenbaum
University of Pennsylvania

Hubert E. Risser
U.S. Geological Survey

Konrad Semrau
Stanford Research Institute

Burkhard Strumpel
Institute for Social Research

T.L. Wright
Chevron Overseas Petroleum, Inc.

COMRATE MEETING - SEPTEMBER 14-15, 1973

Government Observers:

F. Gilman Blake
National Science Foundation

Richard A. Carrigan
National Science Foundation

John D. Morgan, Jr.
U.S. Bureau of Mines

COMRATE MEETING - DECEMBER 16, 1973

Observers:

Bryce Breitenstein
University of Washington

William Dresher
University of Arizona

COMRATE MEETING - DECEMBER 16, 1973

Observers (Continued):

Richard Krablin
The Anaconda Company

Paul Shewmon
National Science Foundation

COMRATE MEETING - JUNE 6-9, 1974

Guests:

Bryce Breitenstein
University of Washington

COMRATE FIELD TRIP - June 11, 1973
(Santa Barbara, California Channel)

Guests:

Bryce Breitenstein
University of Washington

Richard A. Carpenter
Commission on Natural Resources

Ian Kaplan
University of California

T. J. Abshier
Mobil Oil Corporation

E.J. Dickinson
Mobil Oil Corporation

J.A. Forman
Mobil Oil Corporation

W.D. Fritz
Mobil Oil Corporation

T.R. Gamble
Mobil Oil Corporation

D.E. Craggs
Union Oil Company

COMRATE FIELD TRIP - JUNE 11, 1973
(Santa Barbara, California Channel)

Guests (Continued):

R.J. Gillen
Union Oil Company

B.G. Spradlin
Union Oil Company

COMRATE FIELD TRIP - SEPT. 13, 1973
(Bethlehem Steel Corp. Sparrows Point)

Guests:

Edward Johnson
U.S. Bureau of Mines

Courtney Riordan
Environmental Protection Agency

Genevieve Atwood
Environmental Studies Board (NAS)

Edward Johnson
U.S. Bureau of Mines

Robert L. Hoopman
Bethlehem Steel Corporation

Leonard Kesselring
Bethlehem Steel Corporation

H. Halstead
Bethlehem Steel Corporation

J.W. Colbert
Bethlehem Steel Corporation

Gines Flores
Bethlehem Steel Corporation

COMRATE FIELD TRIP - DEC. 14&17, 1973
(Open-pit Mines Vicinity of Tucson and San Manuel Smelter & Mining Operations)

Guests:

Richard Krablin
The Anaconda Company

Robert Lynn
Amax Corporation

T.L. Young
The Magma Copper Company

E.R. Rice
The Magma Copper Company

R. Dugdale
The Magma Copper Company

Fred Matter
University of Arizona

W.R. Ashwill
Ranchers Exploration and Development Corporation

PANEL ON IMPLICATIONS OF MINERAL PRODUCTION FOR HEALTH AND THE ENVIRONMENT

WORKSHOP - MARCH 9, 1973

Conferees:

Blair Bower
Resources for the Future

Elmore Grim
National Environmental Research Center

F.M. Sandoval
U.S. Department of Agriculture

James Craig
Chevron Research Company

Daniel Bienstock
U.S. Bureau of Mines

Paul Yavorsky
U.S. Bureau of Mines

Keith Morgan
Appalachian Lab for Occupational Research

Lester Lave
Carnegie-Mellon University

Observers:

Genevieve Atwood
NAS Energy Board

William Guckert
Surface Mining Reclamation

Richard Hodder
U.S. Department of Agriculture

James Boyer
Bituminous Coal Research, Inc.

Albert Forney
U.S. Bureau of Mines

WORKSHOP - APRIL 22-23, 1974

Conferees:

Bryce Breitenstein
University of Washington

Robert J. Charlson
University of Washington

Joseph Costello
ALFORD

Lester Lave
Carnegie-Mellon University

Keith Morgan
West Virginia University Medical Center

David B. Brooks
Energy, Mines and Resources Canada

Andrew Ford
Thayer School of Engineering, Dartmouth

Eville Gorham
University of Minnesota

George Land
AMAX Coal Company

Roger F. Naill
Thayer School of Engineering, Dartmouth

Williams Reiners
Dartmouth College

Sylvia Milanese
MESA, Department of the Interior

Jack A. Simon
Illinois Geological Survey

Lowell Smith
Washington Environmental Research Center

PANEL ON ESTIMATION OF MINERAL RESERVES AND RESOURCES

WORKSHOP - OCTOBER 12-13, 1973

Conferees:

Harold Bennett
U.S. Bureau of Mines

W. John Cieslewicz
Colorado School of Mines

Willem A. Coster
Occidental Minerals Corporation

Dennis Cox
U.S. Geological Survey

Paul Eimon
Essex International

Ralph Erickson
U.S. Geological Survey

Deverle P. Harris
Pennsylvania State University

John P. Hunt
The Anaconda Company

George McGinn
Mobil Oil Corporation

Charles Meyer
University of California

Pierce D. Parker
AMAX Exploration, Inc.

Walden P. Pratt
U.S. Geological Survey

Paul K. Sims
U.S. Geological Survey

Sheldon Wimpfen
U.S. Bureau of Mines

WORKSHOP - OCTOBER 12-13, 1973

Conferees (Continued):

A.J. Wright
Robertson Research International

WORKSHOP - JUNE 5, 1974

Conferees:

M. King Hubbert
U.S. Geological Survey

Richard L. Jodry
Sun Oil Company

Vincent McKelvey
U.S. Geological Survey

William Mallory
U.S. Geological Survey

T.H. McCulloh
University of Washington

O.B. Shelburne
Mobil Oil Corporation

Alex Steinbergh
Federal Energy Administration

James Winfrey
National Petroleum Council

PANEL ON MINERAL CONSERVATION THROUGH TECHNOLOGY

WORKSHOP - APRIL 10-12, 1974*

Conferees:

Marvin Britton
Corning Glass Works

William J. Cavanaugh
Lone Star Industries

Albert Dietz
Massachusetts Institute of Technology

Marne Dubs
Kennecott Copper Corporation

Julius J. Harwood
Ford Motor Company

W.L. Hawkins
Bell Laboratories

Hans Landsberg
Resources for the Future

Robert L. Loofbourow
Professional Engineer

Darrell D. Porter
E.I. Du Pont Company

Robert Pry
General Electric

T.D. Schlabach
Bell Laboratories

Milton Wadsworth
University of Utah

E.M. Warner
Joy Manufacturing Company

---

*Papers given at this meeting have been reproduced in the Appendix of Source Materials of this panel.

COMRATE SYMPOSIUM ON NATIONAL MINERAL POLICY - AUGUST 23-24, 1973

Conferees:

Robert E. Athay
Department of Commerce

Tom Byrne
Interstate Commerce Commission

Richard Carrigan
National Science Foundation

Gary Cook
Department of Commerce

Albert Crary
National Science Foundation

Floyd L. France
Department of Justice

John H. Hall
Department of the Treasury

James T. Halverson
Federal Trade Commission

Arnold Hoffman
Environmental Protection Agency

John J. Ingersoll
Department of State

Harry LeBovit
General Services Administration

Melville R. Mudge
U.S. Geological Survey

J. Lawrence Muir
Securities and Exchange Commission

Robert D. Nininger
Atomic Energy Commission

Hubert W. Norman
Securities and Exchange Commission

Peter C. Schumaier
Department of Transportation

William L. Springer
Federal Power Commission

Nicholas C. Tolerico
U.S. Tariff Commission

Thomas A. Henrie
U.S. Bureau of Mines

Paul Howard
Bureau of Land Management

Bruce Grant
Mining Enforcement and Safety Administration

Vincent E. McKelvey
U.S. Geological Survey

William M. Porter
Office of Energy Conservation, DoI

John B. Rigg
Minerals, Department of the Interior

John F. Robinson
Development and Budget, DoI

Stephen Wakefield
Energy and Minerals, DoI

John D. Morgan, Jr.
U.S. Bureau of Mines

Paul Meadows
U.S. Bureau of Mines

James Owens
Department of Commerce

Mike Bozzelli
Department of Commerce

Preston Cloud, Jr., Former Chairman, COMRATE

ANNEX III
TABLE OF CONTENTS OF APPENDICES

Appendices to each Panel's report have been prepared as separate documents, containing source material, contributed by panel members or consultants, drawn upon or summarized in the report. The Appendices do not necessarily reflect COMRATE or panel opinion; they are an information resource available on request to readers with a special interest.

APPENDIX TO SECTION I: (REPORT OF PANEL ON MATERIALS CONSERVATION THROUGH TECHNOLOGY)

B. Technology for Materials Supply: Paper Contributed by Panel Member H.H. Kellogg

Chemical Aspects of Solution Mining: Abstract of Paper Presented at Panel Workshop, April 10-11, 1974, by M. Wadsworth, University of Utah

On the Industrial Uses of Space: Paper Contributed by Panel Member T.B. Taylor

C. Pollution Control: Paper Contributed by Panel Member K.G. Semrau

Urban Wastes and Some Related Problems: Paper Contributed by Panel Member D.J. Rose

D. Critical Materials Needs for Current Energy Technology: Paper Contributed by COMRATE Staff Member F. Speer

Materials Requirements for Emerging Energy Technology: Paper Contributed by Panel Member D.J. Rose

E. Materials Conservation Through Substitutes and Product Design: Paper Contributed by Panel Members A.G. Chynoweth and F.P. Huddle, and Staff Member F. Speer

Conservation of Materials in Building: Paper Presented at Panel Workshop, April 10-11, 1974, by A. Dietz, Massachusetts Institute of Technology

Materials Resources and the Automobile Industry: Paper Presented at Panel Workshop, April 10-11, 1974, by J.J. Harwood, Ford Motor Company

Substitution and Recycling in the Communications Industry: Paper Presented at Panel Workshop, April 10-11, 1974, by T.D. Schlabach, Bell Telephone Labs

Plastics - Supplies and Substitutes: Paper Presented at Panel Workshop, April 10-11, 1974, by W.L. Hawkins, Bell Telephone Labs

F. Institutional Aspects of Materials Conservation: Three Papers Contributed by Panel Member F.P. Huddle, 1973-1974

Industrial Symbiosis, Waste Conversion, and Economics of Scale

Some Policy Alternatives for the National Stockpile of Strategic and Critical Materials

Institutional Obstacles to the Acceptance of New Technology in Materials

APPENDIX TO SECTION II: (REPORT OF PANEL ON ESTIMATION OF MINERAL RESERVES AND RESOURCES)

U.S.G.S. Releases Revised U.S. Oil and Gas Resource Estimates; Department of the Interior News Release, March 26, 1974

Synopsis of Procedure: Accelerated National Oil and Gas Resource Evaluation (ANOGRE) W.W. Mallory, U.S.G.S., January 28, 1974

Ratio Between Recoverable Oil per Unit Volume of Sediments for Future Exploratory Drilling to that of the Past for the Conterminous United States: Communication of M. King Hubbert, U.S.G.S., to COMRATE, June 7, 1974

APPENDIX TO SECTION III: (REPORT OF PANEL ON THE IMPLICATIONS OF MINERAL PRODUCTION FOR HEALTH AND THE ENVIRONMENT)

Environmental Impact of Energy Generation: The Sulfur Cycle: Paper Contributed by Panel Member T.D. Brock

Coalworkers' Pneumoconiosis: Its Economic Impact and Prevalence: Paper Prepared for the Panel by J. Costello, ALFORD, U.S. Public Health Service, and W.K.C. Morgan, West Virginia University Medical Center

Letter from W.K.C. Morgan to the Panel, July 22, 1974

Natural Ecosystems and the Industrial Production of Carbon Dioxide: Paper Contributed by Panel Member D.B. Botkin, April 1974

Short-term Health Considerations Regarding the Choice of Coal as a Fuel: Paper Contributed by Panel Member L.B. Lave

APPENDIX TO SECTION IV: (REPORT OF PANEL ON DEMAND FOR FUEL AND MINERAL RESOURCES)

Bibliographic Review of Selected "Demand" Forecasts of Minerals and Commodities, with Commentary on Methodology: Report Prepared for the Panel by Overly Schell Associates, August 1973